ENGINEERING ANALYSIS

ENGINEERING ANALYSIS

A Vector Space Approach

ROBERT J. SCHILLING
Clarkson University

HUA LEE
University of Illinois

A Wiley-Interscience Publication

JOHN WILEY & SONS

New York · Chichester · Brisbane · Toronto · Singapore

Library of Congress Cataloging in Publication Data:

Schilling, Robert J. (Robert Joseph), 1947–
Engineering analysis.

"Based on class notes used at the University of California at Santa Barbara"—
Includes bibliographies and index.
1. Engineering mathematics. 2. Vector spaces.
I. Lee, Hua, 1952– II. Title.
TA330.S35 1987 620'.0042 87-21668
ISBN 0-471-82760-6

Printed in the United States of America

10 9 8 7 6 5 4 3 2 1

To Edgar and Bette Rose

and

To Michelle and Michael

PREFACE

Throughout the history of engineering, various mathematical notions have evolved as useful abstractions. Among these perhaps none is more widespread in use than the concept of linearity. Linear mathematical models, although by no means universally applicable, suffice to describe many important physical phenomena. Moreover, a highly refined theory of vector spaces, linear transformations, and linear systems is available. It is with this theory and certain of its many applications that we shall be concerned.

This book is based on class notes used at the University of California at Santa Barbara. It is designed as a text for a senior or beginning graduate level course in Engineering Analysis. Experience has shown that students with a standard calculus background exhibit little difficulty with the material. The underlying theme of linearity permeates the entire book, and makes for a unified treatment of the topics covered. Indeed, it is the hope of the authors that the reader will soon come to appreciate the many ways in which vector spaces, linear transformations, and linear systems complement one another. Examples of applications are cited throughout the book. However, the principal objective of this text is not the illustration of applications, rather it is the development of a repertoire of analytical tools. The general nature of these mathematical tools should then allow the student to more easily investigate, and more fully appreciate, new theoretical concepts specific to his or her area of specialty.

The style of the presentation of the material is dictated by a few simple underlying principles. Whenever possible, the general case is treated first. Important special cases are then deduced from the general case. This spares the reader of having to *unlearn* a result as various qualifying assumptions are stripped away. A few carefully chosen special cases can, of course, go a long way toward illustrating a general concept. For this reason numerous examples are included throughout the text. A variety of pictorial representations are also used to supplement the discussion. For example, whenever an algebraic concept has a simple geometric interpretation, that interpretation is included in the form of a illustration. In addition, flowcharts are used to document algorithms, and block diagrams are employed to reveal the relationships between the concepts being discussed.

Principal results are stated as logical propositions. Most proofs of propositions are included, particularly when they illustrate an analysis technique that is important in its own right. However, certain results that are germane to the development of the theory require considerable digression in order to develop readable proofs. Rather than wade through an otherwise uninteresting series of preliminary lemmas, the policy instead is to simply refer the reader to an appropriate reference. Numerous exercises are also included throughout the text, and their purpose is to supplement the problems at the end of each chapter, and also to serve as a vehicle for stating useful results of a secondary nature. Definitions appear in the text in two forms, formal and contextual. In both cases the word or phrase being defined is in italic type. In many instances a number of equivalent definitions are available. The policy followed in those cases is a compromise between geometric appeal and algebraic expediency. The geometric path is followed except in those cases where an operational definition, one to be used later for testing purposes, is required.

The development of the material throughout the text proceeds from the general to the particular. This is true both within sections of the book and from one section to the next. The first three chapters are devoted to a discussion of linear or vector spaces. The relationships between the topics covered in this first section of the book are summarized in Fig. 0.1.

In Chapter 1 the reader is introduced to the fundamental notion of an abstract vector space. Various essential concepts such as subspaces, bases, coordinates, and isomorphisms are then considered. The structure of a vector space is then refined in Chapter 2 with the introduction of the norm. Here such topological notions as convergence, continuity, and completeness are considered. Chapter 2 closes with a discussion of contraction mappings and a vector space analysis of successive approximation algorithms. In Chapter 3 the structure of vector spaces is refined still further with the introduction of the inner product. Here the fundamental, geometrically appealing, concept of orthogonality is examined. The notions of orthogonal projections onto subspaces and orthogonal series expansions are developed, and the minimum-norm optimization problem is examined. Chapter 3 closes with an investigation of the classic orthogonal polynomials.

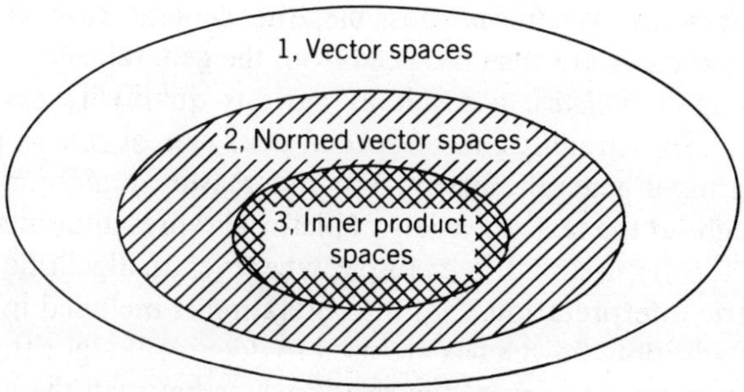

Figure 0.1. Section 1: Vector spaces.

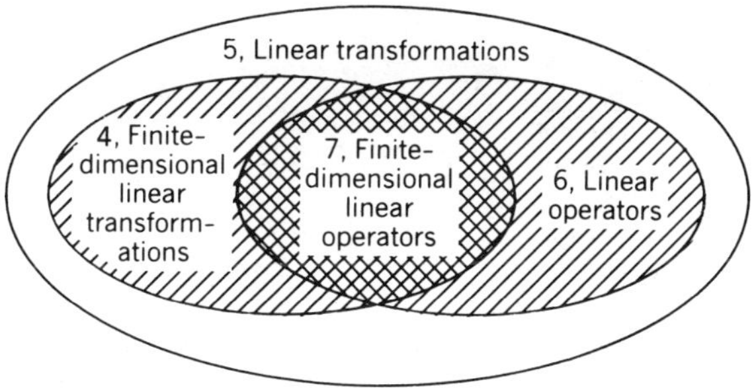

Figure 0.2. Section 2: Linear transformations.

The next four chapters build upon the foundation laid in the first three. In this second section of the book, the discussion focuses on the properties of vectors in one particular space, the space of linear transformations. The relationships between the topics covered in these chapters are summarized in Fig. 0.2.

Again the development of the theory proceeds from the general to the particular, but this time along two distinct lines. In Chapter 4 the reader is introduced to the fundamental notion of a linear transformation between two vector spaces. Essential concepts such as the null space of a transformation, the range space of a transformation, invertible transformations, and bounded transformations are then considered. In Chapter 5 attention is restricted to the finite-dimensional case. Here the important notion of a matrix representation of a linear transformation is examined. Echelon matrices and certain computational algorithms based upon them are then discussed. In Chapter 6 attention is focused upon a particular type of linear transformation, the linear operator. Fundamental operator concepts such as the spectrum, the adjoint operator, normal operators, projections, and sign-definite operators are then considered. Chapter 7 is devoted to transformations which are both finite-dimensional and linear operators. It examines the notion of canonical representations of analytic functions of operators. Included are investigations of the block diagonal representation, the power series representation, the integral representation, and the spectral representation.

The next three chapters illustrate some applications of vector space techniques to engineering analysis problems. In this third section of the book, the focus is on linear systems and orthogonal transformations used for signal processing and estimation. The relationships between the topics covered in these chapters are shown in Fig. 0.3.

In Chapter 8 the discussion is directed toward linear differential systems. Examples of physical systems that can be modeled by differential systems are presented, and the issue of the existence and uniqueness of the solution is addressed. This is followed by a discussion of three representations of linear differential systems: the state space representation, the impulse response representation, and the transfer function representation. Chapter 9 closely parallels

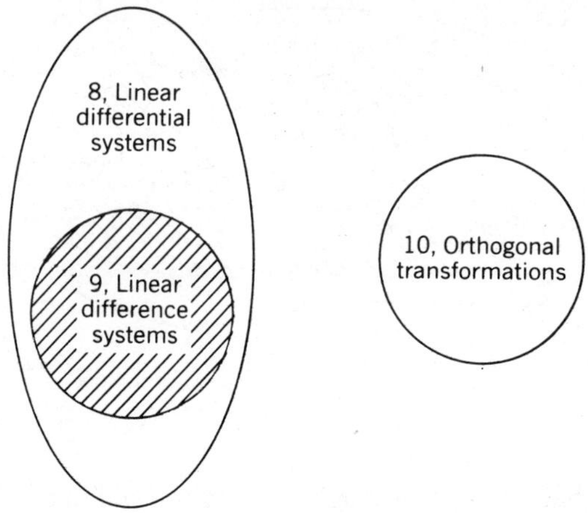

Figure 0.3. Section 3: Linear systems and signal processing.

Chap. 8, but is devoted to discrete-time physical systems and their linear difference system models. The relationship between discrete-time and continuous-time systems is examined. The state space representation, pulse response representation, and transfer function representation of linear difference systems are then developed. Chapter 10 employs a vector space framework to examine and compare fundamental orthogonal transformations used for signal processing and estimation. Here a unified treatment of most of the transforms used in engineering analysis is presented. The sampling theorem is also investigated, and the problem of estimating the amplitude of the signal embedded in noise is examined.

ACKNOWLEDGMENTS

There are many individuals who have contributed to this book. For providing the necessary resources and the proper atmosphere within which to work, we wish to acknowledge the University of California at Santa Barbara, where this project was initiated, and Clarkson University and the University of Illinois, where it was completed. We thank J. LaCoss and S. Whittier for their typing of the manuscript. We are also indebted to the many students who read early drafts of the manuscript and contributed to its revision through questions, comments, and suggestions. Finally, we are grateful to our families, whose patience and support have been indispensable.

R. J. Schilling
Hua Lee

Potsdam, New York
Urbana, Illinois
July 1987.

Contents

1 VECTOR SPACES

Throughout the history of engineering, various mathematical notions have evolved as useful abstractions, and among these there is a particularly fundamental one called the *vector space*. Many engineering problems can be formulated in terms of vector space models, and then solved by using the repertoire of tools associated with the highly refined theory of vector spaces.

1.1. VECTOR SPACES

From the geometric point of view, the most appealing example of a vector space is the space $\mathbf{R}^3$, which is three-dimensional space as we normally think of it. The coordinate axes for the space $\mathbf{R}^3$ are shown in Figure 1.1.

The concept of a vector space is, however, considerably more general (Halmos, 1958). To visualize this, we must first consider the underlying notion of a field of scalars. A *field* is a set $\mathbf{F}$ whose elements, called *scalars*, can be added

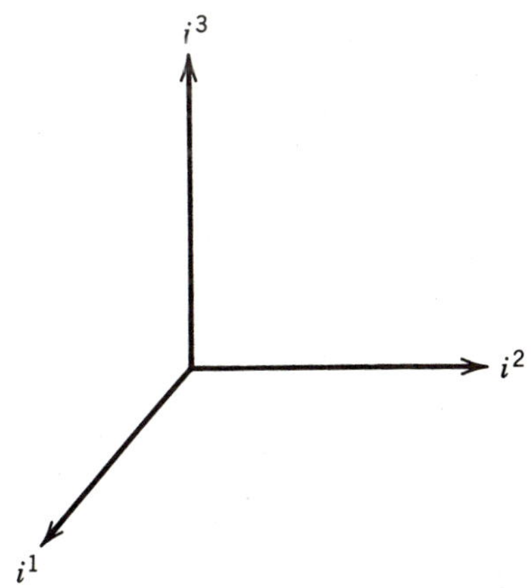

Figure 1.1. The coordinate axes of the space $\mathbf{R}^3$.

and multiplied to produce new elements in **F** (Noble and Deniel, 1977). The fields with which we will be concerned are the field **R** of *real numbers* and the field **C** of *complex numbers.* Other common examples of fields include the field of *binary numbers* $\{0, 1\}$ and the field of *rational polynomials.* A vector space is defined over or with respect to a field **F** in the following manner.

Definition 1.1 (Vector Space). A *vector space* over a field **F** is a set **V** whose elements are called *vectors* together with two operations called *vector addition* and *scalar multiplication* that must satisfy the following axioms.

1. **V** must be *closed* with respect to the two operations. That is, for each pair of vectors $(x, y) \in \mathbf{V}^2$ and each scalar $\alpha \in \mathbf{F}$, the *sum* $x + y$ and the *scalar product* αx must also be vectors in **V**:

$$(x, y) \in \mathbf{V}^2 \quad \Rightarrow x + y \in \mathbf{V},$$
$$(\alpha, x) \in \mathbf{F} \times \mathbf{V} \Rightarrow \alpha x \in \mathbf{V}.$$

2. The two operations must be *associative, commutative,* and *distributive* in the following sense. For each $(x, y, z) \in \mathbf{V}^3$ and each $(\alpha, \beta) \in \mathbf{F}^2$,

$$(x + y) + z = x + (y + z),$$
$$(\alpha\beta)x = \alpha(\beta x),$$
$$x + y = y + x,$$
$$\alpha(x + y) = \alpha x + \alpha y,$$
$$(\alpha + \beta)x = \alpha x + \beta x.$$

3. There must exist *identity elements* $(0, 1) \in \mathbf{V} \times \mathbf{F}$ called the *zero vector* and *unit scalar*, respectively, where

$$x \in \mathbf{V} \Rightarrow x + 0 = x,$$
$$x \in \mathbf{V} \Rightarrow 1x \quad = x.$$

4. For each $x \in \mathbf{V}$ there must be an *additive inverse* element, $-x \in \mathbf{V}$, such that

$$x + (-x) = 0.$$ ■

This definition is, clearly, quite lengthy. We can begin to make the notion more concrete by considering a number of important special cases.

Example 1.1 (The Space $\mathbf{F}^n$). Perhaps the most important example of a vector space is $\mathbf{F}^n$ the space of *column vectors* with *n* components from the field **F**. Let $(\alpha, x, y) \in \mathbf{F} \times \mathbf{F}^n \times \mathbf{F}^n$. We then define vector addition and scalar multiplication on $\mathbf{F^n}$ in terms of the components as

$$(x + y)_k \triangleq x_k + y_k, \qquad 1 \leqslant k \leqslant n$$
$$(\alpha x)_k \triangleq \alpha x_k, \qquad 1 \leqslant k \leqslant n.$$

Here "$\triangleq$" should be read "equals by definition." Note that the right-hand side involves addition and multiplication of scalars. Hence field addition and field multiplication (real or complex) induce vector addition and scalar multiplication as indicated. In view of the definition of vector addition, the zero vector in $\mathbf{F^n}$ is the column vector whose components are all zero. That is,

$$0_k = 0, \qquad 1 \leqslant k \leqslant n.$$

Here 0 is the zero scalar (additive identity) of $\mathbf{F}$. Note that the space $\mathbf{R}^3$ in Fig. 1.1 is a special case of $\mathbf{F^n}$ with $\mathbf{F} = \mathbf{R}$ and $n = 3$. ∎

Example 1.2 (The Space $\mathbf{F^{m \times n}}$). The space $\mathbf{F^{m \times n}}$ consists of the set of all $m \times n$ *matrices* whose components belong to the field $\mathbf{F}$. Hence a vector in $\mathbf{F^{m \times n}}$ is an $m \times n$ matrix. Let $(\alpha, A, B) \in \mathbf{F} \times \mathbf{F^{m \times n}} \times \mathbf{F^{m \times n}}$. We then define vector addition and scalar multiplication as

$$(A + B)_{kj} \triangleq A_{kj} \pm B_{kj}, \qquad 1 \leqslant k \leqslant m, \quad 1 \leqslant j \leqslant n,$$
$$(\alpha A)_{kj} \triangleq \alpha A_{kj}, \qquad 1 \leqslant k \leqslant m, \quad 1 \leqslant j \leqslant n.$$

Again, the right-hand side refers to field addition and field multiplication. The zero vector in this case is the $m \times n$ matrix of zeros,

$$0_{kj} = 0, \qquad 1 \leqslant k \leqslant m, \quad 1 \leqslant j \leqslant n.$$ ∎

Example 1.3 (The Space $\mathbf{F(T, V)}$). Let T be a subset of the field $\mathbf{F}$ and let $\mathbf{V}$ be a vector space over $\mathbf{F}$. The space $\mathbf{F(T, V)}$ then consists of the set of all *functions mapping T into V*:

$$\mathbf{F(T, V)} \triangleq \{f\colon T \to \mathbf{V}\}.$$

Let $(\alpha, f, g) \in \mathbf{F} \times \mathbf{F(T, V)}^2$. We then define vector addition and scalar multiplication on $\mathbf{F(T, V)}$ as

$$(f + g)(t) \triangleq f(t) + g(t), \qquad t \in T,$$
$$(\alpha f)(t) \triangleq \alpha f(t), \qquad t \in T.$$

Note that the right-hand side refers to vector addition and scalar multiplication on the space $\mathbf{V}$ that has already been defined. The zero vector in this case is the function that is *identically zero* on T,

$$0(t) = 0_v, \qquad t \in T.$$

Here 0_v is the zero vector of **V**. As we shall see, the vector space **F(T, V)** is fundamentally different from the previous examples. We call it an example of a *function space.* ∎

Example 1.4 (The Space V × W). Let **V** and **W** be vector spaces over the same field **F**. The space **V** × **W** consists of the set of all *ordered pairs* (v, w) such that $v \in \mathbf{V}$ and $w \in \mathbf{W}$. That is,

$$\mathbf{V} \times \mathbf{W} \triangleq \{(v, w) : v \in \mathbf{V},\ w \in \mathbf{W}\}.$$

Let $(\alpha, (v, w), (p, q)) \in \mathbf{F} \times (\mathbf{V} \times \mathbf{W})^2$. Vector addition and scalar multiplication on **V** × **W** are defined as

$$(v, w) + (p, q) \triangleq (v + p, w + q),$$
$$\alpha(v, w) \triangleq (\alpha v, \alpha w).$$

Note that the right-hand side refers to vector addition and scalar multiplication on **V** and **W**. The zero vector in this case is the ordered pair of zero vectors,

$$0 = (0_v, 0_w).$$

Here 0_v and 0_w denote the zero vectors of **V** and **W**, respectively. We call **V** × **W** the *product space* of **V** and **W**. Note that the notion of a product space can be extended to include more than two spaces. Indeed $\mathbf{F}^n$ can be regarded as the product space of **F** taken n times:

$$\mathbf{F}^n = \mathbf{F} \times \mathbf{F} \times \cdots \times \mathbf{F} \quad (n \text{ times}).$$

Similarly, when we write $(\alpha, x, y) \in \mathbf{F} \times \mathbf{V}^2$, we can take this to mean that (α, x, y) is a vector in the product space $\mathbf{F} \times \mathbf{V}^2$. ∎

1.2. SUBSPACES

In order to fully appreciate the utility of the notion of vector spaces, we must start to refine the concept somewhat. We begin with the idea of one space being contained in another.

Definition 2.1 (Subspace). A subset **U** of a vector space **V** is a *subspace of* **V** ⇔ **U** is itself a vector space using the vector addition and scalar multiplication defined on **V**. ∎

Thus, as the terminology suggests, a subspace is a vector space within a vector space. Again, we illustrate this concept with a number of examples.

Example 2.1 (Zero Subspace). The set $\mathbf{V}$ and the set $\{\mathbf{0}\}$ are the largest and the smallest subspaces of $\mathbf{V}$, respectively. We call $\{\mathbf{0}\}$ the *zero subspace.* ■

Example 2.2 (Functions Mapping 0 Into 0). The set of all functions mapping $\mathbf{F}$ into $\mathbf{V}$ such that $f(0) = 0$ is a subspace of $\mathbf{F}(\mathbf{F}, \mathbf{V})$:

$$\mathbf{U} = \{f \in \mathbf{F}(\mathbf{F}, \mathbf{V}) : f(0) = 0\}.$$ ■

Fortunately, we do not have to test the entire list of axioms in Def. 1.1 to verify whether or not a given subset $\mathbf{U}$ qualifies as a subspace. Indeed, the following simple criterion is sufficient.

Proposition 2.1 (Subspace). A nonempty subset $\mathbf{U}$ is a subspace of a vector space $\mathbf{V}$ if

$$(\alpha, x, y) \in \mathbf{F} \times \mathbf{U}^2 \Rightarrow \alpha x + y \in \mathbf{U}.$$

Proof. The associative, commutative, and distributive laws hold on $\mathbf{U}$ because $\mathbf{U}$ is a subset of $\mathbf{V}$ and $\mathbf{V}$ is a vector space. It remains to show that $\mathbf{U}$ is closed and has the required identity and inverse elements. Since $\mathbf{U}$ is nonempty, there exists at least one $x \in \mathbf{U}$. Thus,

$$\begin{aligned} 0 &= -x + x \\ &= (-1)x + x \\ &\in \mathbf{U}. \end{aligned}$$

Next because the zero vector is in $\mathbf{U}$, for each $x \in \mathbf{U}$, we have

$$\begin{aligned} -x &= -x + 0 \\ &= (-1)x + 0 \\ &\in \mathbf{U}. \end{aligned}$$

Finally, let $(\alpha, x, y) \in \mathbf{F} \times \mathbf{U}^2$. Then $\mathbf{U}$ is closed relative to vector addition and scalar multiplication as

$$\begin{aligned} \alpha x &= \alpha x + 0 \in \mathbf{U}, \\ x + y &= (1)x + y \in \mathbf{U}. \end{aligned}$$ ■

We refer to the weighted sum $\alpha x + y$ as an example of a *linear combination* of x and y. Proposition 2.1 can be summarized in words by saying that a nonempty subset **U** is a subspace of a vector space **V** if it is closed with respect to the operation of performing linear combinations. Because a subspace is itself a vector space, the utility of Prop. 2.1 lies in the fact that it provides us with a simple scheme for establishing new vector spaces. A few general vector spaces can be defined as in Section 1.1. More refined subspaces of them can then be established using Prop. 2.1. The following example illustrates this technique.

Example 2.3 (Odd Functions). Let $\mathbf{F_o(R, R)}$ be the set of all *odd functions* mapping **R** into **R**:

$$\mathbf{F_o(R, R)} \triangleq \{f \in \mathbf{F(R, R)} : f(-t) = -f(t)\}.$$

Then $\mathbf{F_o(R, R)}$ is nonempty since $f(t) = t$ is a vector in $\mathbf{F_o(R, R)}$. Next, let $(\alpha, f, g) \in \mathbf{R} \times \mathbf{F_o(R, R)}^2$. From Exam. 1.3,

$$\begin{aligned}(\alpha f + g)(-t) &= \alpha f(-t) + g(-t)\\ &= -\alpha f(t) - g(t)\\ &= -(\alpha f + g)(t).\end{aligned}$$

Thus $\alpha f + g \in \mathbf{F_o(R, R)}$. It follows from Prop. 2.1 that $\mathbf{F_o(R, R)}$ is a subspace of $\mathbf{F(R, R)}$. Therefore the odd functions mapping **R** into **R** form a vector space. ■

Exercise 2.1 (Intersection Property). Show that if $\mathbf{U_1}$ and $\mathbf{U_2}$ are subspaces of **V**, then the *intersection* of $\mathbf{U_1}$ and $\mathbf{U_2}$, denoted $\mathbf{U_1} \cap \mathbf{U_2}$, is a subspace of **V**. Here,

$$\mathbf{U_1} \cap \mathbf{U_2} \triangleq \{x : x \in \mathbf{U_1} \text{ and } x \in \mathbf{U_2}\}.$$ ■

1.3. DIRECT SUMS

It is often useful to decompose a vector space into a number of subspaces. One such decomposition is the direct sum decomposition.

Definition 3.1 (Direct Sum). A vector space **V** is the *direct sum* of the subspaces $\mathbf{U_1}$ and $\mathbf{U_2}$ $\Leftrightarrow$ for each $x \in \mathbf{V}$ there exist unique $(x^1, x^2) \in \mathbf{U_1} \times \mathbf{U_2}$ such that $x = x^1 + x^2$. In this case we write

$$\mathbf{V} = \mathbf{U_1} \oplus \mathbf{U_2}.$$ ■

Direct sum decompositions often take the form of a subspace and its dual or complement. Examples of direct sum decompositions include the following.

Example 3.1 (Symmetric/Skew-Symmetric Matrices). Let $A \in \mathbf{F}^{\mathbf{n}\times\mathbf{n}}$. Then A^{T}, the *transpose* of A, is obtained by interchanging the rows and columns of A. Consider the subspaces of *symmetric* matrices, and *skew-symmetric* matrices defined, respectively, as

$$\mathbf{F}_{\mathbf{sy}}^{\mathbf{n}\times\mathbf{n}} \triangleq \{A \in \mathbf{F}^{\mathbf{n}\times\mathbf{n}} : A^{\mathrm{T}} = A\},$$
$$\mathbf{F}_{\mathbf{sk}}^{\mathbf{n}\times\mathbf{n}} \triangleq \{A \in \mathbf{F}^{\mathbf{n}\times\mathbf{n}} : A^{\mathrm{T}} = -A\}.$$

The space $\mathbf{F}^{\mathbf{n}\times\mathbf{n}}$ is the direct sum of $\mathbf{F}_{\mathbf{sy}}^{\mathbf{n}\times\mathbf{n}}$ and $\mathbf{F}_{\mathbf{sk}}^{\mathbf{n}\times\mathbf{n}}$. Indeed, for each $A \in \mathbf{F}^{\mathbf{n}\times\mathbf{n}}$, let

$$A_1 = (A + A^{\mathrm{T}})/2,$$
$$A_2 = (A - A^{\mathrm{T}})/2.$$

Then $(A_1, A_2) \in \mathbf{F}_{\mathbf{sy}}^{\mathbf{n}\times\mathbf{n}} \times \mathbf{F}_{\mathbf{sk}}^{\mathbf{n}\times\mathbf{n}}$, and $A = A_1 + A_2$. ■

Example 3.2 (Even–Odd Functions). Let $\mathbf{V}$ be a vector space over $\mathbf{R}$ and consider the space $\mathbf{F(R, V)}$ of functions mapping the field $\mathbf{R}$ into $\mathbf{V}$. Let $\mathbf{F_e(R, V)}$ and $\mathbf{F_o(R, V)}$ denote the subspaces of *even* and *odd functions*, respectively. That is,

$$\mathbf{F_e(R, V)} \triangleq \{f \subset \mathbf{F(R, V)} : f(-t) - f(t)\},$$
$$\mathbf{F_o(R, V)} \triangleq \{f \in \mathbf{F(R, V)} : f(-t) = -f(t)\}.$$

Then $\mathbf{F(R, V)}$ is the direct sum of $\mathbf{F_e(R, V)}$ and $\mathbf{F_o(R, V)}$. Here, for each $f \in \mathbf{F(R, V)}$, let

$$f^1(t) - [f(t) + f(-t)]/2,$$
$$f^2(t) - [f(t) - f(\ \ t)]/2.$$

Then, $(f^1, f^2) \in \mathbf{F_e(R, V)} \times \mathbf{F_o(R, V)}$ and $f = f^1 + f^2$. ■

1.4. CONVEXITY

In engineering applications, some results that are applicable to subspaces can be made more widely applicable by extending them to convex sets.

Definition 4.1 (Convex Set). A subset S of a vector space $\mathbf{V}$ is *convex* $\Leftrightarrow$ for each $(x^0, x^1) \in S^2$:

$$0 \leqslant \alpha \leqslant 1 \Rightarrow (1 - \alpha)x^0 + \alpha x^1 \in S.$$ ■

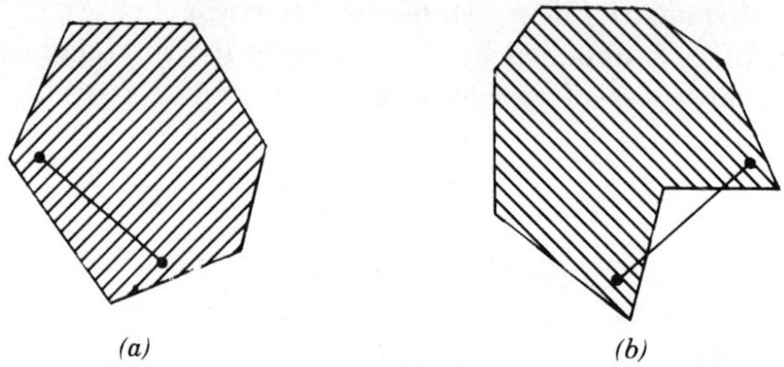

Figure 1.2. (*a*) Convex and (*b*) nonconvex sets in $\mathbf{R}^2$.

The definition of convexity has a simple geometric interpretation. As the parameter α varies from 0 to 1, the vector $(1 - \alpha)x^0 + \alpha x^1$ traverses a straight line segment from x^0 to x^1. Thus a set is convex if and only if, for each pair of points in the set, the line segment connecting them is also in the set. Refer to Figure 1.2 for examples of convex and nonconvex sets in the space $\mathbf{R}^2$.

One of the fundamental properties of convex sets is that the intersection of an arbitrary number of convex sets is itself a convex set as can be seen from the following exercise.

Exercise 4.1 (Intersection Property). Let $\{S_k\}$ be a collection of convex subsets of a vector space $\mathbf{V}$ and let S be the subset obtained by taking the intersection of $\{S_k\}$. Show that $S = \cap S_k$ is a convex set. ■

If S is an arbitrary subset of a vector space $\mathbf{V}$, we can always associate with S a special convex set called the convex hull of S. In particular, we can take the intersection of all convex sets that contain S to produce the *smallest* convex set containing S.

Definition 4.2 (Convex Hull). Let S be a subset of a vector space $\mathbf{V}$. Then the *convex hull* of S is the smallest convex subset of $\mathbf{V}$ that contains S. ■

A sketch of a set S in $\mathbf{R}^2$ and its convex hull is shown in Figure 1.3. If the set S is already convex, then it is equal to its convex hull. This is the case, for example,

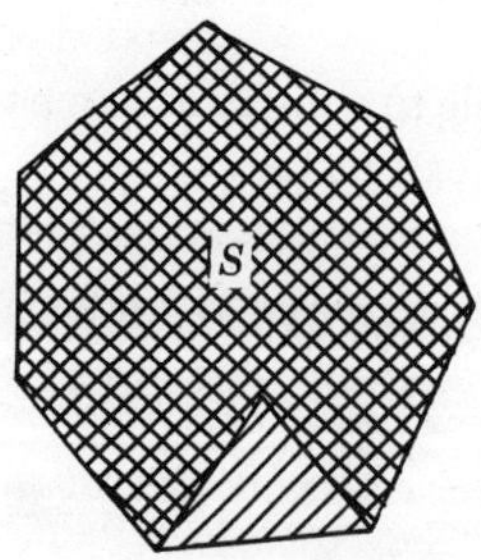

Figure 1.3. A set S in $\mathbf{R}^2$ and its convex hull.

when S is a subspace. Indeed, in view of the closure property, every subspace is convex although the converse is not true. In engineering applications, convex sets often arise from collections of linear inequality constraints, as can be seen from the following exercise.

Exercise 4.2 (Convex Set). Let $A \in \mathbf{F}^{\mathbf{m} \times \mathbf{n}}$ and let $b \in \mathbf{F}^{\mathbf{m}}$. Consider the following subset S of $\mathbf{F}^{\mathbf{n}}$. Show that S is a convex set

$$S = \{x \in \mathbf{F}^{\mathbf{n}} : Ax \leqslant b\}. \qquad \blacksquare$$

Exercise 4.3. Let S be a convex subset of a vector space $\mathbf{V}$ and let $\{x^1, x^2, \ldots, x^m\}$ be vectors in S. Show that if $a \in \mathbf{F}^{\mathbf{m}}$ with $a_k \geqslant 0$ then,

$$\sum_{k=1}^{m} a_k = 1 \Rightarrow \sum_{k=1}^{m} a_k x^k \in S. \qquad \blacksquare$$

1.5. SPAN AND DIMENSION

Next we consider properties of subsets of $\mathbf{V}$ that are not subspaces. Typically these subsets consist of a finite number of vectors or perhaps a countable sequence of vectors.

Definition 5.1 (Linear Combinations). Let $\{x^1, x^2, \ldots, x^m\}$ be vectors in $\mathbf{V}$. Then a *linear combination* of $\{x^1, x^2, \ldots, x^m\}$ is any vector x that can be expressed using a set of weighting coefficients $a \in \mathbf{F}^{\mathbf{m}}$,

$$x = \sum_{k=1}^{m} a_k x^k. \qquad \blacksquare$$

Thus a linear combination is a *weighted sum* where the weighting coefficients are taken from the field over which the space is defined. Suppose S is a subset of a space $\mathbf{V}$ and x^1 and x^2 are linear combinations of vectors in S. Then for each $\alpha \in \mathbf{F}$, the vector $\alpha x^1 + x^2$ is also a linear combination of vectors in $\mathbf{S}$. It follows (Prop. 2.1) that the set of all linear combinations of vectors in S forms a subspace of $\mathbf{V}$. In view of its importance, we give this subspace a special name.

Definition 5.2 (Span). A vector x belongs to the subspace *spanned* by a subset U of $\mathbf{V} \Leftrightarrow x$ is a linear combination of vectors in U. The space spanned by the vectors in U is denoted **span(U)**. $\blacksquare$

The subspace spanned by U is sometimes referred to as the subspace *generated* by U. In certain cases, the subspace spanned by a given set of vectors is the entire space $\mathbf{V}$. This situation is of particular importance as it allows us to determine the dimension of the space.

Definition 5.3 (Dimension). A space **V** is *finite-dimensional* ⇔ it is spanned by a finite number of vectors. Otherwise, **V** is *infinite-dimensional.* The minimum number of vectors needed to span **V** is called the *dimension* of **V**. ■

In the remainder of this chapter we focus on additional properties of finite-dimensional vector spaces. In Chapters 2 and 3 these properties are generalized to include infinite-dimensional spaces.

1.6. LINEAR INDEPENDENCE

Recall that the dimension of a space is the minimum number of vectors needed to span the space. In order to determine the minimum number of vectors, we introduce the notion of linear independence.

Definition 6.1 (Linear Independence). Let $X = \{x^1, x^2, \ldots, x^m\}$ be a subset of a vector space **V**. Then X is a *linearly independent* set ⇔ the following holds. Otherwise X is a *linearly dependent* set

$$\sum_{k=1}^{m} a_k x^k = 0 \Rightarrow a = 0. \qquad ■$$

Thus $\{x^1, x^2, \ldots, x^m\}$ form a linearly independent set if and only if the only linear combination of $\{x^1, x^2, \ldots, x^m\}$ that equals the zero vector is the linear combination all of whose coefficients are zero. Definition 6.1 is an opertional definition of linear independence in that it can be used as a test. An alternative formultion of the notion of linear independence, one that is perhaps more easily interpreted, is summarized in the following exercise.

Exercise 6.1 (Linear Combinations). Show that $\{x^1, x^2, \ldots, x^m\}$ are linearly dependent ⇔ at least one vector in the set can be expressed as a linear combination of the other vectors. ■

Exercise 6.2 (Linear Dependence). Show that any set of vectors containing a linearly dependent subset is itself linearly dependent. Show that any set of vectors containing the zero vector is linearly dependent. ■

Exercise 6.3 (Linear Independence). Show that every subset of a linearly independent set is itself linearly independent. ■

Recall that the dimension of a finite-dimensional space is defined as the minimum number of vectors needed to span the space. It could instead have been defined as the maximum number of linearly independent vectors in the space as can be seen from the following result.

Proposition 6.1 (Linear Independence and Dimension). Linearly independent subsets of an n-dimensional space V contain at most n vectors.

Proof. Let $\{x^1, x^2, \ldots, x^m\}$ be a linearly independent set in $\mathbf{V}$ and suppose $m > n$. Since $\mathbf{V}$ is n-dimensional, there exist vectors $\{y^1, y^2, \ldots, y^n\}$ that span $\mathbf{V}$. Hence

$$x^1 = \sum_{k=1}^{n} a_k y^k,$$

for some $a \in \mathbf{F^n}$. Moreover, $a \neq 0$ since $x^1 \neq 0$. Reindex $\{y^1, y^2, \ldots, y^n\}$, if necessary, such that $a_1 \neq 0$. Then

$$y^1 = \left(x^1 - \sum_{k=2}^{n} a_k y^k\right) \bigg/ a_1.$$

Hence $\{x^1, y^2, y^3, \ldots, y^n\}$ span $\mathbf{V}$. Next, using induction, suppose $\{x_1, \ldots, x^{k-1}, y^k, \ldots, y^n\}$ span $\mathbf{V}$. It follows that

$$x^k = \sum_{j=1}^{k-1} a_j x^j + \sum_{j=k}^{n} a_j y^j,$$

for some $a \in \mathbf{F^n}$. In this case, $a_j \neq 0$ for some $j \geqslant k$ since $\{x^1, x^2, \ldots, x^k\}$ are linearly independent. Reindex $\{y^k, y^{k+1}, \ldots, y^n\}$, if necessary, such that $a_k \neq 0$. Then

$$y^k = \left(x^k - \sum_{j=1}^{k-1} a_j x^j - \sum_{j=k+1}^{n} a_j y^j\right) \bigg/ a_k.$$

Hence $\{x^1, \ldots, x^k, y^{k+1}, \ldots, y^n\}$ span $\mathbf{V}$. By induction, $\{x^1, x^2, \ldots, x^n\}$ span $\mathbf{V}$. If $m > n$, then

$$x^m = \sum_{k=1}^{n} a_k x^k,$$

for some $a \in \mathbf{F^n}$ with $a \neq 0$. But from Definition 6.1 this is impossible since $\{x^1, x^2, \ldots, x^m\}$ are linearly independent. Thus it must be the case that $m \leqslant n$. ■

1.7. BASES

The concepts of span and linear independence can now be combined. In doing so we introduce the fundamental notion of a basis for a vector space.

Definition 7.1 (Basis). A *basis* for a vector space $\mathbf{V}$ is any linearly independent set of vectors that span $\mathbf{V}$. ■

The conceptual value of bases will become apparent as we proceed. For the time being, let us consider a number of examples.

Example 7.1 (Standard Basis for $\mathbf{F}^n$). Let $\mathbf{V} = \mathbf{F}^n$. Then the columns $\mathbf{I} = \{i^1, i^2, \ldots, i^n\}$ of the $n \times n$ *identity matrix* I are linearly independent and span $\mathbf{F}^n$. Hence they form a basis for $\mathbf{F}^n$. This basis is sufficiently important that it is given a special name, the *standard basis* for $\mathbf{F}^n$.

Example 7.2 (Standard Basis for $\mathbf{P}^n(\mathbf{T})$). Let T be a subset of $\mathbf{F}$ and let $\mathbf{P}^n(\mathbf{T})$ be the space of *nth degree polynomials mapping T into $\mathbf{F}$*. Thus vectors in $\mathbf{P}^n(\mathbf{T})$ are functions of the following form for $a \in \mathbf{F}^{n+1}$:

$$f(t) = \sum_{k=1}^{n+1} a_k t^{k-1}.$$

The following *elementary polynominals*, $\mathbf{B} = \{1, t, t^2, \ldots, t^n\}$, are linearly independent and span the space $\mathbf{P}^n(\mathbf{T})$. Hence the elementary polynomials constitute a basis for $\mathbf{P}^n(\mathbf{T})$. This particular basis is important enough to refer to it as the *standard basis* for $\mathbf{P}^n(\mathbf{T})$. Notice that the dimension of $\mathbf{P}^n(\mathbf{T})$ is $n + 1$.

Every nonzero finite-dimensional space $\mathbf{V}$ has an infinite number of bases. However, all bases for $\mathbf{V}$ are equivalent in the sense that they contain the same number of vectors. This can be seen from the following result that is left as an exercise for the reader.

Exercise 7.1 (Basis and Dimension). Show that every basis for an n-dimensional space $\mathbf{V}$ contains exactly n vectors. ■

1.8. COORDINATES

In engineering applications we often do not work with vectors themselves. Instead, we work with coordinates of the vectors. To introduce the notion of coordinates, let $\{x^1, x^2, \ldots, x^n\}$ be a basis for the space $\mathbf{V}$. Then, as $\{x^1, x^2, \ldots, x^n\}$ span $\mathbf{V}$, for each $x \in \mathbf{V}$ there exists an $a \in \mathbf{F}^n$ such that

$$x = \sum_{k=1}^{n} a_k x^k.$$

Moreover, this vector a is *unique*. Indeed, suppose there exists some $b \in \mathbf{F}^n$ with $b \neq a$ where

$$x = \sum_{k=1}^{n} b_k x^k.$$

Using the fact that $0 = x - x$, we can then show that there exists a nonzero linear combination of $\{x^1, x^2, \ldots, x^n\}$ that sums to the zero vector. But this contradicts the fact that $\{x^1, x^2, \ldots, x^n\}$ are linearly independent. Hence we must have $b = a$. In view of this uniqueness, it makes sense to consider the following definition.

Definition 8.1 (Coordinates). Let $\mathbf{B} = \{x^1, x^2, \ldots, x^n\}$ be a basis for $\mathbf{V}$. Then for each $x \in \mathbf{V}$ the unique $a \in \mathbf{F^n}$ such that

$$x = \sum_{k=1}^{n} a_k x^k,$$

is called the *vector of coordinates of* x relative to **B**. In this case we write, $a \triangleq [x]^B$. ■

Hence each basis induces its own *system of coordinates.* Notice that the coordinates of n-dimensional vectors are themselves vectors in the space $\mathbf{F^n}$. The following are examples of bases and the coordinate systems they induce.

Example 8.1 (Coordinates in $\mathbf{R^2}$). Let $\mathbf{V} = \mathbf{R^2}$ and let $\mathbf{B} = \{x^1, x^2\}$ where $x^1 = [1, 2]^T$ and $x^2 = [1, -2]^T$. Next let $x = [2, 0]^T$ be a vector in $\mathbf{V}$. Then from Def. 8.1 we have

$$a_1 + a_2 = 2,$$
$$2a_1 - 2a_2 = 0.$$

Solving these equations for the vector a then yields $[x]^B - [1, 1]^T$. Thc solution is pictured in Figure 1.4. ■

Example 8.2 (Coordinates in $\mathbf{P^4(R)}$). Let $\mathbf{V} = \mathbf{P^4(R)}$ and let $\mathbf{B} = \{1, t, t^2, t^3\}$ be the basis of clcmentary polynomials. Next let $p(t) = 2 - 3t + t^2 - 4t^4$ be a

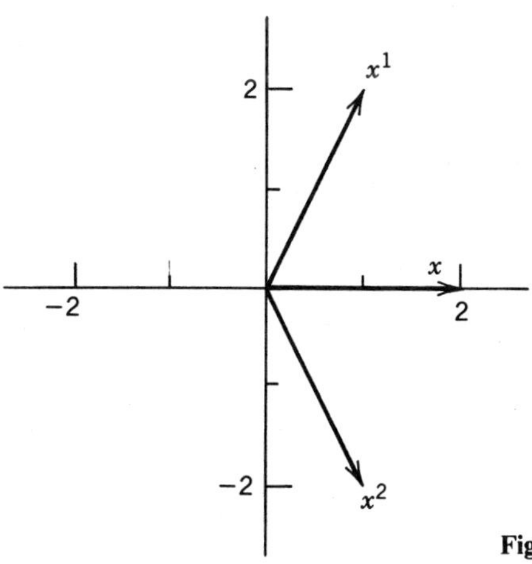

Figure 1.4. The coordinates of x relative to $\{x^1, x^2\}$.

vector in **V**. Then from Def. 8.1 it is readily seen that the coordinates of p are

$$[p]^{\mathbf{B}} = [2, -3, 1, 0, -4]^{\mathrm{T}}.$$

Note that the coordinates of the polynomial p with respect to the standard basis of elementary polynomials are simply the coefficients of p. This is the only basis of $\mathbf{P}^n(\mathbf{R})$ for which this is true. ■

1.9. ISOMORPHISMS

The space $\mathbf{F}^n$ is of special significance. In order to appreciate why, we first consider the notion of a vector space isomorphism.

Definition 9.1 (Isomorphism). An *isomorphism* between a vector space **V** and a vector space **W** is any function f mapping **V** into **W** such that

1. $f(\alpha x + y) = \alpha f(x) + f(y)$,
2. $f(x) = f(y) \Rightarrow x = y$,
3. $\{f(x): x \in V\} = W$.

We refer to properties 1, 2, and 3 by saying f is *linear*, *one-to-one*, and *onto*, respectively. If an isomorphism exists between **V** and **W**, we say **V** and **W** are *isomorphic*. ■

If we describe $f(x)$ by saying that it is the *image of* x *under* f, then the three fundamental properties of an isomorphism can be interpreted as follows. The linear property says that the image of a linear combination must equal the linear combination of the images. The one-to-one property, which is also called the *injective* property, says that no two vectors in **V** can have the same image in **W**. Finally, the onto property,which is also called the *surjective* property, says that every vector in **W** is the image of some vector in **V**. Functions that are both injective and surjective are called *bijective* functions. The importance of the space $\mathbf{F}^n$ lies in the following fundamental result.

Proposition 9.1 (Isomorphism). The space $\mathbf{F}^n$ is isomorphic to every n-dimensional space **V**.

Proof. Let $\{x^1, x^2, \ldots, x^n\}$ be a basis for **V**, and consider the following function f mapping $\mathbf{F}^n$ into **V**:

$$f(a) \triangleq \sum_{k=1}^{n} a_k x^k.$$

We will show that f is linear, one-to-one, and onto. Let $(\alpha, a, b) \in \mathbf{F} \times \mathbf{F^n} \times \mathbf{F^n}$. Then

$$\begin{aligned} f(\alpha a + b) &= \sum_{k=1}^{n} (\alpha a + b)_k x^k \\ &= \sum_{k=1}^{n} (\alpha a_k + b_k) x^k \\ &= \sum_{k=1}^{n} \alpha a_k x^k + \sum_{k=1}^{n} b_k x_k \\ &= \alpha f(a) + f(b). \end{aligned}$$

Hence f is linear. Suppose there exist $(a, b) \in \mathbf{F^n} \times \mathbf{F^n}$ with $b \neq a$ such that $f(b) = f(a)$. Now,

$$\begin{aligned} 0 &= f(a) - f(b) \\ &= \sum_{k=1}^{n} a_k x^k - \sum_{k=1}^{n} b_k x^k \\ &= \sum_{k=1}^{n} (a_k - b_k) x^k, \\ &= \sum_{k=1}^{n} (a - b)_k x^k. \end{aligned}$$

But this is impossible since $\{x^1, x^2, \ldots, x^n\}$ are linearly independent. Hence f is one-to-one. Finally, since $\{x^1, x^2, \ldots, x^n\}$ span $\mathbf{V}$, for each $x \in \mathbf{V}$ there exists an $a \in \mathbf{F^n}$ such that

$$\begin{aligned} f(a) &= \sum_{k=1}^{n} a_k x^k \\ &= x. \end{aligned}$$

Hence f is onto. Therefore f is an isomorphism, and $\mathbf{F^n}$ and $\mathbf{V}$ are isomorphic. ■

Refer to Figure 1.5 for a pictorial description of Prop. 9.1. The significance of Prop. 9.1 lies in the fact that the space $\mathbf{F^n}$ is, in a sense, equivalent to all n-dimensional vector spaces. As such, rather than work directly with a vector x in an n-dimensional space $\mathbf{V}$, we can choose a basis $\mathbf{B}$ instead and then work with the *coordinates* of the vector. That is, we can work with the isomorphic image $[x]^{\mathbf{B}}$, a vector in $\mathbf{F^n}$.

In view of the fact that the space $\mathbf{F^n}$ has a special significance among finite-dimensional spaces, let us examine this space in more detail. In particular, there is a basis for $\mathbf{F^n}$ which has properties that are shared by no other basis.

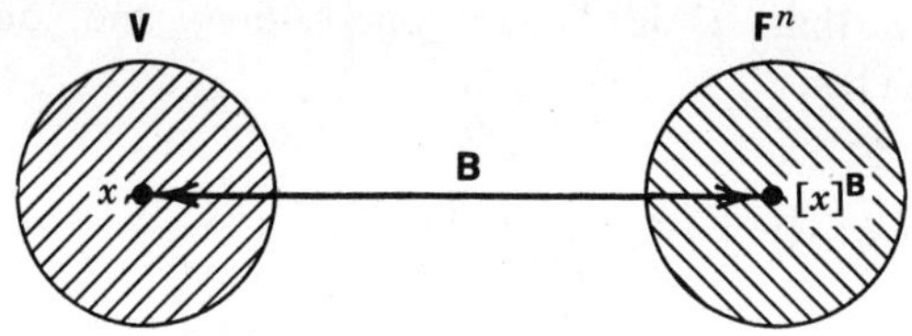

Figure 1.5. A vector in **V** and its coordinates in $\mathbf{F}^n$.

Definition 9.2 (Standard Basis for $\mathbf{F}^n$). The *standard basis for* $\mathbf{F}^n$ is defined $\mathbf{I} = \{i^1, i^2, \ldots, i^n\}$ where i^k is the kth column of the $n \times n$ identity matrix I.

■

The importance of the standard basis for $\mathbf{F}^n$ lies in the nature of the coordinate system it induces. The standard basis is the *implied* basis as can be seen from the following result.

Proposition 9.2 (Standard Basis for $\mathbf{F}^n$). If $x \in \mathbf{F}^n$ and $\mathbf{I}$ is the standard basis for $\mathbf{F}^n$, then $[x]^{\mathbf{I}} = x$.

Proof. Let x be a vector in $\mathbf{F}^n$ and let $a = [x]^{\mathbf{I}}$. Then for $1 \leqslant j \leqslant n$,

$$\begin{aligned} x_j &= \left(\sum_{k=1}^{n} a_k i^k\right)_j \\ &= \sum_{k=1}^{n} a_k i_j^k \\ &= \sum_{k=1}^{n} a_k I_{jk} \\ &= a_j. \end{aligned}$$

■

Thus the *components* of a vector in $\mathbf{F}^n$ are its *coordinates* relative to the standard basis. For all the other bases in $\mathbf{F}^n$ the coordinates of a vector are not the same as the components of the vector. Consequently, given a vector in $\mathbf{F}^n$, if the basis is not specified explicitly, the assumed implicit basis is the standard basis $\mathbf{I}$.

1.10. COORDINATE TRANSFORMATIONS

Many complex engineering problems can be solved by recasting the problems in coordinate systems within which the solutions are more readily apparent. Later we will see some examples of this technique. For the time being, consider the following question. Given the coordinates of a vector relative to one basis, what are its coordinates relative to another basis?

Proposition 10.1 (Coordinate Transformations). Let **V** be an n-dimensional vector space and let $\mathbf{B} = \{b^1, b^2, \ldots, b^n\}$ and $\mathbf{C} = \{c^1, c^2, \ldots, c^n\}$ be two bases for **V**. Let P be the $n \times n$ matrix whose kth column is defined as $p^k \triangleq [b^k]^{\mathbf{C}}$ for $1 \leqslant k \leqslant n$. Then for each vector $x \in \mathbf{V}$, $[x]^{\mathbf{C}} = P[x]^{\mathbf{B}}$.

Proof. Let b and c be the coordinates of x relative to **B** and **C**, respectively. That is $b = [x]^{\mathbf{B}}$ and $c = [x]^{\mathbf{C}}$. Now

$$\begin{aligned}
\sum_{k=1}^{n} c_k c^k &= x \\
&= \sum_{k=1}^{n} b_k b^k \\
&= \sum_{k=1}^{n} b_k \left(\sum_{j=1}^{n} P_{jk} c^j \right) \\
&= \sum_{j=1}^{n} \left(\sum_{k=1}^{n} P_{jk} b_k \right) c^j \\
&= \sum_{j=1}^{n} (Pb)_j c^j \\
&= \sum_{k=1}^{n} (Pb)_k c^k.
\end{aligned}$$

It follows that $c = Pb$ or $[x]^{\mathbf{C}} = P[x]^{\mathbf{B}}$. ■

Refer to Figure 1.6, which depicts the relationship between a vector x in **V** and its various coordinates in $\mathbf{F}^n$.

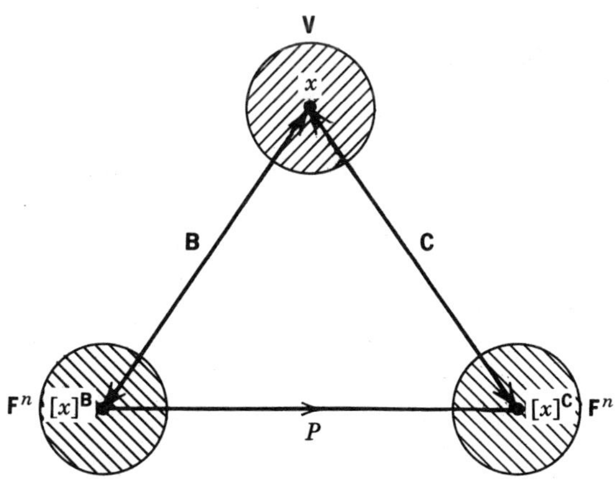

Figure 1.6. The coordinates transformation matrix P.

Exercise 10.1 (Inverse Coordinate Transformation). Use Prop. 10.1 to show that the coordinate transformation matrix P is invertible in which case $[x]^{\mathbf{B}} = P^{-1}[x]^{\mathbf{C}}$. Find a simple expression for the kth column of P^{-1}. ■

An important special case of the coordinate transformation problem involves finding the coordinates of a vector in $\mathbf{F}^n$ relative to some basis $\mathbf{B}$ given the components of the vector. This is tantamount to transforming from the standard basis $\mathbf{I}$ to the desired basis $\mathbf{B}$.

Proposition 10.2 (Coordinates in $\mathbf{F}^n$). Let $\mathbf{B} = \{b^1, b^2, \ldots, b^n\}$ be a basis for $\mathbf{F}^n$, and let B be the $n \times n$ matrix whose kth column is b^k. Then for each vector $x \in \mathbf{F}^n$,

$$[x]^{\mathbf{B}} = B^{-1}x.$$

Proof. Let x be a vector in $\mathbf{F}^n$ and let b be the coordinates of x relative to the basis $\mathbf{B}$. That is, $b = [x]^{\mathbf{B}}$. Then

$$\begin{aligned} x &= \sum_{k=1}^{n} b_k b^k \\ &= Bb \\ &= B[x]^{\mathbf{B}}. \end{aligned}$$

Now B transforms the coordinates of x relative to $\mathbf{B}$ into the components of x. But the components of x are the coordinates of x relative to the standard basis $\mathbf{I}$. Thus B is a coordinate transformation matrix. From Exer. 10.1 it follows that B is invertible. Hence $[x]^{\mathbf{B}} = B^{-1}x$. ■

As an example of an application of Prop. 10.2, consider the following coordinate transformation in the space $\mathbf{R}^2$.

Example 10.1 (Coordinates in $\mathbf{R}^2$). Let $\mathbf{V} = \mathbf{R}^2$ and let $\mathbf{B} = \{b^1, b^2\}$ be a basis where $b^1 = [1, 2]^T$ and $b^2 = [3, 4]^T$. Next, let $x = [-3, 1]^T$ be a vector in $\mathbf{R}^2$. Then from Prop. 10.2

$$\begin{aligned} [x]^{\mathbf{B}} &= B^{-1}x \\ &= \begin{bmatrix} 1 & 3 \\ 2 & 4 \end{bmatrix}^{-1} \begin{bmatrix} -3 \\ 1 \end{bmatrix} \\ &= \begin{bmatrix} -2 & \frac{3}{2} \\ 1 & -\frac{1}{2} \end{bmatrix} \begin{bmatrix} -3 \\ 1 \end{bmatrix} \\ &= \begin{bmatrix} \frac{15}{2} \\ -\frac{7}{2} \end{bmatrix} \end{aligned}$$

■

PROBLEMS

1. Let $\mathbf{V} = \mathbf{F(R, R)}$. Which of the following are subspaces of $\mathbf{V}$? Support your answers by applying Prop. 2.1 or by finding a counterexample to Def. 1.1.

   (a) The set of *periodic* functions of period 2π,

   $$U = \{f \in \mathbf{V} : f(t + 2\pi) = f(t)\}$$

   (b) The set of *nondecreasing* functions,

   $$U = \{f \in \mathbf{V} : s > t \Rightarrow f(s) \geqslant f(t)\}.$$

   (c) The set of functions with *support* S,

   $$U = \{f \in \mathbf{V} : f(t) = 0,\ t \notin S\}.$$

2. Let $\mathbf{V} = \mathbf{F}^{\mathbf{n} \times \mathbf{n}}$. Which of the following are subspaces of $\mathbf{V}$? Support your answers by applying Prop. 2.1 or by finding a counterexample to Def. 1.1.

   (a) The *upper triangular* matrices,

   $$U = \{A \in \mathbf{V} : A_{kj} = 0,\ k > j\}.$$

   (b) The matrices which *commute* with $B \in \mathbf{F}^{\mathbf{n} \times \mathbf{n}}$:

   $$U = \{A \in \mathbf{V} : \mathbf{AB} = \mathbf{BA}\}.$$

   (c) The *nonsingular* matrices. Here $\det(A)$ is the determinant of A,

   $$U = \{A \in \mathbf{V} : \det(A) \neq 0\}.$$

3. Let S be a subset of $\mathbf{F}$ and let $\mathbf{C(S)}$ be the set of continuous functions mapping S into $\mathbf{F}$. Show that $\mathbf{C(S)}$ is a vector space.

4. Consider the following vectors in $\mathbf{R}^3$, where T denotes the transpose,

   $$x^1 = [1, 1, 0]^{\mathrm{T}},$$
   $$x^2 = [0, 1, 1]^{\mathrm{T}},$$
   $$x^3 = [2, 1, -1]^{\mathrm{T}}.$$

   (a) Are $\{x^1, x^2, x^3\}$ linearly independent? If not, express one vector as a linear combination of the others.

   (b) Do $\{x^1, x^2, x^3\}$ span $\mathbf{R}^3$? If not, find a vector in $\mathbf{R}^3$ which is not in the space spanned by $\{x^1, x^2, x^3\}$.

   (c) Do $\{x^1, x^2, x^3\}$ form a basis for $\mathbf{R}^3$? If not, why not?

5. Consider the space $\mathbf{R}^{2\times 3}$.

(a) Find a basis **B** for $\mathbf{R}^{2\times 3}$.
(b) What is the dimension of $\mathbf{R}^{2\times 3}$?
(c) Find $[A]^{\mathbf{B}}$ if

$$A = \begin{bmatrix} 1 & 2 & 3 \\ 2 & 3 & 4 \end{bmatrix}.$$

6. Let $\mathbf{V} = \mathbf{R}^2$ and consider the following two bases for **V**. Let $\mathbf{B} = \{b^1, b^2\}$ where $b^1 = [1, -1]^T$ and $b^2 = [2, 0]^T$, and let $\mathbf{C} = \{c^1, c^2\}$ where $c^1 = [1, 2]^T$ and $c^2 = [2, 1]^T$. Finally, let $x = [-1, 3]^T$.

(a) Find $[x]^{\mathbf{B}}$
(b) Find $[x]^{\mathbf{C}}$
(c) Find P such that $[x]^{\mathbf{C}} = P[x].^{\mathbf{B}}$
(d) Find Q such that $[x]^{\mathbf{B}} = Q[x].^{\mathbf{C}}$
(e) Verify that $PQ = QP = I$.

7. Let S be a subset of **F** and let **P(S)** be the set of *polynomials* mapping S into **F**. Finally, let $\mathbf{P^n(S)}$ be the *polynomials of degree n* mapping S into **F**.

(a) Show that **P(S)** is a vector space.
(b) Show that $\mathbf{P^n(S)}$ is a subspace of **P(S)** for each $n \geqslant 0$.

8. Let $\mathbf{P}^{\mathbf{mn}-1}\mathbf{(F)}$ be the space of $(mn - 1)$-degree polynomials mapping **F** into **F**, and let $\mathbf{F}^{\mathbf{m}\times\mathbf{n}}$ be the space of $m \times n$ matrices. Show that $\mathbf{P}^{\mathbf{mn}-1}\mathbf{(F)}$ and $\mathbf{F}^{\mathbf{m}\times\mathbf{n}}$ are isomorphic.

9. Let $\mathbf{V} = \mathbf{F}^{\mathbf{n}\times\mathbf{n}}$, and let $\mathbf{U}_1$, $\mathbf{U}_2$, and $\mathbf{U}_3$ be the following subspaces of **V**. Show that $\mathbf{V} = \mathbf{U}_1 \oplus \mathbf{U}_2 \oplus \mathbf{U}_3$:

$$\mathbf{U}_1 \triangleq \{A \in \mathbf{V} : A_{kj} = 0,\ k > j\},$$
$$\mathbf{U}_2 \triangleq \{A \in \mathbf{V} : A_{kj} = 0,\ k \neq j\},$$
$$\mathbf{U}_3 \triangleq \{A \in \mathbf{V} : A_{kj} = 0,\ k < j\}.$$

REFERENCES

Halmos, P. R., *Finite Dimensional Vector Spaces*, Van Nostrand Reinhold, Princeton, N.J., 1958.
Noble, B., and J. W. Deniel, *Applied Linear Algebra*, McGraw-Hill, New York, 1977.

2 NORMED VECTOR SPACES

The structure of vector spaces can be refined considerably with the introduction of a measure of the *distance* between two vectors. Once the notion of distance or length is developed, the important concepts of convergence, continuity, and completeness can then be investigated.

2.1. NORMS

In order to develop a measure of the distance between two vectors, we begin by considering the concept of the *length* or *size* of a vector.

Definition 1.1 (Norm). A *norm* on a vector space **V** is a function denoted $\|\cdot\|$ mapping **V** into **R** having the following properties. For each $(\alpha, x, y) \in \mathbf{F} \times \mathbf{V}^2$,

1. $\|x\| \geqslant 0$,
2. $\|x\| = 0 \Leftrightarrow x = 0$,
3. $\|\alpha x\| = |\alpha| \, \|x\|$,
4. $\|x + y\| \leqslant \|x\| + \|y\|$.

A vector space **V** taken together with an appropriately defined norm is called a *normed vector space.* ■

Thus a norm is a nonnegative function mapping **V** into **R** which is zero only at the origin of **V**. The following are examples of normed vector spaces.

Example 1.1 (The Normed Space $\mathbf{F}^n_p$). For the space $\mathbf{F}^n$, let $1 \leqslant p < \infty$. Then the following is a valid norm on $\mathbf{F}^n$:

$$\|x\|_p \triangleq \left(\sum_{k=1}^{n} |x_k|^p \right)^{1/p}.$$

The special case $p = 2$ is of particular significance as we shall see in Chapter 3. We refer to $\|x\|_2$ as the *Euclidean norm* of the vector x. ■

Example 1.2 (The Normed Space $\mathbf{F}^n_\infty$). For the space $\mathbf{F}^n$, the following is also a valid norm:

$$\|x\|_\infty = \max_{k=1}^{n}\{|x_k|\}.$$ ■

The p-norms defined in Exam. 1.1 are called the family of *Holder norms* on $\mathbf{F}^n$. The ∞-norm in Exam. 1.2 is actually a limiting special case of the Holder norms, as can be seen from the following exercise.

Exercise 1.1 (Holder Norms). Let $x \in \mathbf{F}^n$. Show that $\|x\|_p \to \|x\|_\infty$ as $p \to \infty$. ■

We refer to property 4 in Definition 1.1 as the *triangle inequality* property. When $\mathbf{V} = \mathbf{R}^2_2$, it says that no side of a triangle can be longer than the sum of the lengths of the other two sides. Refer to Figure 2.1 for a pictorial representation of this geometric interpretation of the triangle inequality.

Whereas the triangle inequality puts an upper bound on the sum of two vectors, it is also possible to put a lower bound on the difference of two vectors. In particular, consider the following exercise.

Exercise 1.2 (Norm of Difference). Let $\mathbf{V}$ be a normed vector space, and let $(x, y) \in \mathbf{V}^2$. Show that

$$|(\|x\| - \|y\|)| \leqslant \|x - y\|.$$ ■

The previous examples of norms are defined for finite-dimensional vector spaces. Note that in both cases the norm reduces to the absolute value when the

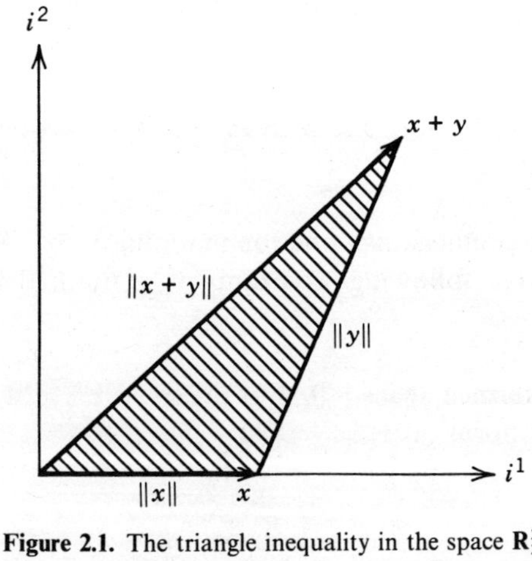

Figure 2.1. The triangle inequality in the space $\mathbf{R}^2_2$.

dimension is $n = 1$. It is also possible to define norms over infinite-dimensional vector spaces as the following example illustrates.

Example 1.3 (The Normed Space $\mathbf{C}_\infty$(T)). Let $T = [\alpha, \beta]$ be a *bounded interval* in **R**. For the space **C(T)** of *continuous functions* mapping T into **R**, the following is a valid norm:

$$\|f\|_\infty = \max_{t \in T}\{|f(t)|\}. \qquad \blacksquare$$

2.2. NATURAL METRICS

A norm is actually a special case of a more general function called a metric. A metric is a function that measures the distance between two points in a set. More specifically, we define a metric as follows.

Definition 2.1 (Metric). A *metric* on a set S is a function denoted d mapping $S \times S$ into **R** having the following properties. For each $(x, y, z) \in S^3$,

1. $d(x, y) \geqslant 0$,
2. $d(x, y) = 0 \Leftrightarrow x = y$,
3. $d(x, y) = d(y, x)$,
4. $d(x, y) \leqslant d(x, z) + d(z, y)$.

A set S taken together with an appropriately defined metric is called a *metric space*. ■

Note that the definitions of a norm and a metric are quite similar. Indeed, the two concepts are closely related, as can be seen from the following result.

Proposition 2.1 (Natural Metric). Let **V** be a normed vector space. Then the following is a valid metric on **V**. We call it the *natural metric* induced by the norm

$$d(x, y) = \|x - y\|.$$

Proof. From the definition of d it is clear that 1 and 2 of Def. 1.1 ⇒ 1 and 2 of Def. 2.1. Next, using 3 of Def. 1.1,

$$\begin{aligned} d(x, y) &= \|x - y\| \\ &= \|(-1)(y - x)\| \\ &= |(-1)|\,\|y - x\| \\ &= \|y - x\| \\ &= d(y, x). \end{aligned}$$

Finally, using the triangle inequality property of the norm we have

$$\begin{aligned} d(x, y) &= \|x - y\| \\ &= \|x - z + z - y\| \\ &\leq \|x - z\| + \|z - y\| \\ &= d(x, z) + d(z, y). \end{aligned}$$

■

Thus the norm of a vector x is really the distance between x and the zero vector 0 as measured by the natural metric. That is,

$$\|x\| = d(x, 0)$$

It follows that the norm of a vector is a measure of the *length* or *size* of the vector. In this sense the norm is analogous to the absolute value function, which measures the size of a scalar.

2.3. CONVERGENCE AND CONTINUITY

The utility of the norm lies in the fact that it allows us to define convergence in a vector space **V** in terms of convergence on the real line **R**.

Definition 3.1 (Convergence). Let $\{x^k\}$ be a sequence of vectors in a normed vector space **V**. Then $\{x^k\}$ *converges* $\Leftrightarrow$ there exists a vector $x \in \mathbf{V}$ called the *limit* of the sequence such that

$$\|x^k - x\| \to 0 \quad \text{as} \quad k \to \infty.$$

In this case we often write $x^k \to x$ as $k \to \infty$ where the norm is understood or left implicit. ■

Note that the limit x must *itself* belong to **V** if $\{x^k\}$ is to qualify as a convergent sequence of vectors. The following example illustrates how convergence in a vector space reduces to convergence on the real line.

Example 3.1. Let $\mathbf{V} = \mathbf{R}_2^3$ where the Euclidean norm ($p = 2$) is used. Consider the following sequence in **V**. For $k = 1, 2, \ldots$, let

$$x_j^k = \frac{jk - 1}{k}, \qquad 1 \leq j \leq 3.$$

As a candidate for the limit consider $x = [1, 2, 3]^T$. Then from Exam. 1.1,

$$\begin{aligned}\|x^k - x\|_2^2 &= \sum_{j=1}^{3} |x_j^k - x|^2 \\ &= \sum_{j=1}^{3} \left|\frac{jk-1}{k} - j\right|^2 \\ &= \sum_{j=1}^{3} \left|\frac{-1}{k}\right|^2 \\ &= \frac{3}{k^2}.\end{aligned}$$

It follows that $x^k \to x$ as $k \to \infty$. ■

An important special case of a convergent sequence arises when the vectors in the sequence are partial sums. The limit vector is then called a convergent series.

Definition 3.2 (Convergent Series). An infinite series, $\sum_{k=1}^{\infty} x^k$, *converges* in a normed vector space $\mathbf{V} \Leftrightarrow$ there exists an $x \in \mathbf{V}$ such that x is the limit of the sequence of partial sums, $s^k = \sum_{j=1}^{k} x^j$. In this case we write

$$\sum_{k=1}^{\infty} x^k = x.$$ ■

Next consider the notion of continuity in a normed vector space. The continuity of a function mapping one normed vector space into another is defined with respect to the norms in the following manner.

Definition 3.3 (Continuity). Let **V** and **W** be normed vector spaces and let $f: \mathbf{V} \to \mathbf{W}$. Then for each $x \in \mathbf{V}$, the function f is *continuous at* $x \Leftrightarrow$ for each $\varepsilon > 0$ there exists a $\delta > 0$ such that

$$\|y - x\|_V < \delta \Rightarrow \|f(y) - f(x)\|_W < \varepsilon.$$

More generally, the function f is *continuous* $\Leftrightarrow f$ is continuous at every point in **V**. ■

Exercise 3.1 (The space C(V, W)). Let **V** and **W** be normed vector spaces, and let **C(V, W)** be the set of *continuous functions* mapping **V** into **W**. Show that **C(V, W)** is a vector space. (*Hint*: Show that it is a subspace of **F(V, W)**). ■

The concepts of convergence and continuity are interrelated as can be seen from the following exercise.

Exercise 3.2. Let $\mathbf{V}$ and $\mathbf{W}$ be normed vector spaces, and let $f:\mathbf{V}\to\mathbf{W}$ be continuous. Show that

$$x^k \to x \quad \text{as} \quad k\to\infty \Rightarrow f(x^k)\to f(x) \quad \text{as} \quad k\to\infty. \qquad ■$$

2.4. OPEN AND CLOSED SETS

The notions of open and closed sets can also be considered within the framework of normed vector spaces. For a concise formulation, it is helpful to first consider an important special case of an open set, namely, the open ball.

Definition 4.1 (Open Ball). Let $\mathbf{V}$ be a normed vector space, and let $x\in\mathbf{V}$. Then for each $\varepsilon > 0$, the *open ball* of radius ε centered at x is denoted $B(x, \varepsilon)$ and is defined as

$$B(x,\varepsilon) \triangleq \{y\in\mathbf{V} : \|y - x\| < \varepsilon\} \qquad ■$$

Thus $B(x, \varepsilon)$ is the set of all vectors in $\mathbf{V}$ whose distance from x is less than ε, as measured by the natural metric. The name *ball* arises from the fact that in the space $\mathbf{R}_2^3$ the set $B(0, 1)$ is in fact a sphere of unit radius centered at the origin. For the space $\mathbf{R}_2^2$ the ball reduces to a disk as shown in Figure 2.2.

Depending upon the nature of the norm used, the open ball sometimes does not represent a geometric ball at all but instead takes the form of a diamond or a square. Consider, in particular, the following exercise.

Exercise 4.1 (Open Ball). Sketch the open unit ball $B(0, 1)$ in the normed vector spaces $\mathbf{R}_1^2$, $\mathbf{R}_2^2$, and $\mathbf{R}_\infty^2$. ■

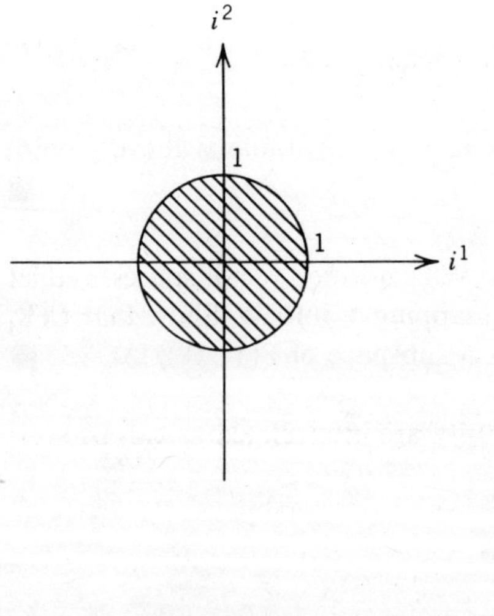

Figure 2.2. The unit ball $B(0, 1)$ in the space $\mathbf{R}_2^2$.

Let S be a subset of a normed vector space $\mathbf{V}$. Every subset S of $\mathbf{V}$ has an interior $\mathring{S}$, a boundary δS, and a closure $\tilde{S}$ associated with it. Each of these sets can be defined in terms of an open ball. We begin with the interior of the set S.

Definition 4.2 (Interior). Let S be a subset of $\mathbf{V}$. A point $x \in \mathbf{V}$ is an *interior point* of $S \Leftrightarrow$ there exists an $\varepsilon > 0$ such that $B(x, \varepsilon) \subseteq S$. The set of all interior points of S is called the *interior of* S and is denoted $\mathring{S}$. ∎

Note that in general $\mathring{S} \subseteq S$. When the interior of a set is equal to the set, we say the set is open.

Definition 4.3 (Open Set). A subset S of a normed vector space $\mathbf{V}$ is *open* $\Leftrightarrow \mathring{S} = S$. ∎

Clearly the open ball $B(x, \varepsilon)$ is an open set as the name suggests. In a similar manner we can define closed sets in terms of something called the closure of the set. Here the notation $\varnothing$ denotes the *empty set*, the set containing no elements.

Definition 4.4 (Closure). Let S be a subset of $\mathbf{V}$. A point $x \in \mathbf{V}$ is a *closure point* of $S \Leftrightarrow$ for each $\varepsilon > 0$, $B(x, \varepsilon) \cap S \neq \varnothing$. The set of all closure points of S is called the *closure of* S and is denoted $\tilde{S}$. ∎

Note that in general $S \subseteq \tilde{S}$. When the closure of a set is equal to the set, we say the set is closed.

Definition 4.5 (Closed Set). A subset S of a normed vector space $\mathbf{V}$ is *closed* $\Leftrightarrow \tilde{S} = S$. ∎

Some sets are open, some are closed, some are both, and some are neither as can be seen from the following exercise.

Exercise 4.2. Let $\mathbf{V} = \mathbf{R}^2$. Find an example of subsets S of $\mathbf{V}$ that are open, closed, both open and closed, and neither open nor closed. ∎

The distinction between open and closed sets has a simple geometric interpretation when the set has a nonempty boundary. We define the boundary of a set as follows.

Definition 4.6 (Boundary). Let S be a subset of $\mathbf{V}$. A point $x \in \mathbf{V}$ is a *boundary point* of $S \Leftrightarrow x \in \tilde{S}$ and $x \notin \mathring{S}$. The set of all boundary points of S is called the *boundary of* S and is denoted δS. ∎

Not every set has a nonempty boundary. However, if $\delta S \neq \varnothing$ then S is open if and only if it does not contain *any* of its boundary. Similarly, S is closed if and only if it does contain *all* of its boundary. A more concise way to characterize the boundary δS is to use the notion of the complement of one set relative to another. If S is a subset of $\mathbf{V}$, then the *complement of* S with respect to $\mathbf{V}$ is denoted $\mathbf{V} - S$ and is defined as the set of all points in $\mathbf{V}$ that are not in S. That is,

$$\mathbf{V} - S \triangleq \{x \in \mathbf{V} : x \notin S\}.$$

The boundary δS then has a simple interpretation. It is the complement of the interior of S relative to the closure of S. That is,

$$\delta S = \bar{S} - \mathring{S}.$$

Exercise 4.3 (Open and Closed). Show that the entire normed vector space $\mathbf{V}$ and the empty set $\varnothing$ are each both open and closed. Thus they have no boundary. ■

Exercise 4.4 (Complement). Let S be a subset of a normed vector space $\mathbf{V}$. Show that S is open $\Leftrightarrow$ its complement $\mathbf{V} - S$ is closed. ■

Exercise 4.5 (Boundary). Let S be a subset of a normed vector space $\mathbf{V}$. Show that the boundary of S is equal to the boundary of its complement. ■

As was the case with convex sets, the *intersection* of an arbitrary collection of closed sets is a closed set. The dual of this statement also holds, namely, the *union* of an arbitrary collection of open sets is an open set. The importance of closed sets lies in the fact that they contain all of their limit points.

Proposition 4.1 (Closed Sets). Let S be a subset of a normed vector space $\mathbf{V}$. Then S is closed $\Leftrightarrow$ S contains the limits of all of its convergent sequences.

Proof. Let $\{x^k\}$ be a convergent sequence in S and let x be its limit point. Then x is a closure point of S. Hence if S is closed, then $x \in S$. Suppose S is not closed. Then there exists a closure point x of S which is not in S. Let $\{x^k\}$ be any sequence in S such that

$$\|x^k - x\| < \frac{1}{k}, \qquad k \geqslant 1.$$

Then $\{x^k\}$ converges to x, but $x \notin S$. ■

The concept of closure can be combined with the notion of the span of a set. The combination allows us to introduce a useful new concept called the density of a subset.

Definition 4.7 (Density). A subset S of a normed vector space $\mathbf{V}$ is *dense* in $\mathbf{V} \Leftrightarrow$ the closure of the subspace spanned by S is $\mathbf{V}$. ■

Recall that the span of a set S is the set of all linear combinations of vectors in S. Thus if S is dense in $\mathbf{V}$, then each vector in $\mathbf{V}$ can be represented as the limit of a convergent sequence of linear combinations of vectors in S. Typically, the limit of a convergent sequence of linear combinations of vectors is an infinite series of vectors.

Exercise 4.6. Let $\mathbf{V}$ be a finite-dimensional normed vector space, and let $\mathbf{X}$ be any basis for $\mathbf{V}$. Show that $\mathbf{X}$ is dense in $\mathbf{V}$. ■

Exercise 4.7. Let $\mathbf{V}$ be the vector space of even periodic functions of period 2π that can be represented by a Fourier series. Show that the sequence $\{\cos(kt)\}$ for $k \geqslant 0$ is dense in $\mathbf{V}$. ■

2.5. BANACH SPACES

The concept of convergence is an important one since many engineering problems are solved by constructing a sequence of increasingly accurate *approximate* solutions $\{x^k\}$. If the sequence converges, the limit x is then an exact solution. To apply this iterative solution technique, it is helpful to have access to a simple test for convergence. Unfortunately, the definition of convergence is not entirely satisfactory for this purpose. Indeed, in order to successfully apply the definition, one must know a priori the limit point x. To avoid this practical shortcoming, we introduce the following alternative classification of sequences.

Definition 5.1 (Cauchy Sequence). A sequence $\{x^k\}$ in a normed vector space $\mathbf{V}$ is a *Cauchy sequence* $\Leftrightarrow$ for each $\varepsilon > 0$ there exists an $N > 0$ such that

$$(m, n) \geqslant (N, N) \Rightarrow \|x^m - x^n\| < \varepsilon.$$ ■

Note that the above classification of sequences makes no mention of a limit point. Hence it can be applied using only knowledge of how the general terms of the sequence are formulated. Convergent sequences include Cauchy sequences as a special case as is evident from the following exercise.

Exercise 5.1. Let $\mathbf{V}$ be a normed vector space. Show that every convergent sequence in $\mathbf{V}$ is a Cauchy sequence. ■

Although every convergent sequence is a Cauchy sequence, the converse is not true. Indeed, consider the following counterexample.

Example 5.1. Let $\mathbf{V} = \mathbf{P}_1([0, 1])$, the space of polynomials mapping $[0, 1]$ into $\mathbf{R}$ with the norm defined as follows:

$$\|p\|_1 = \int_0^1 |p(t)|\, dt.$$

As a sequence of vectors in **V** consider $\{p^k\}$ where p^k is the following polynomial:

$$p^k(t) = \sum_{j=1}^{k} \frac{t^{j-1}}{(j-1)!}.$$

First we show that $\{p^k\}$ is a Cauchy sequence in **V**. For convenience, suppose $m > n$. Then

$$\begin{aligned}
\| p^m - p^n \|_1 &= \int_0^1 |p^m(t) - p^n(t)| \, dt \\
&= \int_0^1 \left| \sum_{j=n+1}^{m} \frac{t^{j-1}}{(j-1)!} \right| dt \\
&= \int_0^1 \sum_{j=n+1}^{m} \frac{t^{j-1}}{(j-1)!} \, dt \\
&= \sum_{j=n+1}^{m} \int_0^1 \frac{t^{j-1}}{(j-1)!} \, dt \\
&= \sum_{j=n+1}^{m} \frac{1}{j!} \\
&\leqslant \sum_{j=n+1}^{\infty} \frac{1}{j!} \\
&= \sum_{j=0}^{\infty} \frac{1}{j!} - \sum_{j=0}^{n} \frac{1}{j!} \\
&= e - \sum_{j=0}^{n} \frac{1}{j!}.
\end{aligned}$$

Clearly for each $\varepsilon > 0$ there exists an $N > 0$ such that $(m, n) \geqslant (N, N) \Rightarrow \| p^m - p^n \|_1 < \varepsilon$. Thus $\{p^k\}$ is a Cauchy sequence in $\mathbf{V} = \mathbf{P}_1([0, 1])$. However $\{p^k\}$ does not converge in **V** because the limit point $p(t) = e^t$ is *not* a polynomial and is therefore not a vector in **V**. ■

Although this counterexample shows that the Cauchy criterion is generally weaker than the convergence criterion, there are many normed vector spaces within which the Cauchy criterion is equivalent to the convergence criterion. Indeed, these spaces are of sufficient importance that they are given a special name. They are called complete normed vector spaces or Banach spaces.

Definition 5.2 (Banach Space). A normed vector space **V** is *complete* $\Leftrightarrow$ every Cauchy sequence converges in **V**. A complete normed vector space is called a *Banach space*. ■

Clearly, the space $\mathbf{P_1([0,1])}$ is not a Banach space. However, the following important spaces are complete and are therefore examples of Banach spaces.

Example 5.2 (The Banach Space $\mathbf{C_\infty(T)}$). The space $\mathbf{C_\infty(T)}$ of continuous functions mapping T into $\mathbf{R}$ defined in Exam. 1.3 is a Banach space. ■

Example 5.3 (The Banach Space $\mathbf{F^n_p}$). The spaces $\mathbf{F^n_p}$ for $1 \leqslant p \leqslant \infty$ defined in Exam. 1.1 and Exam. 1.2 are Banach spaces. ■

Exercise 5.2. Let $\mathbf{V}$ and $\mathbf{W}$ be Banach spaces with norms $\|\cdot\|_{\mathbf{V}}$ and $\|\cdot\|_{\mathbf{W}}$, respectively. Show that $\mathbf{V} \times \mathbf{W}$ is also a Banach space with $\|(v, w)\| = \|v\|_V + \|w\|_W$. ■

The fact that $\mathbf{F^n_p}$ and spaces like it are Banach spaces is a consequence of the following fundamental result.

Proposition 5.1 (Banach Spaces). Every finite-dimensional normed vector space $\mathbf{V}$ is a Banach space.

Proof. We proceed by induction. Suppose $\mathbf{V}$ is one-dimensional. Let $\{b^1\}$ be a basis for $\mathbf{V}$. Given a sequence $\{x^k\}$ in $\mathbf{V}$, for each x^k we have $x^k = a_k b^1$ for some $a_k \subset \mathbf{F}$. Hence $\{x^k\}$ converges in $\mathbf{V} \leftrightarrow \{a_k\}$ converges in $\mathbf{F}$. But every field $\mathbf{F}$ is complete in the sense of Def. 5.2. Hence every one-dimensional normed vector space $\mathbf{V}$ is complete. Next, suppose every $(n-1)$-dimensional normed vector space is complete. Let $\mathbf{V}$ be n-dimensional, and let $\mathbf{B} = \{b^1, b^2, \ldots, b^n\}$ be a basis for $\mathbf{V}$. Next, let $1 \leqslant m \leqslant n$, and let $\mathbf{V_m}$ be the $(n-1)$-dimensional subspace of $\mathbf{V}$ spanned by $\{b^1, \ldots, b^{m-1}, b^{m+1}, \ldots, b^n\}$. Define

$$\beta_m = \inf\{\|x - b^m\|,\ x \in \mathbf{V_m}\}.$$

Here inf represents the *infimum* or greatest lower bound. Clearly, $\beta_m \geqslant 0$. Suppose $\beta_m = 0$. Then there exists a sequence $\{x^k\}$ in $\mathbf{V_m}$ such that $x^k \to b^m$ as $k \to \infty$. Thus $\{x^k\}$ converges in $\mathbf{V}$. Hence $\{x^k\}$ is a Cauchy sequence. But $\{x^k\}$ is a subset of $\mathbf{V_m}$, and the space $\mathbf{V_m}$ is complete. Therefore x^k converges in $\mathbf{V_m}$, in which case $b^m \in \mathbf{V_m}$. But this contradicts the fact that the basis B is a linearly independent set of vectors. Thus $\beta_m > 0$ for $1 \leqslant m \leqslant n$. Let $\beta > 0$ be defined as

$$\beta = \min_{m=1}^{n}\{\beta_m\}.$$

Next, let $\{x^k\}$ be a Cauchy sequence in $\mathbf{V}$. Then for each x^k there exists a set of coordinates $a^k \in \mathbf{F^n}$ such that

$$x^k = \sum_{i=1}^{n} a_i^k b^i.$$

Next, let $k > j$, and let $1 \leqslant m \leqslant n$. Then, using the definition of β, we have the following inequality:

$$\begin{aligned}
\|x^k - x^j\| &= \left\| \sum_{i=1}^{n} a_i^k b^i - \sum_{i=1}^{n} a_i^j b^i \right\| \\
&= \left\| \sum_{i=1}^{n} (a_i^k - a_i^j) b^i \right\| \\
&= \left\| (a_m^k - a_m^j) b^m + \sum_{i \neq m} (a_i^k - a_i^j) b^i \right\| \\
&= \left\| (a_m^k - a_m^j) \left(b^m - \sum_{i \neq m} \frac{a_i^k - a_i^j}{a_m^j - a_m^k} b^i \right) \right\| \\
&\geqslant (a_m^k - a_m^j)\beta.
\end{aligned}$$

Since $\{x^k\}$ is a Cauchy sequence, it follows that $\{a_m^k\}$ is a Cauchy sequence in $\mathbf{F}$ for each m. Since $\mathbf{F}$ is complete, there exists an $a \in \mathbf{F}^n$ such that $a^k \to a$ as $k \to \infty$. Let x be the vector in $\mathbf{V}$ whose coordinates are a. That is,

$$x = \sum_{j=1}^{n} a_j b^j.$$

Clearly, $x \in \mathbf{V}$. Furthermore, the Cauchy sequence $\{x^k\}$ converges to x, for

$$\begin{aligned}
\|x^k - x\| &= \left\| \sum_{j=1}^{n} a_j^k b^j - \sum_{j=1}^{n} a_j b^j \right\| \\
&= \left\| \sum_{j=1}^{n} (a_j^k - a_j) b^j \right\| \\
&\leqslant \sum_{j=1}^{n} \|(a_j^k - a_j) b^j\| \\
&= \sum_{j=1}^{n} (|a_j^k - a_j|) \|b^j\|
\end{aligned}$$

It follows that $\mathbf{V}$ contains the limits of all its Cauchy sequences. Hence $\mathbf{V}$ is a Banach space. ■

If we examine subspaces of Banach spaces, we find that the complete property is equivalent to the closed property as can be seen from the following exercise.

Exercise 5.3 (Closure and Completeness). Let $\mathbf{U}$ be a subspace of a Banach space $\mathbf{V}$. Show that $\mathbf{U}$ is a Banach space $\Leftrightarrow$ $\mathbf{U}$ is closed. ■

2.6. SUMMABLE SEQUENCES

An important class of Banach spaces can be generated from what are called the summable sequences. Let $\ell_{\mathbf{p}}$ be the set of all sequences $\{a_k\}$ of scalars in $\mathbf{F}$ which are bounded in the following sense

$$\sum_{k=1}^{\infty} |a_k|^p < \infty.$$

Define vector addition and scalar multiplication on $\ell_{\mathbf{p}}$ as follows. Let $(\alpha, a, b) \in \mathbf{F} \times \ell_{\mathbf{p}}^2$. Then for $1 \leqslant k < \infty$,

$$(a + b)_k \triangleq a_k + b_k,$$

$$(\alpha a)_k \triangleq \alpha a_k.$$

The zero vector in this case is the sequence of zero scalars. That is, $0_k = 0_F$ for $1 \leqslant k < \infty$. It is clear from the definition of scalar multiplication that $(\alpha, a) \in \mathbf{F} \times \ell_{\mathbf{p}} \Rightarrow \alpha a \in \ell_{\mathbf{p}}$. Hence the set $\ell_{\mathbf{p}}$ is closed with respect to scalar multiplication. In Prop. 6.3 we show that $\ell_{\mathbf{p}}$ is also closed with respect to vector addition and is therefore a vector space. We call $\ell_{\mathbf{p}}$ the space of *summable sequences.* As a candidate for a norm on $\ell_{\mathbf{p}}$, consider

$$\|a\|_p = \left(\sum_{k=1}^{\infty} |a_k|^p\right)^{1/p}.$$

To show that $\|\cdot\|_p$ is indeed a valid norm, we must show that it satisfies the four fundamental properties of a norm. Let $(\alpha, a) \in \mathbf{F} \times \ell_{\mathbf{p}}$. Clearly, $\|a\|_p \geqslant 0$, and also $\|a\|_p = 0 \Leftrightarrow a = 0$. To verify the third property we have

$$\begin{aligned}
\|\alpha a\|_p &= \left(\sum_{k=1}^{\infty} |\alpha a_k|^p\right)^{1/p} \\
&= \left(\sum_{k=1}^{\infty} |\alpha|^p |a_k|^p\right)^{1/p} \\
&= \left(|\alpha|^p \sum_{k=1}^{\infty} |a_k|^p\right)^{1/p} \\
&= |\alpha| \left(\sum_{k=1}^{\infty} |a_k|^p\right)^{1/p} \\
&= |\alpha| \|a\|_p.
\end{aligned}$$

It remains to verify the last property of the norm, the triangle inequality. Inequalities can be very useful in engineering applications as they often allow us to place upper bounds on quantities which we would like to show are in some sense small. For example, the quantity might represent the error in an estimate of the exact solution to a problem. In order to develop the triangle inequality in the space ℓ_p we first digress to develop two preliminary inequalities, each of which is useful in its own right.

Proposition 6.1 (Scalar Inequality). Let $\alpha \geqslant 0$ and $\beta \geqslant 0$, and let $0 \leqslant \lambda \leqslant 1$. Then

$$\alpha^\lambda \beta^{1-\lambda} \leqslant \lambda\alpha + (1-\lambda)\beta.$$

Proof. Direct substitution verifies that the inequality holds when $\lambda = 0$ and when $\lambda = 1$. Let $\lambda \in (0, 1)$, and consider the following function $f: [0, \infty) \to R$:

$$f(t) = t^\lambda - \lambda(t+1) - 1.$$

First, we will show that $f(t) \leqslant 0$. The first derivative of $f(t)$ is $f^{(1)}(t) = \lambda(t^{\lambda-1} - 1)$. Thus

$$\begin{aligned} f^{(1)}(t) = 0 &\Leftrightarrow t^{\lambda-1} = 1 \\ &\Leftrightarrow (\lambda - 1)\ln(t) = 0 \\ &\Leftrightarrow \ln(t) = 0 \\ &\Leftrightarrow t = 1. \end{aligned}$$

Next, the second derivative of $f(t)$ is $f^{(2)}(t) = \lambda(\lambda-1)t^{\lambda-2}$. Thus $f^{(2)}(t) < 0$ for $t > 0$. It follows that $f(t) \leqslant f(1) = 0$. Hence from the definition for $f(t)$ we have

$$t^\lambda \leqslant \lambda(t-1) + 1.$$

Next, let $\alpha \geqslant 0$, let $\beta > 0$, and let $t = \alpha/\beta$. Then, substituting $t = \alpha/\beta$ into the last inequality yields

$$\left(\frac{\alpha}{\beta}\right)^\lambda \leqslant \lambda\left(\frac{\alpha}{\beta} - 1\right) + 1.$$

If we now multiply both sides of this inequality by $\beta > 0$, we get the inequality in Prop. 6.1. Note that the inequality in Prop. 6.1 is also valid when $\beta = 0$ as can be verified by direct substitution. ∎

Proposition 6.1 is useful as a tool for establishing other inequalities. The first example of its use is in establishing the following family of inequalities known as the *Holder inequalities* in the space ℓ_p.

Proposition 6.2 (Holder Inequality in ℓ_p). Let $(p, q) \in (1, \infty)^2$ with $1/p + 1/q = 1$. Then for each $(a, b) \in \ell_p \times \ell_q$,

$$\sum_{k=1}^{\infty} |a_k b_k| \leqslant \|a\|_p \|b\|_q.$$

Proof. Because the inequality in Prop. 6.2 clearly holds if $a = 0$ or if $b = 0$, suppose $a \neq 0$ and $b \neq 0$. For $1 \leqslant k < \infty$ let $\alpha = (|a_k|/\|a\|_p)^p$, let $\beta = (|\beta_k|/\|b\|_q)^q$, and let $\lambda = 1/p$ where p and q are as in Prop. 6.2. Then applying Prop. 6.1,

$$\frac{|a_k|}{\|a\|_p}\left(\frac{|b_k|}{\|b\|_q}\right)^{q(1-1/p)} \leqslant \frac{1}{p}\left(\frac{|a_k|}{\|a\|_p}\right)^p + \left(1 - \frac{1}{p}\right)\left(\frac{|b_k|}{\|b\|_q}\right)^q.$$

Since $1/p + 1/q$, this simplifies to

$$\frac{|a_k b_k|}{\|a\|_p \|b\|_q} \leqslant \frac{1}{p}\left(\frac{|a_k|}{\|a\|_p}\right)^p + \frac{1}{q}\left(\frac{|b_k|}{\|b\|_q}\right)^q.$$

We can now sum over $1 \leqslant k < \infty$. Let S represent the resulting left-hand side. That is, $S = (\sum_{k=1}^{\infty} |a_k b_k|)/(\|a\|_p \|b\|_q)$. Then

$$\begin{aligned} S &\leqslant \frac{1}{p}\sum_{k=1}^{\infty}\left(\frac{|a_k|}{\|a\|_p}\right)^p + \frac{1}{q}\sum_{k=1}^{\infty}\left(\frac{|b_k|}{\|b\|_q}\right)^q \\ &= \frac{1}{p\|a\|_p^p}\sum_{k=1}^{\infty}|a_k|^p + \frac{1}{q\|b\|_q^q}\sum_{k=1}^{\infty}|b_k|^q \\ &= \frac{1}{p} + \frac{1}{q} \\ &= 1. \end{aligned}$$

■

In view of Prop. 6.2, we are now in a position to establish the triangle inequality for the ℓ_p norm. Note that this also verifies that the set ℓ_p is closed with respect to vector addition.

Proposition 6.3 (Minkowski). Let $p \in [1, \infty)$, and let $(a, b) \in \ell_p \times \ell_p$. Then $a + b \in \ell_p$. Moreover,

$$\|a + b\|_p \leqslant \|a\|_p + \|b\|_p.$$

Proof. First consider the special case $p = 1$. Using the triangle inequality in **F**, we have

$$\|a + b\|_1 = \sum_{k=1}^{\infty} |a_k + b_k|$$

$$\leqslant \sum_{k=1}^{\infty} (|a_k| + |b_k|)$$

$$= \sum_{k=1}^{\infty} |a_k| + \sum_{k=1}^{\infty} |b_k|$$

$$= \|a\|_1 + \|b\|_1.$$

Next, let $p \in (1, \infty)$, and let $\binom{p}{j}$ be the *binomial coefficient.* That is, $\binom{p}{j} = p!/(j!(p-j)!)$. Then for $(a, b) \in \ell_{\mathbf{p}}^2$,

$$\begin{aligned}
\|a + b\|_p^p &= \sum_{k=1}^{\infty} |a_k + b_k|^p \\
&\leqslant \sum_{k=1}^{\infty} (|a_k| + |b_k|)^p \\
&= \sum_{k=1}^{\infty} \sum_{j=0}^{p} \binom{p}{j} |a_k|^j |b_k|^{p-j} \\
&= \sum_{j=0}^{p} \binom{p}{j} \sum_{k=1}^{\infty} |a_k|^j |b_k|^{p-j} \\
&= \|a\|_p^p + \|b\|_p^p + \sum_{j=1}^{p-1} \binom{p}{j} \sum_{k=1}^{\infty} |a_k|^j |b_k|^{p-j}.
\end{aligned}$$

Next, for $1 \leqslant j \leqslant p-1$ define the sequences $a_k^j = |a_k|^j$ and $b_k^j = |b_k|^{p-j}$ for $1 \leqslant k < \infty$. Then $a \in \ell_{\mathbf{p}} \Rightarrow a^j \in \ell_{\mathbf{p/j}}$. Similarly, $b \in \ell_{\mathbf{p}} \Rightarrow b^j \in \ell_{\mathbf{p/(p-j)}}$. Thus, using the Holder inequality of Prop. 6.2,

$$\begin{aligned}
\|a + b\|_p^p &\leqslant \|a\|_p^p + \|b\|_p^p + \sum_{j=1}^{p-1} \binom{p}{j} \sum_{k=1}^{\infty} |a_k^j b_k^j| \\
&\leqslant \|a\|_p^p + \|b\|_p^p + \sum_{j=1}^{p-1} \binom{p}{j} \|a^j\|_{p/j} \|b^j\|_{p/(p-j)} \\
&= \|a\|_p^p + \|b\|_p^p + \sum_{j=1}^{p-1} \binom{p}{j} \|a^j\|_p^j \|b^j\|_p^{p-j} \\
&= \sum_{j=0}^{p} \binom{p}{j} \|a^j\|_p^j \|b^j\|_p^{p-j} \\
&= (\|a\|_p + \|b\|_p)^p.
\end{aligned}$$

■

Thus $\ell_{\mathbf{p}}$ is a valid normed vector space for $1 \leqslant p < \infty$. An important limiting special case of the family of $\ell_{\mathbf{p}}$ spaces is the space ℓ_∞, which is the space of sequences that are bounded in the sense that $\|a\|_\infty < \infty$ where

$$\|a\|_\infty = \sup_{k=1}^{\infty} \{|a_k|\}.$$

Here sup denotes the *supremum* or smallest upper bound. We call ℓ_∞ the space of *uniformly bounded sequences.* Note that we cannot use the max operator in the definition of $\|\cdot\|_\infty$. Indeed, since there are an infinite number of terms, it is possible that a maximum will never be achieved even when the sequence is bounded. The following exercise illustrates this point with a simple example.

Exercise 6.1. Consider the sequence $\{a_k\}$ where $a_k = 1 - 1/k$ for $k \geqslant 1$. Show that $\|a\|_\infty = 1$, but that $\max_{k=1}^{\infty}\{|a_k|\}$ does not exist. ■

Exercise 6.2. Consider the sequence $\{a_k\}$ where $a_k = 1$ for $k \geqslant 1$. Show that $a \in \ell_\infty - \ell_1$. ■

The family of normed vector spaces ℓ_p for $p \in [1, \infty]$ is important because each member of this family is a complete normed vector space as can be seen from the following result. The interested reader can refer to Luenberger (1969) for a proof.

Proposition 6.4 (The Banach Space ℓ_p). For each $p \in [1, \infty]$ the normed vector space ℓ_p is a Banach space. ■

In view of Exer. 6.2, it is evident that the space ℓ_∞ is larger than the space ℓ_1. More generally, there is a simple strict inclusion relationship between the ℓ_p spaces as can be seen from the following result.

Proposition 6.5 (ℓ_p Inclusion). Let $a \in \ell_1$. Then $\in \ell_p$ for $1 \leqslant p \leqslant \infty$. Moreover,

$$\|a\|_\infty \leqslant \|a\|_p \leqslant \|a\|_1.$$

Proof. Let $a \in \ell_1$, and let $n \geqslant 1$. Then for each $p \in [1, \infty)$ we have

$$\begin{aligned}\sum_{k=1}^{n} |a_k|^p &\leqslant \left(\sum_{k=1}^{n} |a_k|\right)^p \\ &\leqslant \left(\sum_{k=1}^{\infty} |a_k|\right)^p \\ &= \|a\|_1^p.\end{aligned}$$

Because this inequality holds for all $1 \leqslant n < \infty$, let $n \to \infty$ and then take the $1/p$ power of both sides. Then $\|a\|_p \leqslant \|a\|_1$. Hence ℓ_1 is a subspace of ℓ_p. Next, suppose $a \in \ell_p$. Then for each $k \geqslant 1$:

$$\begin{aligned}|a_k| &\leqslant (|a_k|^p)^{1/p} \\ &\leqslant \left(\sum_{j=1}^{\infty} |a_j|^p\right)^{1/p} \\ &= \|a\|_p.\end{aligned}$$

Thus $\|a\|_\infty \leqslant \|a\|_p$. ■

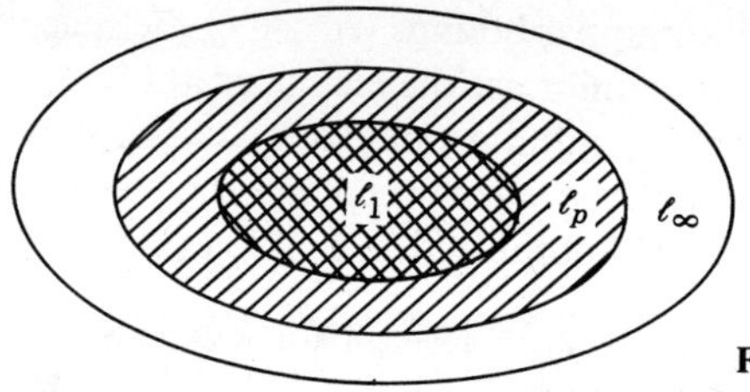

Figure 2.3. Inclusion relations for the ℓ_p spaces.

Refer to Figure 2.3 for a pictorial summary of the strict inclusion properties of the ℓ_p spaces. Notice that ℓ_1 is a subspace of ℓ_p and ℓ_p is a subspace of ℓ_∞ for $1 \leqslant p \leqslant \infty$.

2.7. INTEGRABLE FUNCTIONS

A second important class of Banach spaces can be generated from the Lebesgue-integrable functions. To that end let $T = [\alpha, \beta]$ be an interval in $\mathbf{R}$, and let $w\colon T \to (0, \infty)$ be a positive weighting function. For each $p \in [1, \infty)$ consider the set $\mathbf{L_p(T, w)}$ of functions mapping T into $\mathbf{F}$ which are Lebesgue-integrable and bounded in the following sense.

$$\int_T |f(t)|^p w(t)\, dt < \infty.$$

We define vector addition and scalar multiplication on $\mathbf{L_p(T, w)}$ as follows. Let $(\alpha, f, g) \in \mathbf{F} \times \mathbf{L_p^2(T, w)}$. Then for $t \in T$,

$$(f + g)(t) \triangleq f(t) + g(t),$$
$$(\alpha f)(t) \triangleq \alpha f(t).$$

The zero vector in this case is the function which is identically zero on T. That is, $0(t) = \mathbf{0_F}$ for $t \in T$. It is clear from the definition of scalar multiplication that $(\alpha, f) \in \mathbf{F} \times \mathbf{L_p(T, w)} \Rightarrow \alpha f \in \mathbf{L_p(T, w)}$. Hence the set $\mathbf{L_p(T, w)}$ is closed with respect to scalar multiplication. Later we show that it is also closed with respect to vector addition and is therefore a vector space. We call $\mathbf{L_p(T, w)}$ the space of *Lebesgue-integrable* functions. As a candidate for a norm on $\mathbf{L_p(T, w)}$, consider,

$$\|f\|_p = \left(\int_T |f(t)|^p w(t)\, dt\right)^{1/p}.$$

To show that $\|\cdot\|_p$ is a valid norm, we must again show that it satisfies the four fundamental properties of a norm. Clearly, for each $f \in \mathbf{L_p(T, w)}$ we have $\|f\|_p \geqslant 0$. However, it is *not* true that $\|f\|_p = 0 \Leftrightarrow f = 0$. Indeed, $\|f\|_p = 0$ if and only if f is identically equal to zero *except* possibly on a subset of T of *Lebesgue measure zero* (for example, a countable set of points in T). In order to circumvent this problem we introduce the notion of equivalence classes of functions.

Definition 7.1 (Equivalence Class). For each $f \in \mathbf{L_p(T, w)}$ the *equivalence class* generated by f is denoted $E(f)$ and defined

$$E(f) = \{g \in \mathbf{L_p(T, w)} : g(t) = f(t) \text{ almost everywhere}\}.$$

Here *almost everywhere* is everywhere except possibly on a set of Lebesgue measure zero. ■

Thus the equivalence class of a function f is the set of all functions which are equal to f everywhere except possibly on a set of Lebesgue measure zero. Sets of Lebesgue measure zero include, for example, sets of isolated points. Integrating a function over a set of measure zero contributes nothing to the integral. Hence $\|f - g\| = 0$ for all f and g belonging to the same equivalence class. As such, if we take as vectors in $\mathbf{L_p(T, w)}$ not functions but *equivalence classes* of functions, then

$$\|f\|_p = 0 \Leftrightarrow f = 0,$$

as required by the second property of the norm. To establish the third property of the norm, let $(\alpha, f) \in \mathbf{F} \times \mathbf{L_p(T, w)}$. Then,

$$\begin{aligned}
\|\alpha f\|_p &= \left(\int_T |\alpha f(t)|^p w(t)\, dt\right)^{1/p} \\
&= \left(\int_T |\alpha|^p |f(t)|^p w(t)\, dt\right)^{1/p} \\
&= \left(|\alpha|^p \int_T |f(t)|^p w(t)\, dt\right)^{1/p} \\
&= |\alpha| \left(\int_T |f(t)|^p w(t)\, dt\right)^{1/p} \\
&= |\alpha| \|f\|_p.
\end{aligned}$$

It remains to show that $\|\cdot\|_p$ satisfies the triangle inequality. To do this, we must first digress to establish the integrable function version of the Holder inequality.

Proposition 7.1 (Holder Inequality in $\mathbf{L_p(T, w)}$). Let $(p, q) \in (1, \infty)^2$ with $1/p + 1/q = 1$. Then for each $(f, g) \in \mathbf{L_p(T, w)} \times \mathbf{L_q(T, w)}$

$$\int_T |f(t)g(t)| w(t)\, dt \leqslant \|f\|_p \|g\|_q.$$

Proof. Because direct substitution verifies that the inequality holds when $f = 0$ and when $g = 0$, suppose $f \neq 0$ and $g \neq 0$. Let $\alpha = (|f(t)|/\|f\|_p)^p$, let $\beta = (|g(t)|/\|g\|_q)^q$, and let $\lambda = 1/p$ where p and q are as in Prop 7.1. Then using the scalar inequality of Prop. 6.1,

$$\begin{aligned}\frac{|f(t)g(t)|}{\|f\|_p\|g\|_q} &= \frac{|f(t)|}{\|f\|_p}\left(\frac{|g(t)|}{\|g\|_q}\right)^{q(1-1/p)} \\ &\leqslant \frac{1}{p}\left(\frac{|f(t)|}{\|f\|_p}\right)^p + \left(1 - \frac{1}{p}\right)\left(\frac{|g(t)|}{\|g\|_q}\right)^q \\ &\leqslant \frac{1}{p}\left(\frac{|f(t)|}{\|f\|_p}\right)^p + \frac{1}{q}\left(\frac{|g(t)|}{\|g\|_q}\right)^q.\end{aligned}$$

If we now multiply both sides by $w(t) > 0$ and integrate over T, this yields

$$\begin{aligned}\frac{\int_T |f(t)g(t)|w(t)\,dt}{\|f\|_p\|g\|_q} &\leqslant \frac{1}{p} + \frac{1}{q} \\ &= 1.\end{aligned}$$

■

The triangle inequality can now be established in $\mathbf{L_p(T, w)}$. Note that this also verifies that the set $\mathbf{L_p(T, w)}$ is closed with respect to vector addition.

Proposition 7.2 (Minkowski). Let $p \in [1, \infty)$, and let $(f, g) \in \mathbf{L_p^2(T, w)}$. Then $f + g \in \mathbf{L_p(T, w)}$. Moreover,

$$\|f + g\|_p \leqslant \|f\|_p + \|g\|_p.$$

Proof. For $p = 1$, the inequality in Prop. 7.2 follows directly from the triangle inequality in $\mathbf{F}$. Let $p \in (1, \infty)$, and let $\binom{p}{k}$ be the binomial coefficient. Then,

$$\begin{aligned}\|f + g\|_p^p &= \int_T |f(t) + g(t)|^p w(t)\,dt \\ &\leqslant \int_T (|f(t)| + |g(t)|)^p w(t)\,dt \\ &= \int_T \sum_{k=0}^{p} \binom{p}{k} |f(t)|^k |g(t)|^{p-k} w(t)\,dt \\ &= \sum_{k=0}^{p} \binom{p}{k} \int_T |f(t)|^k |g(t)|^{p-k} w(t)\,dt \\ &= \|f\|_p^p + \|g\|_p^p + \sum_{k=1}^{p-1} \binom{p}{k} \int_T |f(t)|^k |g(t)|^{p-k} w(t)\,dt.\end{aligned}$$

Next, for $1 \leqslant k \leqslant p-1$ define $f^k(t) = |f(t)|^k$ and $g^k(t) = |g(t)|^{p-k}$. Then $f \in \mathbf{L_p(T, w)} \Rightarrow f^k \in \mathbf{L_{p/k}(T, w)}$. Similarly, $g \in \mathbf{L_p(T, w)} \Rightarrow g^k \in \mathbf{L_{p/(p-k)}(T, w)}$. Thus, using the Holder inequality of Prop. 7.1,

$$\begin{aligned}
\|f+g\|_p^p &= \|f\|_p^p + \|g\|_p^p + \sum_{k=1}^{p-1} \binom{p}{k} \int_T |f^k(t) g^k(t)| w(t)\, dt \\
&\leqslant \|f\|_p^p + \|g\|_p^p + \sum_{k=1}^{p-1} \binom{p}{k} \|f^k\|_{p/k} \|g^k\|_{p/(p-k)} \\
&\leqslant \|f\|_p^p + \|g\|_p^p + \sum_{k=1}^{p-1} \binom{p}{k} \|f^k\|_p^k \|g^k\|_p^{p-k} \\
&\leqslant \sum_{k=0}^{p} \binom{p}{k} \|f^k\|_p^k \|g^k\|_p^{p-k} \\
&= (\|f\|_p + \|g\|_p)^p.
\end{aligned}$$

■

Thus $\mathbf{L_p(T, w)}$ is a normed vector space for each $p \in [1, \infty)$. An important *limiting case* of the family of $\mathbf{L_p(T, w)}$ spaces is the space $\mathbf{L_\infty(T, w)}$ of functions which are bounded in the following sense. If $\mathbf{E(f)}$ is the equivalence class generated by f, then $f \in \mathbf{L_\infty(T, w)}$ if $\|f\|_\infty < \infty$ where

$$\|f\|_\infty = \inf_{g \in E(f)} (\sup_{t \in T} \{|g(t)| w(t)\}).$$

We call this the *essential supremum* of f. Note, for example, that if $f(t) = 0$ almost everywhere in T, then $\|f\|_\infty = 0$ because the function that is identically zero in T belongs to the same equivalence class as f.

Exercise 7.1. Consider the function $f(t) = e^{-t}$ with $T = [0, \infty)$ and $w(t) = 1$. Show that $f \in \mathbf{L_p(T, w)}$ for every $p \in [1, \infty]$. ■

The family of $\mathbf{L_p(T, w)}$ spaces is an important one because each member of this family is a complete normed vector space, as can be seen from the following result. The interested reader is referred to Royden (1963) for a proof.

Proposition 7.3 (The Banach Space $\mathbf{L_p(T, w)}$). For each $p \in [1, \infty]$, the normed vector space $\mathbf{L_p(T, w)}$ is a Banach space. ■

It should be pointed out that if the *Lebesgue integral* is replaced by the more familiar *Riemann integral*, then the resulting space $\mathbf{R_p(T, w)}$ is *not* complete and is therefore not a Banach space. Thus there are Cauchy sequences of Riemann-integrable functions which do not converge to Riemann-integrable functions. This is the motivation for using the Lebesgue integral for the $\mathbf{L_p(T, w)}$ spaces.

A common special case of the space $\mathbf{L_p(T, w)}$ arises when the weighting function is $w(t) = 1$ for $t \in T$. This is referred to as *uniform weighting*. In this case we simply write $\mathbf{L_p(T)}$ where the absence of an explicit weighting function indicates uniform weighting is assumed.

As was the case with the spaces of summable sequences, the spaces of Lebesgue-integrable functions satisfy some simple inclusion relations (Desoer and Vidyasagar, 1975) as is evident from the following result.

Proposition 7.4 ($\mathbf{L_p(T, w)}$ Inclusion). The vector space $\mathbf{L_1(T, w)} \cap \mathbf{L_\infty(T, w)}$ is a subspace of $\mathbf{L_p(T, w)}$ for $1 \leqslant p \leqslant \infty$.

Proof. Let $f \in \mathbf{L_1(T, w)}$, and let S be the subset of T for which $f(t) \geqslant 1$. Because $f \in \mathbf{L_1(T, w)}$, we have $\mu(S) < \infty$ where μ denotes the Lebesgue measure of S. If $f \in \mathbf{L_\infty(T, w)}$

$$\int_S |f(t)|^p w(t)\, dt < \|f\|_\infty^p\, \mu(S)$$

Then

$$\begin{aligned}
\|f\|_p^p &= \int_T |f(t)|^p w(t)\, dt \\
&= \int_S |f(t)|^p w(t)\, dt + \int_{T-S} |f(t)|^p w(t)\, dt \\
&\leqslant \int_S |f(t)|^p w(t)\, dt + \int_{T-S} |f(t)| w(t)\, dt \\
&\leqslant \int_S |f(t)|^p w(t)\, dt + \int_T |f(t)| w(t)\, dt \\
&= \int_S \|f\|_\infty^p\, \mu(S) + \|f\|_1 \\
&< \infty.
\end{aligned}$$

■

Refer to Figure 2.4 for a pictorial summary of the inclusion relations for the $\mathbf{L_p}$ spaces where it can be seen that $\mathbf{L_1(T, w)} \cap \mathbf{L_\infty(T, w)}$ is a subspace of $\mathbf{L_p(T, w)}$ for each $1 \leqslant p \leqslant \infty$.

Exercise 7.2. Let $T = [0, \infty)$. Find an example of a function which is in $\mathbf{L_\infty(T)} - \mathbf{L_1(T)}$. ■

Exercise 7.3. Let $T = [0, \infty)$. Find an example of a function which is in $\mathbf{L_1(T)} - \mathbf{L_\infty(T)}$. ■

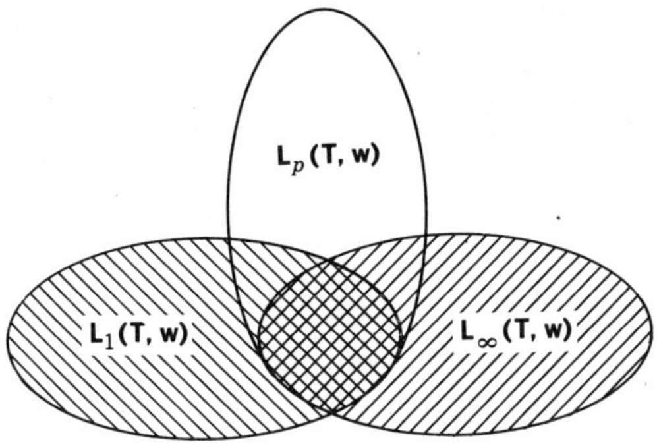

Figure 2.4. Inclusion relations for the $\mathbf{L}_p(\mathbf{T}, \mathbf{w})$ spaces.

2.8. CONTRACTION MAPPINGS

Many engineering problems can be solved by constructing a sequence of increasingly accurate approximate solutions. This is called the method of *successive approximation.* It has a natural formulation within the framework of normed vector spaces where the approximate solutions are represented by a sequence of vectors. Under certain rather general conditions, the sequence of vectors can be shown to converge and, moreover, bounds on the *rate* of convergence can be established. To see this, we begin by considering a special class of functions that map a normed vector space $\mathbf{V}$ into itself. We use the notation Tx for the image of the vector x under the operator T. Note that x may itself be a function if the space $\mathbf{V}$ is a function space.

Definition 8.1 (Contraction). Let $\mathbf{V}$ be a normed vector space and let $T: \mathbf{V} \to \mathbf{V}$. Then T is a *contraction mapping* on $\mathbf{V} \Leftrightarrow$ there exists a $\rho < 1$ called a *contraction constant* such that for each $(x, y) \in \mathbf{V}^2$:

$$\|Tx - Ty\| \leqslant \rho \|x - y\|$$ ■

The term *contraction* arises from the requirement that distance between any pair of points in $\mathbf{V}$ must shrink or contract when those points are operated on by the operator T. Thus the distance between the images of the points, $\|Tx - Ty\|$, is always smaller than the distance between the points, $\|x - y\|$. Refer to Figure 2.5 which shows how the unit ball $B(x, 1)$ gets mapped into the ball $B(Tx, \rho)$ where ρ is the contraction constant.

Exercise 8.1 (Contraction). Let T be a contraction mapping on a normed vector space $\mathbf{V}$. Show that T is continuous. ■

If the space $\mathbf{V}$ is a Banach space, then each contraction mapping on $\mathbf{V}$ has a unique point associated with it which remains fixed when operated on by T.

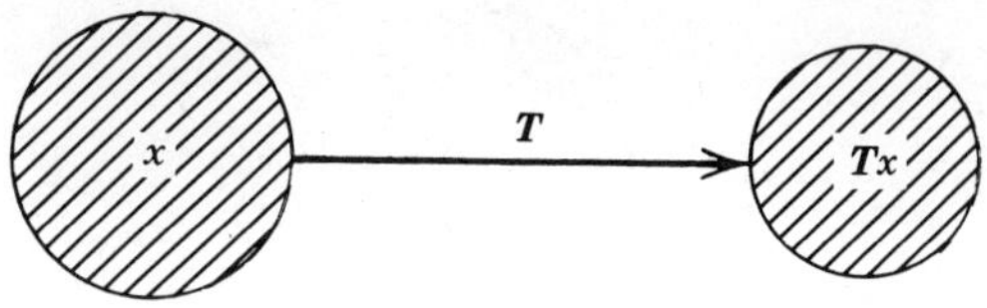

Figure 2.5. Image of unit ball under contraction mapping.

Proposition 8.1 (Fixed Point Theorem). Let $\mathbf{T}$ be a contraction mapping on a Banach space $\mathbf{V}$. Then there exists a unique $x^* \in \mathbf{V}$ called the *fixed point* of $\mathbf{T}$ such that

$$\mathbf{T}x^* = x^*$$

Proof. Let $x^0 \in \mathbf{V}$ and let $x^k = \mathbf{T}x^{k-1}$ for $k \geqslant 1$. Next, let k and j be positive integers with $k > j$. Then, using Def. 8.1, the triangle inequality, and the limit of a *geometric series* of scalars, we have

$$\begin{aligned}
\|x^k - x^j\| &= \|\mathbf{T}^k x^0 - \mathbf{T}^j x^0\| \\
&\leqslant \rho^j \|\mathbf{T}^{k-j} x^0 - x^0\| \\
&= \rho^j \|x^{k-j} - x^0\| \\
&\leqslant \rho^j (\|x^{k-j} - x^{k-j-1}\| + \cdots + \|x^1 - x^0\|) \\
&\leqslant \rho^j (1 + \rho + \cdots + \rho^{k-j-1}) \|x^1 - x^0\| \\
&= \rho^j \left(\sum_{i=0}^{k-j-1} \rho^i \right) \|x^1 - x^0\| \\
&\leqslant \rho^j \left(\sum_{i=0}^{\infty} \rho^i \right) \|x^1 - x^0\| \\
&= \frac{\rho^j}{1-\rho} \|x^1 - x^0\|.
\end{aligned}$$

Thus $\{x^k\}$ is a Cauchy sequence in $\mathbf{V}$. Since $\mathbf{V}$ is a complete normed vector space, the sequence $\{x^k\}$ converges to a point $x^* \in \mathbf{V}$. Using the continuity of $\mathbf{T}$ from Exer. 8.1, we then have

$$\begin{aligned}
\mathbf{T}x^* &= \mathbf{T} \lim_{k \to \infty} \{x^k\} \\
&= \lim_{k \to \infty} \{\mathbf{T}x^k\} \\
&= \lim_{k \to \infty} \{x^{k+1}\} \\
&= x^*.
\end{aligned}$$

Thus the limit x^* is a fixed point of T. Suppose there exists another fixed point $y \in \mathbf{V}$. Then from Def. 8.1,

$$\begin{aligned} \|y - x^*\| &= \|Ty - Tx^*\| \\ &\leqslant \rho\|y - x^*\|. \end{aligned}$$

Because $\rho < 1$, the only way this can hold is if $y = x^*$. Hence the fixed point x^* is unique. ■

Thus every contraction mapping on a Banach space **V** has a unique fixed point. Although this may appear to be a rather specialized result, it is in fact quite general and potentially very useful in engineering applications. Its generality arises from the fact that the operator T can be *nonlinear* and the space **V** can be any complete normed vector space. The utility of the fixed point theorem lies in the fact that it can be used to establish the convergence of a general class of successive approximation algorithms. Consider, in particular, the following generic successive approximation algorithm where **V** is a normed vector space and T is an operator which maps **V** into **V**.

Algorithm 8.1 (Successive Approximation)

1. Pick $x^0 \in \mathbf{V}$ and set $k = 1$.
2. Compute $x^k = Tx^{k-1}$.
3. Set $k = k + 1$.
4. Go to 2. ■

If **V** is a Banach space and T is a contraction mapping, then this successive approximation algorithm will converge to the unique fixed point of T. Moreover, we can put an upper bound on the rate of convergence in terms of the contraction constant ρ as can be seen from the following result.

Proposition 8.2 (Successive Approximation). Let **V** be a Banach space, and let T be a contraction mapping on **V**. Then the sequence $\{x^k\}$ constructed by Alg. 8.1 converges to the fixed point x^* of T, and the rate of convergence is

$$\|x^k - x^*\| \leqslant \frac{\rho^k}{1-\rho} \|Tx^0 - x^0\|.$$

Proof. Let k and j be positive integers with $j > k$. Then, using Def. 8.1 and the triangle inequality,

$$\begin{aligned} \|x^j - x^k\| &= \|T^j x^0 - T^k x^0\| \\ &\leqslant \rho^k \|T^{j-k} x^0 - x^0\| \end{aligned}$$

$$
\begin{aligned}
&= \rho^k \|x^{j-k} - x^0\| \\
&\leqslant \rho^k(\|x^{j-k} - x^{j-k-1}\| + \cdots + \|x^1 - x^0\|) \\
&\leqslant \rho^k(1 + \rho + \cdots + \rho^{j-k-1})\|x^1 - x^0\| \\
&= \rho^k \left(\sum_{i=0}^{j-k-1} \rho^i \right) \|x^1 - x^0\| \\
&\leqslant \rho^k \left(\sum_{i=0}^{\infty} \rho^i \right) \|x^1 - x^0\| \\
&= \frac{\rho^k}{1-\rho} \|x^1 - x^0\|.
\end{aligned}
$$

Let $x^* \in \mathbf{V}$ be the fixed point of $\boldsymbol{T}$. Then, using the triangle inequality,

$$
\begin{aligned}
\|x^k - x^*\| &= \|x^k - x^j + x^j - x^*\| \\
&\leqslant \|x^k - x^j\| + \|x^j - x^*\| \\
&= \|x^j - x^k\| + \|x^j - x^*\| \\
&\leqslant \frac{\rho^k}{1-\rho} \|x^1 - x^0\| + \|x^j - x^*\| \\
&= \frac{\rho^k}{1-\rho} \|x^1 - x^0\| + \|\boldsymbol{T}^j x^0 - \boldsymbol{T}^j x^*\| \\
&\leqslant \frac{\rho^k}{1-\rho} \|x^1 - x^0\| + \rho^j \|x^0 - x^*\|.
\end{aligned}
$$

This holds for all $j > k$. Letting $j \to \infty$, the second term vanishes, and we have

$$
\begin{aligned}
\|x^k - x^*\| &\leqslant \frac{\rho^k}{1-\rho} \|x^1 - x^0\| \\
&= \frac{\rho^k}{1-\rho} \|\boldsymbol{T}x^0 - x^0\|.
\end{aligned}
$$

■

Many algorithms can be cast in the general form of Alg. 8.1 by appropriately defining the normed space **V** and the operator $\boldsymbol{T}$. To show that these algorithms converge, it is sufficient to show that **V** is a Banach space and that $\boldsymbol{T}$ is a contraction. Note that $\boldsymbol{T}$ can be both nonlinear and infinite-dimensional. The objective is to define $\boldsymbol{T}$ in such a way that the fixed point is a solution to the problem.

In Chapter 8 we examine an illustration of the fixed point theorem in which $\boldsymbol{T}$ is an integral operator and its fixed point $\mathbf{x}^*$ is the unique solution to a general system of nonlinear differential equations. In the interim consider the following exercise that shows that we can also solve systems of linear algebraic equations by the successive approximation technique.

Exercise 8.2 (Successive Approximation). Let $\mathbf{V} = \mathbf{F}_\infty^n$ and let T be the following nonlinear mapping from $\mathbf{V}$ into $\mathbf{V}$:

$$Tx = b - (A - I)x.$$

Here $A \in \mathbf{F}^{n \times n}$ is a diagonally dominant matrix which has been normalized. That is,

$$|A_{kk}| = 1, \qquad 1 \leqslant k \leqslant n,$$

$$\sum_{k \neq j} |A_{kj}| < 1, \qquad 1 \leqslant j \leqslant n.$$

Show that T is a contraction and that Alg. 8.1 converges, in this case, to the unique solution of the linear algebraic system $Ax = b$. ■

The fixed point theorem can be generalized in a number of ways which make it more widely applicable (Vidyasagar, 1978). For example, it is only necessary that T^k be a contraction for $k > N$ for some fixed integer N. Furthermore, under certain conditions it is not necessary that T be a contraction on the entire Banach space $\mathbf{V}$, as can be seen from the following exercise.

Exercise 8.2 (Contraction Mapping). Let $\mathbf{V}$ be a Banach space, and let $T:\mathbf{V} \to \mathbf{V}$. Suppose there exists a closed ball $S = \{x : \|x - z\| \leqslant \alpha\}$ which is invariant under T in the sense that $x \in S \Rightarrow Tx \in S$. Suppose that the restriction of T to the closed ball S is a contraction. That is, for each $(x, y) \in S^2$,

$$\|Tx - Ty\| \leqslant \rho \|x - y\|,$$

for some $\rho < 1$. Show that T has a unique fixed point $x^* \in S$ and show that if $x^0 \in S$, then $x^k = T^k x^0$ converges to x^* with

$$\|x^k - x^*\| \leqslant \frac{\rho^k}{1 - \rho} \|Tx^0 - x^0\|.$$ ■

PROBLEMS

1. Show that the norm for $\mathbf{F}_p^n$ proposed in Exam. 1.1 is a valid norm for each $p \in [1, \infty)$. (*Hint*: Show that $\mathbf{F}_p^n$ is, in effect, a subspace of $\mathbf{l}_p$).
2. Show that the norm for $\mathbf{F}_\infty^n$ proposed in Exam. 1.2 is a valid norm. Do this by direct verification of the four fundamental properties.

3. Show that the set $\mathbf{C}_\infty(\mathbf{T})$ defined in Exam. 1.3 is a vector space.

4. Consider the space $\mathbf{F}^{\mathbf{m}\times\mathbf{n}}$.

(a) Find a norm for $\mathbf{F}^{\mathbf{m}\times\mathbf{n}}$.

(b) Show that it is a valid norm.

5. Let $\mathbf{C}_1([0, 1])$ be the subspace of $\mathbf{L}_1([0, 1])$ consisting of continuous functions. Show that $\mathbf{C}_1([0, 1])$ is *not* a Banach space by constructing a Cauchy sequence in $\mathbf{C}_1([0, 1])$ which converges to a function f which is not continuous. Sketch the sequence of functions. (*Hint*: Try $x_k(t) = t^k$).

6. Let S be a subset of a normed vector space $\mathbf{V}$ with δS nonempty.

(a) Show that S is open $\Leftrightarrow S \cap \delta S = \varnothing$.

(b) Show that S is closed $\Leftrightarrow S \cap \delta S = \delta S$.

7. Show that the union of a finite number of closed sets is a closed set.

8. Show that the intersection of a finite number of open sets is an open set.

9. Show that all Cauchy sequences are bounded.

10. Let $\ell_{\mathbf{p}}^{\mathbf{n}}$ be the space of all sequences $\{x^k\}$ of vectors in $\mathbf{F}^{\mathbf{n}}$ such that $\|\{x^k\}\| < \infty$ where

$$\|\{x^k\}\| = \left(\sum_{k=1}^{\infty} \|x^k\|_p\right)^{1/p}.$$

Here $\|\cdot\|_p$ is the norm on $\mathbf{F}_{\mathbf{p}}^{\mathbf{n}}$ from Exam. 1.1. Show that $\ell_{\mathbf{p}}^{\mathbf{n}}$ is a Banach space for $1 \leqslant n < \infty$ and $p \in [1, \infty)$.

11. Let $\mathbf{L}_{\mathbf{p}}^{\mathbf{n}}(\mathbf{T}, \mathbf{w})$ be the space of all functions f mapping T into $\mathbf{F}^{\mathbf{n}}$ such that $\|f\| < \infty$ where,

$$\|f\| = \left(\int_T \|f(t)\|_p^p w(t)\, dt\right)^{1/p}.$$

Show that $\mathbf{L}_{\mathbf{p}}^{\mathbf{n}}(\mathbf{T}, \mathbf{w})$ is a Banach space for $1 \leqslant n < \infty$ and $p \in [1, \infty)$.

12. Let $\mathbf{V}$ be a normed vector space, and let $f(x) = \|x\|$. Show that $f{:}\mathbf{V} \to \mathbf{R}$ is continuous.

13. Let $f{:}\mathbf{R} \to \mathbf{R}$ be continuously differentiable. Suppose that $t \in [0, 1] \Rightarrow f(t) \in [0, 1]$, and suppose that on $[0, 1]$ we have $|f^{(1)}(t)| \leqslant \rho$ for some $\rho < 1$. Show that f is a contraction on $[0, 1]$, and show that if $x^0 \in [0, 1]$ and $x^k = f(x^{k-1})$, then $\{x^k\}$ converges to the solution of $f(x) = x$.

REFERENCES

Desoer, C. A. and M. Vidyasagar, *Feedback Systems: Input-Output Properties*, Academic, New York, 1975.

Luenberger, D. G., *Optimization by Vector Space Methods*, Wiley, New York, 1969.

Royden, H. L., *Real Analysis*, Macmillan, New York, 1963.

Vidyasagar, M., *Nonlinear Systems Analysis*, Prentice-Hall, Englewood Cliffs, NJ, 1978.

3 INNER PRODUCT SPACES

The structure of vector spaces can be refined further with the introduction of a measure of the *orientation* between two vectors. Once the notion of orientation is developed, vector space formulations based on the powerful concepts of orthogonality and projections can then be explored.

3.1. INNER PRODUCTS

In order to develop a measure of the orientation between two vectors we begin by considering something called the inner product of a pair of vectors.

Definition 1.1 (Inner Product). An *inner product* on a vector space **V** is a function denoted $\langle \cdot, \cdot \rangle$ mapping $\mathbf{V} \times \mathbf{V}$ into **F** having the following properties. For each $(\alpha, x, y, z) \in \mathbf{F} \times \mathbf{V}^3$:

1. $\overline{\langle x, y \rangle} = \langle y, x \rangle$,
2. $\langle x, x \rangle > 0, x \neq 0$,
3. $\langle \alpha x, y \rangle = \alpha \langle x, y \rangle$,
4. $\langle x + y, z \rangle = \langle x, z \rangle + \langle y, z \rangle$.

A vector space **V** taken together with an appropriately defined inner product is called an *inner product space.* ■

Here the overbar denotes the *complex conjugate.* Thus from property 1 we find that the inner product of a vector with itself is always real, even when the field **F** is complex. This property, together with the others, allows us to develop a simple relationship between the inner product and the norm. Before exploring this relationship, we first examine some important examples of inner product spaces.

Example 1.1 (The Inner Product Space $\mathbf{F^n}$). Let $A \in \mathbf{F^{n \times n}}$ be any $n \times n$ nonsingular matrix. Define the *Hermitian transpose* of A as the transpose of the complex conjugate of A. That is, $A^{\mathrm{H}} = \bar{A}^{\mathrm{T}}$. Next, let $\mathbf{V} = \mathbf{F^n}$. Then the following is a valid inner product on $\mathbf{V}$:

$$\langle x, y \rangle = y^{\mathrm{H}} A^{\mathrm{H}} A x.$$

The most important special case occurs when $A = I$, in which case the inner product is also called the *dot product*. ∎

Example 1.2 (The Inner Product Space $\mathbf{F^{n \times n}}$). Let $A \in \mathbf{F^{n \times n}}$ be any $n \times n$ matrix. Define the *trace* of A as the sum of the diagonal elements of A. That is, $\mathrm{tr}(A) \triangleq A_{11} + A_{22} + \cdots + A_{nn}$. Next, let $\mathbf{V} = \mathbf{F^{n \times n}}$. Then the following is a valid inner product on $\mathbf{V}$:

$$\langle A, B \rangle = \mathrm{tr}(AB^{\mathrm{H}}).$$ ∎

Example 1.3 (The Inner Product Space P(T)). Let T be a bounded interval in $\mathbf{R}$, and let $\mathbf{V} = \mathbf{P(T)}$, the space of polynomials mapping T into $\mathbf{F}$. Then the following is a valid inner product on $\mathbf{V}$:

$$\langle f, g \rangle = \int_T f(t)\overline{g(t)}\, dt.$$ ∎

Starting with the four fundamental properties of the inner product, a number of additional useful properties can be derived. Some of these secondary properties are summarized in the following exercises.

Exercise 1.1 (Linearity). Let $\mathbf{V}$ be any inner product space. Show that the inner product is *linear* in its first argument and *conjugate linear* in its second argument. That is, show that

$$\langle a_1 x^1 + a_2 x^2, y \rangle = a_1 \langle x^1, y \rangle + a_2 \langle x^2, y \rangle,$$

$$\langle x, b_1 y^1 + b_2 y^2 \rangle = \bar{b}_1 \langle x, y^1 \rangle + \bar{b}_2 \langle x, y^2 \rangle.$$ ∎

Exercise 1.2. Let $\mathbf{V}$ be any inner product space. Show that $\langle x, x \rangle = 0 \Leftrightarrow x = 0$. ∎

Exercise 1.3. Let $\mathbf{V}$ be any inner product space. Show that $\langle x, z \rangle = \langle y, z \rangle$ for each $z \in \mathbf{V} \Leftrightarrow x = y$. ∎

3.2. STANDARD INNER PRODUCT

Recall that the space $\mathbf{F^n}$ is of special significance among finite-dimensional spaces because it is isomorphic to every n-dimensional space. In view of this property, it is worthwhile to examine inner products on $\mathbf{F^n}$ in more detail. As it turns out, each inner product on $\mathbf{F^n}$ has associated with it a special matrix called a weighting matrix, as can be seen from the following result.

Proposition 2.1 (Inner Products). Let $\langle \cdot, \cdot \rangle$ be any inner product on $\mathbf{F^n}$, and let $W \in \mathbf{F^{n \times n}}$ be defined $W_{kj} = \langle i^j, i^k \rangle$ for $1 \leqslant k, j \leqslant n$ where i^k is the kth column of the identity matrix I. Then for each $(x, y) \in \mathbf{F^n} \times \mathbf{F^n}$:

$$\langle x, y \rangle = y^{\mathrm{H}} W x.$$

Proof. Let $(x, y) \in \mathbf{F^n} \times \mathbf{F^n}$. Then x and y can be represented in terms of the standard basis $\mathbf{I} = \{i^1, i^2, \ldots, i^n\}$

$$x = \sum_{k=1}^{n} x_k i^k,$$

$$y = \sum_{k=1}^{n} y_k i^k.$$

Using the fundamental properties of the inner product and Exer. 1.1 we have,

$$\begin{aligned}
\langle x, y \rangle &= \left\langle \sum_{k=1}^{n} x_k i^k, \sum_{j=1}^{n} y_j i^j \right\rangle \\
&= \sum_{k=1}^{n} x_k \left\langle i^k, \sum_{j=1}^{n} y_j i^j \right\rangle \\
&= \sum_{k=1}^{n} x_k \sum_{j=1}^{n} \bar{y}_j \langle i^k, i^j \rangle \\
&= \sum_{k=1}^{n} \sum_{j=1}^{n} x_k \bar{y}_j \langle i^k, i^j \rangle \\
&= \sum_{k=1}^{n} \sum_{j=1}^{n} x_k \bar{y}_j W_{jk} \\
&= \sum_{j=1}^{n} \bar{y}_j \sum_{k=1}^{n} W_{jk} x_k \\
&= \sum_{j=1}^{n} \bar{y}_j (Wx)_j \\
&= y^{\mathrm{H}} W x.
\end{aligned}$$

■

Thus every inner product on $\mathbf{F}^n$ has a matrix W associated with it called a *weighting matrix*. Weighting matrices have a number of special properties which set them apart from matrices in general, as can be seen from the following exercise.

Exercise 2.1 (Weighting Matrices). Let W be the inner product weighting matrix in Prop. 2.1. Show that W has the following properties:

1. $W^H = W$,
2. $x^H W x > 0, \; x \neq 0$,
3. $\det(W) > 0$. ■

In view of Prop. 2.1, there are an unlimited number of inner products that can be defined on $\mathbf{F}^n$. However, among these there is one particular inner product that is of sufficient importance that we give it a special name. We call it the standard inner product on $\mathbf{F}^n$.

Definition 2.1 (Standard Inner Product). For the space $\mathbf{F}^n$, the *standard inner product* is defined as

$$\langle x, y \rangle = y^H x. \qquad ■$$

Thus the standard inner product is the simplest of all possible inner products on $\mathbf{F}^n$. It is obtained by using the weighting matrix $W = I$. The standard inner product in $\mathbf{F}^n$ is sometimes called the *dot product.*

3.3. NATURAL NORMS

Just as every norm induces a natural metric, every inner product induces a natural norm. Let $\mathbf{V}$ be an inner product space, and consider the following candidate for a norm on $\mathbf{V}$:

$$\|x\| = (\langle x, x \rangle)^{1/2}.$$

To show that this is in fact a valid norm, we must show that it satisfies the four fundamental properties of a norm listed in Def. 2.1.1. From property 2 of the inner product we have $\|x\| \geqslant 0$, while from Exer. 1.2 we have $\|x\| = 0 \Leftrightarrow x = 0$. To verify the third property of the norm, let $(\alpha, x) \in \mathbf{F} \times \mathbf{V}$. Then using Exer. 1.1,

$$\begin{aligned} \|\alpha x\| &= (\langle \alpha x, \alpha x \rangle)^{1/2} \\ &= (\alpha \langle x, \alpha x \rangle)^{1/2} \\ &= (\alpha \bar{\alpha} \langle x, x \rangle)^{1/2} \end{aligned}$$

$$= (|\alpha|^2\langle x, x\rangle)^{1/2}$$
$$= |\alpha|(\langle x, x\rangle)^{1/2}$$
$$= |\alpha|\,\|x\|.$$

The verification of the last property of the norm, the triangle inequality, is somewhat more involved. First we examine the following fundamental relationship between inner products and natural norms.

Proposition 3.1 (Cauchy–Schwarz Inequality). Let $\mathbf{V}$ be an inner product space, and let $\|x\| = (\langle x, x\rangle)^{1/2}$. Then for each $(x, y) \in \mathbf{V} \times \mathbf{V}$:

$$|\langle x, y\rangle| \leqslant \|x\|\,\|y\|.$$

Proof. If $x = 0$, then using property 3 of the inner product it follows that $|\langle x, y\rangle| = \|x\|\,\|y\|$. Suppose $x \neq 0$. Let $z = y - (\langle y, x\rangle/\|x\|^2)x$. Then using the properties of the inner product

$$\langle z, x\rangle = \left\langle y - \frac{\langle y, x\rangle}{\|x\|^2}\, x, x\right\rangle$$
$$= \langle y, x\rangle - \frac{\langle y, x\rangle}{\|x\|^2}\langle x, x\rangle$$
$$= \langle y, x\rangle - \frac{\langle y, x\rangle}{\|x\|^2}\|x\|^2$$
$$= \langle y, x\rangle - \langle y, x\rangle$$
$$= 0.$$

Next using this result and the properties of the inner product we conclude that

$$\|z\|^2 = \langle z, z\rangle$$
$$= \left\langle z, y - \frac{\langle y, x\rangle}{\|x\|^2}\, x\right\rangle$$
$$= \langle z, y\rangle$$
$$= \left\langle y - \frac{\langle y, x\rangle}{\|x\|^2}\, x, y\right\rangle$$
$$= \langle y, y\rangle - \frac{\langle y, x\rangle}{\|x\|^2}\langle x, y\rangle$$
$$= \langle y, y\rangle - \frac{\overline{\langle x, y\rangle}}{\|x\|^2}\langle x, y\rangle$$

$$= \langle y, y\rangle - \frac{\overline{\langle x, y\rangle}\langle x, y\rangle}{\|x\|^2}$$

$$= \langle y, y\rangle - \frac{|\langle x, y\rangle|^2}{\|x\|^2}$$

$$= \|y\|^2 - \frac{|\langle x, y\rangle|^2}{\|x\|^2}.$$

Clearly, $\|z\|^2 \geqslant 0$. Thus multiplying both sides of the last equation by $\|x\|^2 > 0$ yields

$$0 \leqslant \|x\|^2\|y\|^2 - |\langle x, y\rangle|^2.$$

Finally, rearranging this inequality and taking the square root of both sides yields

$$|\langle x, y\rangle| \leqslant \|x\|\,\|y\|. \qquad \blacksquare$$

We refer to the inequality in Prop. 3.1 as the *Cauchy–Schwarz inequality*. Given the Cauchy–Schwarz inequality, we are now in a position to make the following claim.

Proposition 3.2 (Natural Norm). Let **V** be an inner product space. Then the following is a valid norm on **V**. We call it the *natural norm* induced by the inner product

$$\|x\| = (\langle x, x\rangle)^{1/2}.$$

Proof. Since the first three properties of the norm have already been established, it suffices to show that $\|x\|$ satisfies the triangle inequality. Using the Cauchy–Schwarz inequality of Prop. 3.1 and the properties of the inner product,

$$\begin{aligned}
\|x + y\|^2 &= \langle x + y, x + y\rangle \\
&= \langle x, x\rangle + \langle x, y\rangle + \langle y, x\rangle + \langle y, y\rangle \\
&= \|x\|^2 + \langle x, y\rangle + \overline{\langle x, y\rangle} + \|y\|^2 \\
&= \|x\|^2 + 2\,\mathrm{Re}(\langle x, y\rangle) + \|y\|^2 \\
&\leqslant \|x\|^2 + 2|\langle x, y\rangle| + \|y\|^2 \\
&\leqslant \|x\|^2 + 2\|x\|\,\|y\| + \|y\|^2 \\
&= (\|x\| + \|y\|)^2.
\end{aligned} \qquad \blacksquare$$

Thus every inner product space induces a normed vector space by means of the natural norm. While the natural norm is obtained from the inner product, it

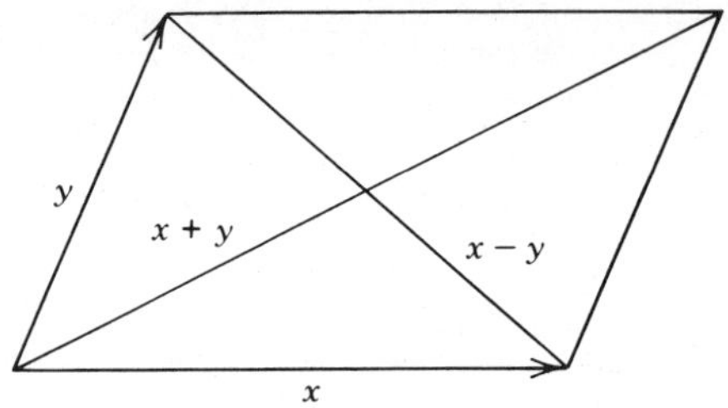

Figure 3.1. The parallelogram law in $\mathbf{R}^2$.

is also possible to reconstruct the inner product from the natural norm. This result is known as the *polarization identity*, and its verification for the case $\mathbf{F} = \mathbf{R}$ is left as an exercise.

Exercise 3.2 (Polarization Identity). Let $\mathbf{V}$ be a real inner product space ($\mathbf{F} = \mathbf{R}$). Show that the inner product can be expressed in terms of its natural norm as

$$\langle x, y \rangle = \tfrac{1}{4}\|x + y\|^2 - \tfrac{1}{4}\|x - y\|^2 \qquad \blacksquare$$

A version of the polarization identity valid for complex inner product spaces can be found in the problems at the end of this chapter. There is another simple relationship that distinguishes natural norms from norms in general. It is known as the parallelogram law.

Exercise 3.3 (Parallelogram Law). Let $\mathbf{V}$ be an inner product space, and let $\|x\|$ be the natural norm on $\mathbf{V}$. Show that

$$\|x + y\|^2 + \|x - y\|^2 = 2\|x\|^2 + 2\|y\|^2. \qquad \blacksquare$$

The term *parallelogram law* arises from the fact that Exer. 3.3 has a simple geometric interpretation when the inner product space is the space $\mathbf{R}^2$. Here if the vectors x and y represent two adjacent sides of a parallelogram, then $x + y$ and $x - y$ represent the two diagonals as shown in Figure 3.1. Thus the parallelogram law is an algebraic generalization of the geometric statement that the sum of the squares of the four sides of a parallelogram equals the sum of the squares of the two diagonals.

3.4. HILBERT SPACES

Because every inner product space induces a normed vector space, the properties associated with normed vector spaces apply to inner product spaces as well. In particular, we can classify inner product spaces in terms of their completeness.

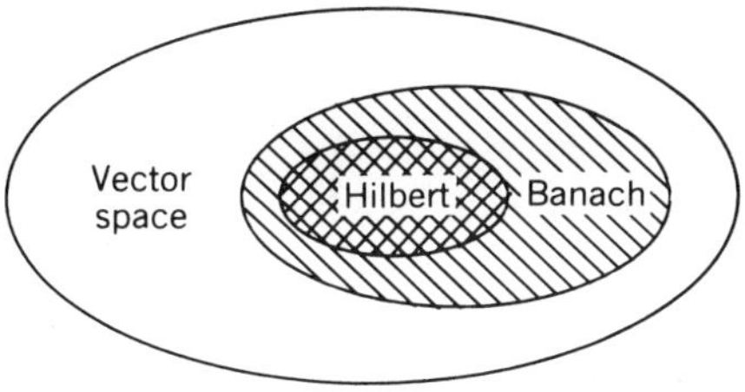

Figure 3.2. General inclusion properties of vector spaces.

Definition 4.1 (Hilbert Space). An inner product space is *complete* $\Leftrightarrow$ the normed vector space it induces is complete. A complete inner product space is called a *Hilbert space.* ■

Thus Hilbert spaces are those inner product spaces that induce Banach spaces with the natural norm. Refer to Figure 3.2 for a pictorial description of the relationship among vector spaces, Banach spaces, and Hilbert spaces. The following are some important examples of Hilbert spaces.

Example 4.1 (The Hilbert Space $\mathbf{F}^n$). For the space $\mathbf{F}^n$, consider the standard inner product

$$\langle x, y \rangle = y^{\mathrm{H}} x.$$

The natural norm in this case is

$$\begin{aligned} \|x\| &= (\langle x, x \rangle)^{1/2} \\ &= (x^{\mathrm{H}} x)^{1/2} \\ &= \left(\sum_{k=1}^{n} x_k \bar{x}_k \right)^{1/2} \\ &= \left(\sum_{k=1}^{n} |x_k|^2 \right)^{1/2} \\ &= \|x\|_2. \end{aligned}$$

Hence the normed vector space induced is $\mathbf{F}_2^n$ as discussed in Exam. 2.1.1. But every finite-dimensional normed vector space is a Banach space. Hence $\mathbf{F}^n$, with the standard inner product, is a Hilbert space. The Hilbert space $\mathbf{F}^n$ is called *Euclidean n-space.* ■

Example 4.2 (The Hilbert Space ℓ_2). For the space ℓ_2 consider the following inner product

$$\langle a, b \rangle = \sum_{k=1}^{\infty} a_k \bar{b}_k.$$

The natural norm in this case is

$$\begin{aligned}\|a\| &= (\langle a, a\rangle)^{1/2}\\ &= \left(\sum_{k=1}^{\infty} a_k\bar{a}_k\right)^{1/2}\\ &= \left(\sum_{k=1}^{\infty} |a_k|^2\right)^{1/2}\\ &= \|a\|_2.\end{aligned}$$

Hence the normed vector space induced is ℓ_2 as discussed in Section 2.5. But the space ℓ_p is a Banach space for $1 \leqslant p \leqslant \infty$. Consequently, ℓ_2 is a Hilbert space. We call the Hilbert space ℓ_2 the space of *square summable sequences.* ■

Example 4.3 (The Hilbert Space $\mathbf{L_2(T, w)}$). For the space $\mathbf{L_2(T, w)}$ consider the following inner product:

$$\langle f, g\rangle = \int_T f(t)\overline{g(t)}w(t)\, dt.$$

The natural norm in this case is

$$\begin{aligned}\|f\| &= (\langle f, f\rangle)^{1/2}\\ &= \left(\int_T f(t)\overline{f(t)}w(t)\, dt\right)^{1/2}\\ &= \left(\int_T |f(t)|^2 w(t)\, dt\right)^{1/2}\\ &= \|f\|_2.\end{aligned}$$

Hence the normed vector space induced is $\mathbf{L_2(T, w)}$, as discussed in Section 2.6. But the space $\mathbf{L_p(T, w)}$ is a Banach space for $1 \leqslant p \leqslant \infty$. Consequently, $\mathbf{L_2(T, w)}$ is a Hilbert space. We call the Hilbert space $\mathbf{L_2(T, w)}$ the space of *square integrable functions.*

Exercise 4.1. Show that for the Hilbert space ℓ_2, the Cauchy–Schwartz inequality is a special case of the Holder inequality in Prop. 2.6.2. ■

Exercise 4.2. Show that for the Hilbert space $\mathbf{L_2(T, w)}$, the Cauchy–Schwartz inequality is a special case of the Holder inequality in Prop. 2.7.1. ■

3.5. ORTHOGONALITY

Next, we consider an important algebraic generalization of the geometric notion of one vector being perpendicular to another. It is this concept in particular that is the key to exploiting the rich structure of inner product spaces.

Definition 5.1 (Orthogonality). Let $X = \{x^1, x^2, \ldots, x^n\}$ be a set of vectors in an inner product space $\mathbf{V}$. Then X is an *orthogonal* set $\Leftrightarrow \langle x^k, x^j \rangle = 0$ for $k \neq j$. Similarly, X is a *normal* set $\Leftrightarrow \|x^k\| = 1$ for $1 \leqslant k \leqslant n$. Finally, X is an *orthonormal* set $\Leftrightarrow X$ is orthogonal and normal. ■

Thus two vectors are orthogonal if and only if the inner product of one with the other is zero. A vector is normal if and only if its norm is unity. The orthogonality of a set of vectors can be summarized concisely in terms of a special matrix called the Gram matrix.

Definition 5.2 (Gram Matrix). Let $X = \{x^1, x^2, \ldots, x^n\}$ be a set of vectors in an inner product space $\mathbf{V}$. Then the *Gram matrix* associated with X is defined

$$G_{kj}(X) \triangleq \langle x^k, x^j \rangle, \qquad 1 \leqslant k, j \leqslant n$$ ■

From Def. 5.1 and Def. 5.2 it is clear that X is orthogonal if and only if $G(X)$ is a *diagonal* matrix. Furthermore, X is orthonormal if and only if $G(X)$ is a special diagonal matrix, the identity matrix I. More generally, it can be shown (e.g., Luenberger, 1969) that the set X is a linearly independent set if and only if the Gram matrix $G(X)$ is *nonsingular*.

Example 5.1 (Orthonormal Set). For the inner product space $\mathbf{F}^n$, the standard basis vectors $\mathbf{I} = \{i^1, i^2, \ldots, i^n\}$ form an orthonormal set. Here i^k denotes the kth column of the $n \times n$ identity matrix I. ■

We can extend the notion of orthogonality to an infinite sequence of vectors in the following manner.

Definition 5.3 (Orthogonal Sequence). The vectors $\{x^k\}$ form an *orthogonal sequence* $\Leftrightarrow$ for each $j \geqslant 1$ the set $X_j = \{x^1, x^2, \ldots, x^j\}$ is orthogonal. ■

There are many examples of orthogonal sequences. In terms of engineering applications, the following illustration of an orthogonal sequence is a familiar one.

Example 5.2 (Orthogonal Sequence). Let $\mathbf{V} = \mathbf{L_2([0, 1])}$. Then the following vectors form an orthogonal sequence in $\mathbf{V}$:

$$f^k(t) = \cos(2\pi k t), \qquad 1 \leqslant k < \infty.$$ ■

The notion of linear independence can also be generalized to infinite sequences in a similar manner.

Definition 5.3 (Linearly Independent Sequence). The vectors $\{x^k\}$ form a *linearly independent sequence* $\Leftrightarrow$ for each $j \geqslant 1$ the set $X_j = \{x^1, x^2, \ldots, x^j\}$ is linearly independent.

Note that the zero vector is orthogonal to all vectors and linearly dependent upon all vectors. If we restrict our attention to nonzero vectors, then orthogonality is a stronger criterion than linear independence, as is evident from the following result.

Proposition 5.1 (Linear Independence). Nonzero orthogonal vectors are linearly independent.

Proof. Let $\{x^k\}$ be an orthogonal sequence of nonzero vectors. Suppose the sequence $\{x^k\}$ is linearly dependent. Then there exists an $n \geqslant 1$ and an $a \in \mathbf{F}^n$ with $a \neq 0$ such that

$$\sum_{k=1}^{n} a_k x^k = 0.$$

Because the vectors $\{x^k\}$ are orthogonal, using Def. 5.1 for $1 \leqslant j \leqslant n$ we have

$$\begin{aligned} 0 &= \langle 0, x^j \rangle \\ &= \left\langle \sum_{k=1}^{n} a_k x^k, x^j \right\rangle \\ &= \sum_{k=1}^{n} a_k \langle x^k, x^j \rangle \\ &= a_j \langle x^j, x^j \rangle \\ &= a_j \|x^j\|^2. \end{aligned}$$

But because the vectors $\{x^k\}$ are nonzero, it follows that $\|x^j\| \neq 0$, which implies that $a = 0$. But this is a contradiction. Hence the vectors $\{x^k\}$ are linearly independent. ■

Given a sequence of nonzero vectors $\{x^k\}$, we can always generate from it a new sequence $\{y^k\}$ which is orthogonal. The orthogonal sequence is generated from the original sequence by forming linear combinations of vectors in accordance with the following algorithm.

Algorithm 5.1 (Gram–Schmidt)

1. Set $k = 1$, and $y^1 = x^1$.
2. Set $k = k + 1$.
3. Compute y^k as follows:

$$y^k = x^k - \sum_{j=1}^{k-1} \frac{\langle x^k, y^j \rangle}{\| y^j \|^2} y^j.$$

4. Go to 2. ■

Note that there is no constraint that requires that the original sequence be linearly independent, only nonzero. However, if the original sequence of vectors is linearly dependent, then at least one of the generated vectors will be the zero vector, which is, orthogonal to all vectors.

Proposition 5.2 (Gram–Schmidt). The vectors $\{ y^k \}$ constructed by applying Alg. 5.1 are orthogonal.

Proof. Let $m \geqslant 1$. It suffices to show that $\langle y^m, y^k \rangle = 0$ for $1 \leqslant k < m$. We proceed by induction. From Alg. 5.1,

$$\begin{aligned} \langle y^2, y^1 \rangle &= \left\langle x^2 - \frac{\langle x^2, y^1 \rangle}{\| y^1 \|^2} y^1, y^1 \right\rangle \\ &= \langle x^2, y^1 \rangle - \frac{\langle x^2, y^1 \rangle}{\| y^1 \|^2} \langle y^1, y^1 \rangle \\ &= \langle x^2, y^1 \rangle - \frac{\langle x^2, y^1 \rangle}{\| y^1 \|^2} \| y^1 \|^2 \\ &= \langle x^2, y^1 \rangle - \langle x^2, y^1 \rangle \\ &= 0 \end{aligned}$$

Hence the first two vectors generated by Alg. 5.1 are orthogonal. Suppose the first $m - 1$ vectors generated are orthogonal. Then from Alg. 5.1 for $1 \leqslant k < m$

$$\begin{aligned} \langle y^m, y^k \rangle &= \left\langle x^m - \sum_{j=1}^{m-1} \frac{\langle x^m, y^j \rangle}{\| y^j \|^2} y^j, y^k \right\rangle \\ &= \langle x^m, y^k \rangle - \sum_{j=1}^{m-1} \frac{\langle x^m, y^j \rangle}{\| y^j \|^2} \langle y^j, y^k \rangle \\ &= \langle x^m, y^k \rangle - \frac{\langle x^m, y^k \rangle}{\| y^k \|^2} \| y^k \|^2 \\ &= \langle x^m, y^k \rangle - \langle x^m, y^k \rangle \\ &= 0. \end{aligned}$$

Thus the first m vectors generated by Alg. 5.1 are orthogonal. By induction the entire sequence $\{y^k\}$ is orthogonal. ■

Algorithm 5.1 can, of course, be applied to finite sequences as well. The following example is a case in point.

Example 5.3 (Gram–Schmidt Orthogonalization). Let $\mathbf{V} = \mathbf{R}^3$, and let $X = [x^1, x^2, x^3]$ be the following matrix:

$$X = \begin{bmatrix} 1 & 1 & -1 \\ 1 & 0 & -1 \\ 0 & -1 & 1 \end{bmatrix}.$$

Applying Alg. 5.1 to the columns of X using the standard inner product in $\mathbf{R}^3$ we have

$$\begin{aligned} y^1 &= x^1 \\ &= [1, 1, 0]^{\mathrm{T}}. \end{aligned}$$

$$\begin{aligned} y^2 &= x^2 - \frac{\langle x^2, y^1 \rangle}{\| y^1 \|^2}\, y^1 \\ &= x^2 - \tfrac{1}{2} y^1 \\ &= [\tfrac{1}{2}, -\tfrac{1}{2}, -1]^{\mathrm{T}}. \end{aligned}$$

$$\begin{aligned} y^3 &= x^3 - \frac{\langle x^3, y^1 \rangle}{\| y^1 \|^2}\, y^1 - \frac{\langle x^3, y^2 \rangle}{\| y^2 \|^2}\, y^2 \\ &= x^3 + y^1 + \tfrac{2}{3} y^2 \\ &= [\tfrac{1}{3}, -\tfrac{1}{3}, \tfrac{1}{3}]^{\mathrm{T}}. \end{aligned}$$

As a check,

$$\langle y^1, y^2 \rangle = \tfrac{1}{2} - \tfrac{1}{2} = 0,$$

$$\langle y^1, y^3 \rangle = \tfrac{1}{3} - \tfrac{1}{3} = 0,$$

$$\langle y^2, y^3 \rangle = \tfrac{1}{6} + \tfrac{1}{6} - \tfrac{1}{3} = 0.$$ ■

One of the most well-known theorems of trigonometry says that for every right triangle the square of the hypotenuse is equal to the sum of the squares of the other two sides. The Pythagorean theorem can also be formulated in a vector space framework as can be seen from the following result.

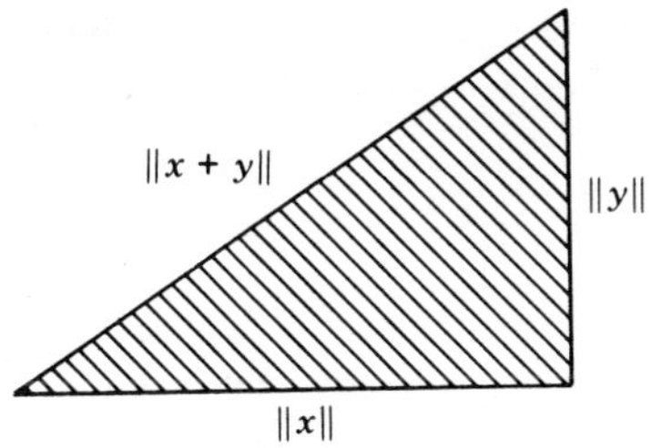

Figure 3.3. Pythagorean theorem in $\mathbf{R}^2$.

Proposition 5.3 (Pythagorean Theorem). Let $\{x^1, x^2, \ldots, x^n\}$ be an orthogonal set in an inner product space $\mathbf{V}$. Then

$$\left\| \sum_{k=1}^{n} x^k \right\|^2 = \sum_{k=1}^{n} \|x^k\|^2.$$

Proof. Using the definition of the natural norm and the orthogonality of the set $\{x^1, x^2, \ldots, x^n\}$ we have

$$\begin{aligned}
\left\| \sum_{k=1}^{n} x^k \right\|^2 &= \left\langle \sum_{k=1}^{n} x^k, \sum_{j=1}^{n} x^j \right\rangle \\
&= \sum_{k=1}^{n} \left\langle x^k, \sum_{j=1}^{n} x^j \right\rangle \\
&= \sum_{k=1}^{n} \sum_{j=1}^{n} \langle x^k, x^j \rangle \\
&= \sum_{k=1}^{n} \langle x^k, x^k \rangle \\
&= \sum_{k=1}^{n} \|x^k\|^2.
\end{aligned}$$ ■

Refer to Figure 3.3 for a geometric interpretation of Prop. 5.3 for the case when $\mathbf{V} = \mathbf{R}^2$ and $n = 2$. This is the classic Pythagorean theorem from trigonometry.

3.6. ORTHOGONAL PROJECTIONS

Given the notion of orthogonality, we are now in a position to decompose a vector into the sum of its projections onto subspaces. To introduce this important concept, we begin by identifying suitable subspaces.

Definition 6.1 (Orthogonal Complement). The *orthogonal complement* of a subset U of an inner product space $\mathbf{V}$ is denoted $\mathbf{U}^\perp$ and is defined

$$\mathbf{U}^\perp = \{x \in \mathbf{V} : y \in U \Rightarrow \langle x, y \rangle = 0\}$$

Thus the orthogonal complement of U is a set denoted $\mathbf{U}^\perp$ consisting of all vectors in $\mathbf{V}$ that are orthogonal to every vector in U. It is of interest to note that even though U can be an arbitrary subset of an inner product space $\mathbf{V}$, its orthogonal complement $\mathbf{U}^\perp$ will always be a subspace of $\mathbf{V}$ and that subspace will be closed.

Exercise 6.1 (Orthogonal Complement). Let U be a subset of an inner product space $\mathbf{V}$. Show that $\mathbf{U}^\perp$ is a closed subspace of $\mathbf{V}$. ■

Since the orthogonal complement of a set U is itself a set $\mathbf{U}^\perp$, it is possible to take the orthogonal complement of the orthogonal complement, which is denoted $\mathbf{U}^{\perp\perp}$. In light of Exer. 6.1, the set $\mathbf{U}^{\perp\perp}$ is a closed subspace of $\mathbf{V}$. When $\mathbf{V}$ is a Hilbert space, it is in fact the closure of the subspace of $\mathbf{V}$ spanned by U, as can be seen from the following exercise.

Exercise 6.2. Let U be a subset of a Hilbert space $\mathbf{V}$. Show that $\mathbf{U}^{\perp\perp}$ is the closure of **span(U)**. ■

In view of Exer. 6.2, if $\mathbf{U}$ is itself a closed subspace, then $\mathbf{U}^{\perp\perp} = \mathbf{U}$. The case when $\mathbf{U}$ is a closed subspace of $\mathbf{V}$ and $\mathbf{V}$ is a Hilbert space is of particular importance. It leads to the following fundamental result known as the projection theorem.

Proposition 6.1 (Projection Theorem). Let $\mathbf{U}$ be a closed subspace of a Hilbert space $\mathbf{V}$. Then $\mathbf{V}$ can be decomposed into the direct sum

$$\mathbf{V} = \mathbf{U} \oplus \mathbf{U}^\perp.$$

Proof. We will only prove the special case when $\mathbf{U}$ is finite-dimensional. For a general proof the reader is referred to Bachman and Narici (1966). Suppose $\mathbf{U}$ is n-dimensional. Let $\mathbf{Y} = \{y^1, y^2, \ldots, y^n\}$ be an orthonormal basis for $\mathbf{U}$. Next, let $x \in \mathbf{V}$ and let $(x^0, x^1) \in \mathbf{V}^2$ be defined as follows:

$$x^0 = \sum_{k=1}^{n} \langle x, y^k \rangle y^k,$$

$$x^1 = x - x^0.$$

Clearly, $x = x^0 + x^1$ by construction. Furthermore, $x^0 \in \mathbf{U}$ since x^0 is a linear combination of the basis vectors $\mathbf{Y}$. Next, we demonstrate that $x^1 \in \mathbf{U}^\perp$ and that the decomposition of x is unique. Now, for each $1 \leqslant k \leqslant n$ because **Y** is an orthonormal set we have

$$\begin{aligned}
\langle x^1, y^k \rangle &= \langle x - x^0, y^k \rangle \\
&= \langle x, y^k \rangle - \langle x^0, y^k \rangle \\
&= \langle x, y^k \rangle - \left\langle \sum_{j=1}^{n} \langle x, y^j \rangle y^j, y^k \right\rangle \\
&= \langle x, y^k \rangle - \sum_{j=1}^{n} \langle x, y^j \rangle \langle y^j, y^k \rangle \\
&= \langle x, y^k \rangle - \langle x, y^k \rangle \langle y^k, y^k \rangle \\
&= \langle x, y^k \rangle - \langle x, y^k \rangle \| y^k \|^2 \\
&= \langle x, y^k \rangle - \langle x, y^k \rangle \\
&= 0.
\end{aligned}$$

Thus x^1 is orthogonal to each basis vector in **Y**. But since every vector in **U** is a linear combination of the basis vectors, it follows that x^1 is orthogonal to every vector in **U**. Hence $x^1 \in \mathbf{U}^{\perp}$. To show that the decomposition of x is unique, suppose there exist $(z^0, z^1) \in \mathbf{U} \times \mathbf{U}^{\perp}$ such that $x = z^0 + z^1$. Then using the Pythagorean theorem, we have

$$\begin{aligned}
0 &= \|x - x\|^2 \\
&= \|x^0 + x^1 - (z^0 + z^1)\|^2 \\
&= \|(x^0 - z^0) + (x^1 - z^1)\|^2 \\
&= \|x^0 - z^0\|^2 + \|x^1 - z^1\|^2.
\end{aligned}$$

It follows that $z^0 = x^0$ and $z^1 = x^1$. Hence the decomposition of x is unique. ■

The projection theorem is a theoretical foundation for many engineering applications, particularly in the field of optimization theory. To apply it, one must show that the subspace **U** is closed and the inner product space **V** is complete. Every finite-dimensional subspace of a Hilbert space **V** is closed. This is a consequence of the results in Chapter 2 because a subspace of a Banach space is closed if it is complete, and all finite-dimensional normed vector spaces are complete. In light of Prop. 6.1, we are now in a position to formally define the term orthogonal projection.

Definition 6.2 (Orthogonal Projection). Let **U** be a closed subspace of a Hilbert space **V**, and let $x \in \mathbf{V}$. Decompose x into the direct sum $x = x^0 + x^1$ where $(x^0, x^1) \in \mathbf{U} \times \mathbf{U}^{\perp}$. Then x^0 is called the *orthogonal projection* of x onto the subspace **U**. ■

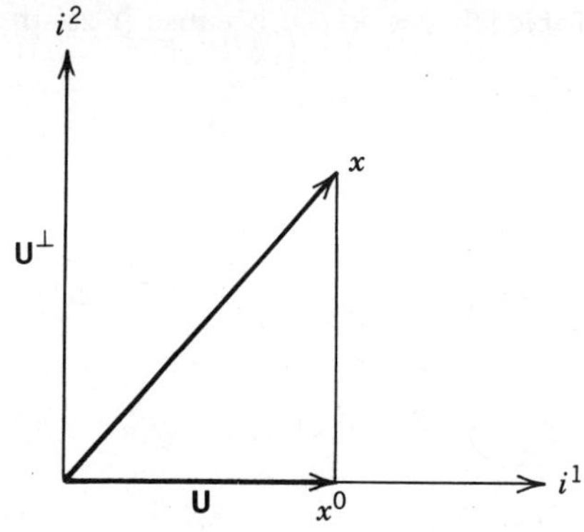

Figure 3.4. An orthogonal projection onto a line in $\mathbf{R}^2$.

Thus the orthogonal projection of x onto $\mathbf{U}$ is the first component of the direct sum decomposition of x induced by the subspace $\mathbf{U}$ and its orthogonal complement $\mathbf{U}^{\perp}$.

Exercise 6.3 (Orthogonal Projections). Show that if x^0 is the orthogonal projection of x onto $\mathbf{U}$, then $x - x^0$ is the orthogonal projection of x onto $\mathbf{U}^{\perp}$. ■

The following is a simple example of a vector x and its orthogonal projections onto the subspaces $\mathbf{U}$ and $\mathbf{U}^{\perp}$.

Example 6.2 (Orthogonal Projection). Let $\mathbf{V} = \mathbf{R}^2$, and let $\mathbf{U}$ be the subspace spanned by the first standard basis vector. That is, $\mathbf{U} = \{\alpha i^1 : \alpha \in \mathbf{F}\}$. Because $\{i^1, i^2\}$ is an orthogonal basis, the orthogonal complement $\mathbf{U}^{\perp}$ is the subspace spanned by the remaining basis vector i^2. That is, $\mathbf{U}^{\perp} = \{\alpha i^2 : \alpha \in \mathbf{F}\}$. For each $x \in \mathbf{R}^2$, we have

$$x = x_1 i^1 + x_2 i^2.$$

Thus $x^0 = x_1 i^1$ is the orthogonal projection of x onto $\mathbf{U}$, while $x - x^0 = x_2 i^2$ is the orthogonal projection of x onto $\mathbf{U}^{\perp}$. Refer to Figure 3.4 for a pictorial description. ■

3.7. ORTHOGONAL EXPANSIONS

An elegant representation of periodic functions is the familiar Fourier series. In this section we show that the Fourier series is actually a special case of a more general formulation called an orthogonal expansion. Roughly speaking, an orthogonal expansion of a vector x is a decomposition of x into the sum of its orthogonal projections onto a complete set of subspaces. We begin by identifying the vectors which span or generate these subspaces.

Definition 7.1 (Complete Orthogonal Set). A sequence $\{x^k\}$ in a Hilbert space **V** is a *complete orthogonal set* in $\mathbf{V} \Leftrightarrow$ it is an orthogonal set and

$$\langle x, x^k \rangle = 0,\ k \geqslant 1 \Rightarrow x = 0. \qquad \blacksquare$$

Thus an orthogonal set is complete if and only if the only vector in **V** that is orthogonal to every member of the set is the zero vector. An equivalent and more traditional way to define the completeness of an orthogonal set X is to say that X is complete if and only if the closure of the subspace spanned by X is the entire space **V**. The formulation in Def. 7.1 is an operational definition in that it can be tested easily. The following simple example is a case in point.

Example 7.1 (Complete Orthogonal Set). For the Hilbert space $\mathbf{l}_2$, let $\{a^k\}$ be the following sequence of orthonormal vectors

$$\begin{aligned} a^1 &= \{1, 0, 0, 0, \ldots\}, \\ a^2 &= \{0, 1, 0, 0, \ldots\}, \\ a^3 &= \{0, 0, 1, 0, \ldots\}. \\ &\vdots \end{aligned}$$

Applying Def. 7.1 we have

$$\begin{aligned} \langle a, a^k \rangle &= \sum_{j=1}^{\infty} a_j \bar{a}_j^k \\ &= \sum_{j=1}^{\infty} a_j I_{kj} \\ &= a_k \end{aligned}$$

Hence $\langle a, a^k \rangle = 0$ for $k \geqslant 1 \Rightarrow a = 0$. Thus $\{a^k\}$ is a complete orthogonal set in $\mathbf{l}_2$. $\blacksquare$

Not every Hilbert space has a countable complete orthogonal set associated with it. Those that do are referred to as separable Hilbert spaces.

Definition 7.2 (Separable Space). An inner product space **V** is *separable* $\Leftrightarrow$ there exists a countable subset U such that the closure of the subspace spanned by U is **V**. $\blacksquare$

Thus a Hilbert space **V** is separable if and only if it contains a countable complete set, a set which is dense in **V**. A countable set is any set which can be placed in a one-to-one correspondence with the nonnegative integers. Thus a sequence is a countable set. For separable Hilbert spaces, a complete orthogonal set of nonzero vectors is the infinite-dimensional equivalent of an orthogonal basis, as can be seen from the following fundamental result.

Proposition 7.1 (Fourier). Let $\{x^k\}$ be a complete orthogonal set of nonzero vectors in a Hilbert space $\mathbf{V}$. Then each $x \in \mathbf{V}$ can be expressed as

$$x = \sum_{k=1}^{\infty} \frac{\langle x, x^k \rangle}{\|x^k\|^2} x^k \qquad \blacksquare$$

The interested reader can refer to Aubin (1979) for a proof of Prop. 7.1. We call the infinite series depicted in Prop. 7.1, an *orthogonal expansion* of x relative to the complete orthogonal set $\{x^k\}$. The coefficients of that expansion are sometimes called the *Fourier coefficients* of x. The name arises from the fact that Fourier's classic series of sines and cosines is an important special case of an orthogonal expansion as seen in the following example.

Example 7.2 (Orthogonal Expansion). Let $\mathbf{V}$ be the subspace of $\mathbf{L}_2([0, 2\pi])$ consisting of periodic functions with a period of 2π. The complex exponentials $\mathbf{X} = \{x^k\}$ form a complete orthogonal set for $\mathbf{V}$. Here if $j = (-1)^{1/2}$, then

$$x^k(t) = \exp(jkt), \qquad -\infty < k < \infty.$$

Hence for each $x \in \mathbf{V}$, from Prop. 7.1 we have

$$\begin{aligned} x(t) &= \sum_{k=-\infty}^{\infty} c_k x^k(t), \\ &= \sum_{k=-\infty}^{\infty} c_k \exp(jkt). \end{aligned}$$

The kth Fourier coefficient is

$$\begin{aligned} c_k &= \frac{\langle x, x^k \rangle}{\|x^k\|^2} \\ &= \frac{\langle x, \exp(jkt) \rangle}{2\pi} \\ &= \frac{1}{2\pi} \int_0^{2\pi} x(t) \exp(-jkt)\, dt. \end{aligned} \qquad \blacksquare$$

Prop. 7.1 applies to finite-dimensional spaces as well. Here a complete orthogonal set is simply an orthogonal basis. The simplest example of a complete orthogonal set for the space $\mathbf{F}^n$ is the standard basis $\mathbf{I}$. Indeed, the Fourier coefficients in this case are simply the components of the vector. The following is another example of a finite-dimensional orthogonal expansion.

Example 7.3 (Finite Orthogonal Expansion). Let $\mathbf{V} = \mathbf{R}^3$, and let $\{x^1, x^2, x^3\}$ be the three orthogonal vectors constructed with the Gram–Schmidt procedure in Exam. 5.3. That is, if $X = [x^1, x^2, x^3]$, then

$$X = \begin{bmatrix} 1 & \frac{1}{2} & \frac{1}{3} \\ 1 & -\frac{1}{2} & -\frac{1}{3} \\ 0 & -1 & \frac{1}{3} \end{bmatrix}.$$

Next, let $x = [1, 0, 2]^T$. Then to expand x in terms of the complete orthogonal set $\{x^1, x^2, x^3\}$, we compute the Fourier coefficients according to Prop. 7.1 as follows:

$$a_1 = \frac{\langle x, x^1 \rangle}{\|x^1\|^2} = \tfrac{1}{2},$$

$$a_2 = \frac{\langle x, x^2 \rangle}{\|x^2\|^2} = -1,$$

$$a_3 = \frac{\langle x, x^3 \rangle}{\|x^3\|^2} = 3.$$

Hence the orthogonal expansion of x in terms of the complete orthogonal set is $x = \frac{1}{2}x^1 - x^2 + 3x^3$. As a check,

$$x = \frac{1}{2}\begin{bmatrix} 1 \\ 1 \\ 0 \end{bmatrix} - \begin{bmatrix} \frac{1}{2} \\ -\frac{1}{2} \\ -1 \end{bmatrix} + 3\begin{bmatrix} \frac{1}{3} \\ -\frac{1}{3} \\ \frac{1}{3} \end{bmatrix} = \begin{bmatrix} 1 \\ 0 \\ 2 \end{bmatrix}.$$ ■

Next we use the notion of an orthogonal expansion to obtain an explicit expression for the orthogonal projection of a vector onto a finite-dimensional subspace $\mathbf{U}$.

Proposition 7.2 (Orthogonal Projection). Let $\mathbf{U}$ be an n-dimensional subspace of a separable Hilbert space $\mathbf{V}$, and let $\{x^1, x^2, \ldots, x^n\}$ be any orthogonal basis for $\mathbf{U}$. Then for each $x \in \mathbf{V}$, the orthogonal projection of x onto $\mathbf{U}$ can be expressed as

$$x^0 = \sum_{k=1}^{n} \frac{\langle x, x^k \rangle}{\|x^k\|^2} x^k.$$

Proof. Let $\{x^{n+1}, x^{n+2}, \ldots\}$ be a complete orthogonal set of nonzero vectors in $\mathbf{U}^\perp$. Then from Prop. 7.1, the sequence $\{x^k\}$ for $k \geqslant 1$ is a complete orthogonal set for $\mathbf{V}$. Hence for each $x \in \mathbf{V}$ from Prop. 7.1 we have

$$x = \sum_{k=1}^{\infty} a_k x^k.$$

Here $a_k = \langle x, x^k\rangle / \|x^k\|^2$ for $k \geqslant 1$. This representation of x can be decomposed as follows:

$$\begin{aligned} x &= \sum_{k=1}^{n} a_k x^k + \sum_{k=n+1}^{\infty} a_k x^k \\ &= x^0 + \sum_{k=n+1}^{\infty} a_k x^k. \end{aligned}$$

Clearly, $x^0 \in \mathbf{U}$. Since the direct sum decomposition of x induced by $\mathbf{U}$ is unique, x^0 is the orthogonal projection of x onto $\mathbf{U} \Leftrightarrow x - x^0 \in \mathbf{U}^{\perp}$. But $x - x^0 \in \mathbf{U}^{\perp}$ because for $1 \leqslant k \leqslant n$,

$$\begin{aligned} \langle x - x^0, x^k\rangle &= \left\langle \sum_{j=n+1}^{\infty} a_j x^j, x^k \right\rangle \\ &= \sum_{j=n+1}^{\infty} a_j \langle x^j, x^k\rangle \\ &= 0. \end{aligned}$$

■

Example 7.4 (Orthogonal Projection). Let $\mathbf{V} = \mathbf{R}^3$, and let U be the subspace of $\mathbf{V}$ spanned by $\{i^1, i^2\}$. Then for each $x \in \mathbf{V}$, the orthogonal projection of x onto $\mathbf{U}$ is

$$\begin{aligned} x &= \sum_{k=1}^{2} \frac{\langle x, i^k\rangle}{\|i^k\|^2} i^k \\ &= \sum_{k=1}^{2} \langle x, i^k\rangle i^k \\ &= \sum_{k=1}^{2} x_k i^k \\ &= [x_1, x_2, 0]^{\mathrm{T}}. \end{aligned}$$

■

The motivation for the use of the term *projection* can be seen in Figure 3.5, which shows the orthogonal projection computed in Exam. 7.4. Here the subspace $\mathbf{U}$ is the plane spanned by the first two standard basis vectors $\{i^1, i^2\}$. The orthogonal projection x^0 of x onto $\mathbf{U}$ corresponds to the shadow the arrow vector x would cast on the $i^1 - i^2$ plane by a light beam projected down from above. Of course the concept of orthogonal projections is much more general since we can speak of projecting sequences and functions onto subspaces as well. Indeed, the beauty of the vector space paradigm is that it allows us to develop powerful algebraic generalizations of simple geometric notions such as these.

In light of Prop. 7.2, we can now develop a geometric interpretation of orthogonal expansions using the notion of projections. Let $\{x^k\}$ be a complete

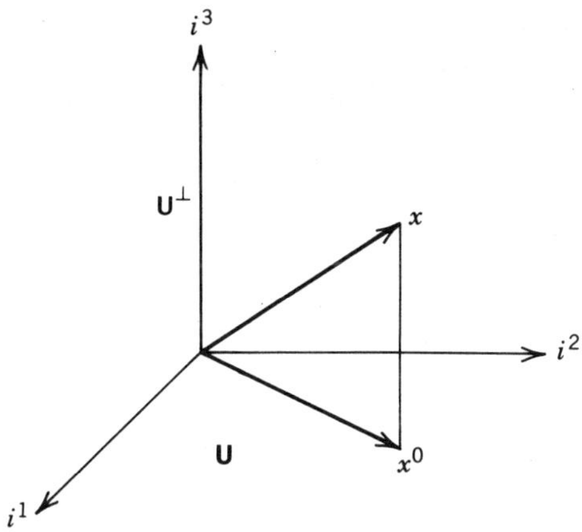

Figure 3.5. An orthogonal projection onto a plane in $\mathbf{R}^3$.

orthogonal set, and let $\mathbf{U_k}$ be the subspace of $\mathbf{V}$ spanned by the kth member of that set x^k. Then from Prop. 7.2 it is clear that the term $(\langle x, x^k\rangle/\|x^k\|^2)x^k$ represents the orthogonal projection of x onto the subspace $\mathbf{U_k}$. Hence the orthogonal expansion of x is simply the decomposition of x into the sum of its projections onto the subspaces spanned by the members of the complete orthogonal set.

Exercise 7.1 (Gram–Schmidt Orthogonalization). Examine the Gram–Schmidt orthogonalization procedure of Alg. 5.1, and explain how the algorithm works in terms of orthogonal projections. ■

If a vector x is projected onto a subspace $\mathbf{U}$, then the length of the orthogonal projection x^0 will never be longer than the length of the original vector x. This simple geometric observation is the idea behind the following general inequality known as Bessel's inequality.

Proposition 7.3 (Bessel Inequality). Let $\{x^k\}$ be an orthonormal sequence in a Hilbert space V. Then for each $x \in V$

$$\sum_{k=1}^{\infty} |\langle x, x^k\rangle|^2 \leqslant \|x\|^2. \quad ■$$

Refer to Aubin (1979) for a proof of Prop. 7.3. The left-hand side of the Bessel inequality can be interpreted as the square of the norm of the orthogonal projection of x onto the closure of the subspace spanned by the orthonormal set. If the set is complete, then the Bessel inequality turns into an equality. The resulting equation in this case is called *Parseval's identity*.

3.8. MINIMUM NORM PROBLEMS

Orthogonal projections can be viewed in still another way. Consider the minimum norm problem. That is, let **U** be a closed subspace of a Hilbert space **V**, and let $x \in \mathbf{V}$. The problem is to find the vector $x^* \in \mathbf{U}$ that is *closest* to x as measured by the natural metric. This problem has many practical manifestations when the vector spaces **V** and **U** are selected appropriately. The solution to the minimum norm problem is given by the following result.

Proposition 8.1 (Minimum Norm). Let **U** be a closed subspace of a Hilbert space **V**, and let x^0 be the orthogonal projection of $x \in \mathbf{V}$ onto **U**. Then for each $y \in \mathbf{U}$

$$\|x - x^0\| \leqslant \|x - y\|.$$

Proof. We consider only the special case when U is finite-dimensional and **V** is separable. The interested reader can refer to Luenberger (1969) for a general proof. Suppose **U** is n-dimensional. Let $\{x^1, x^2, \ldots, x^n\}$ be an orthonormal basis for **U**, and let $\{x^{n+1}, x^{n+2}, \ldots\}$ be a complete orthonormal set for $\mathbf{U}^\perp$. Then from Prop. 6.1 it follows that $\{x^k\}$ for $k \geqslant 1$ is a complete orthonormal set for **V**. Hence for each $(x, y) \in \mathbf{V} \times \mathbf{U}$ we have

$$x = \sum_{k=1}^{\infty} a_k x^k,$$

$$y = \sum_{k=1}^{n} b_k x^k.$$

Here the Fourier coefficients of x and y are $a_k = \langle x, x^k \rangle$ and $b_k = \langle y, x^k \rangle$, respectively. Next, from Prop. 7.2, the orthogonal projection of x onto the subspace **U** can be represented as

$$x^0 = \sum_{k=1}^{n} a_k x^k.$$

Using the Pythagorean theorem we then have

$$\begin{aligned} \|x - x^0\|^2 &= \left\| \sum_{k-1}^{\infty} a_k x^k - \sum_{k=1}^{n} a_k x^k \right\|^2 \\ &= \left\| \sum_{k=n+1}^{\infty} a_k x^k \right\|^2 \\ &= \sum_{k=n+1}^{\infty} \|a_k x^k\|^2 \end{aligned}$$

Again using the Pythagorean theorem and the expressions for x and y we have

$$\begin{aligned}
\|x-y\|^2 &= \left\| \sum_{k=1}^{\infty} a_k x^k - \sum_{k=1}^{n} b_k x^k \right\|^2 \\
&= \left\| \sum_{k=1}^{n} (a_k - b_k) x^k + \sum_{k=n+1}^{\infty} a_k x^k \right\|^2 \\
&= \left\| \sum_{k=1}^{n} (a_k - b_k) x^k \right\|^2 + \left\| \sum_{k=n+1}^{\infty} a_k x^k \right\|^2 \\
&\geqslant \left\| \sum_{k=n+1}^{\infty} a_k x^k \right\|^2 \\
&= \sum_{k=n+1}^{\infty} \|a_k x^k\|^2 \\
&= \|x - x^0\|^2.
\end{aligned}$$

■

Thus the orthogonal projection of a vector $x \in \mathbf{V}$ onto a subspace $\mathbf{U}$ is that vector x^0 in $\mathbf{U}$ which is *nearest* to x as measured by the natural metric. Note that if $x \in \mathbf{U}$, then $x^0 = x$, in which case the distance between them is zero. A pictorial representation of the minimum norm property of orthogonal projections in the space $\mathbf{R}^2$ is shown in Figure 3.6.

Example 8.1. Let $\mathbf{V} = \mathbf{F}^n$, and consider the following optimization problem:

Minimize: $\|x - y\|$

Subject to: $\langle y, c^k \rangle = 0, \qquad 1 \leqslant k \leqslant m.$

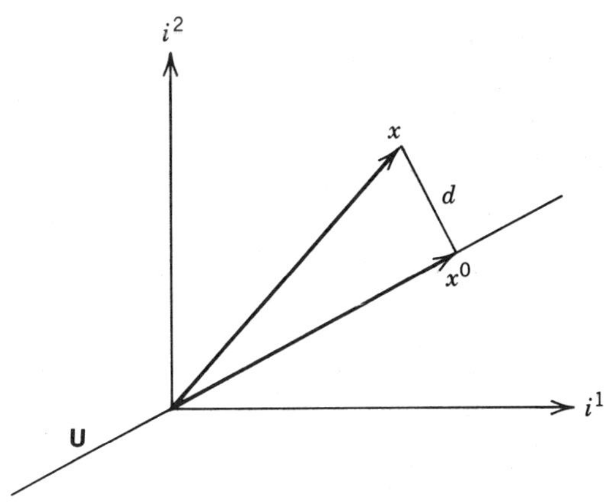

Figure 3.6. The minimum norm property of orthogonal projections.

Here $\{c^1, c^2, \ldots, c^m\}$ are arbitrary vectors in $\mathbf{F}^n$. This is a problem for which Prop. 8.1 is applicable because the set of all vectors satisfying the *equality constraints* is a closed subspace of $\mathbf{V}$. ■

The minimum norm or nearest vector property can be generalized in an important direction. Recall from Chapter 1 that every subspace is a convex set. If we replace the closed subspace $\mathbf{U}$ with a closed convex set S, we get the following generalization of Prop. 8.1.

Proposition 8.2 (Convex Sets). Let $\mathbf{V}$ be a Hilbert space, and let S be a closed convex subset of $\mathbf{V}$. Then for each $x \in \mathbf{V}$ there exists a unique $x^0 \in S$ such $\|x - x^0\| \leqslant \|x - y\|$ for all $y \in S$. Moreover, x^0 is the unique vector with this property $\Leftrightarrow \langle x - x^0, y - x^0 \rangle \leqslant 0$ for each $y \in S$.

A proof of Prop. 8.2 can be found in Luenberger (1969). The utility of Prop. 8.2 lies in the fact that it is quite widely applicable in engineering applications.

Example 8.2. Let $\mathbf{V} = \mathbf{F}^n$ and consider the following more general optimization problem which includes both *equality constraints* and *inequality constraints*:

Minimize: $\|x - y\|$

Subject to: $\langle y, c^k \rangle = 0, \quad 1 \leqslant k \leqslant m.$

$\langle y, d^k \rangle \leqslant 0, \quad 1 \leqslant k \leqslant p.$ ■

Here $\{c^1, c^2, \ldots, c^m\}$ and $\{d^1, d^2, \ldots, d^p\}$ are arbitrary vectors in $\mathbf{F}^n$. The set of all vectors satisfying both sets of constraints is no longer a subspace. However, it is a closed convex set and therefore Prop. 8.2 is applicable. ■

3.9. ISOMORPHISMS REVISITED

Let $\mathbf{V}$ and $\mathbf{W}$ be vector spaces over the same field $\mathbf{F}$. Recall that a vector space isomorphism between $\mathbf{V}$ and $\mathbf{W}$ is a linear, one-to-one function mapping $\mathbf{V}$ onto $\mathbf{W}$. We generalize this notion to inner product spaces as follows.

Definition 9.1 (Inner Product Space Isomorphism). Let $\mathbf{V}$ and $\mathbf{W}$ be inner product spaces defined over the same field $\mathbf{F}$. An *isomorphism* between $\mathbf{V}$ and $\mathbf{W}$ is any function f mapping $\mathbf{V}$ into $\mathbf{W}$ with the following properties:

1. f is a vector space isomorphism,
2. $(x, y) \in \mathbf{V}^2 \Rightarrow \langle f(x), f(y) \rangle_{\mathbf{W}} = \langle x, y \rangle_{\mathbf{V}}$.

In this case we say the inner product spaces $\mathbf{V}$ and $\mathbf{W}$ are *isomorphic*. ■

Thus an inner product space isomorphism is a linear, one-to-one, onto mapping which *preserves the inner product*. The following exercise shows that an inner product space isomorphism also preserves the natural norm.

Exercise 9.1 (Natural Norm). Show that if f is an inner product space isomorphism between **V** and **W**, then for each $x \in \mathbf{V}$

$$\|f(x)\|_{\mathbf{W}} = \|x\|_{\mathbf{V}}. \qquad \blacksquare$$

Isomorphic inner product spaces have the same underlying topology. As such, complete orthogonal sets in one space generate complete orthogonal sets in the other space in the following manner.

Proposition 9.1 (Complete Orthogonal Sets). Let f be an inner product space isomorphism between the Hilbert spaces **V** and **W**. Then $\{x^k\}$ is a complete orthogonal set in $\mathbf{V} \Leftrightarrow \{f(x^k)\}$ is a complete orthogonal set in **W**.

Proof. From Def. 9.1 the set $\{x^k\}$ is orthogonal in $\mathbf{V} \Leftrightarrow$ the set $\{f(x^k)\}$ is orthogonal in **W**. Suppose the set $\{x^k\}$ is not complete. Then from Def. 7.1 there exists an $x \in \mathbf{V}$ with $x \neq 0$ such that

$$\langle x, x^k \rangle_{\mathbf{V}} = 0, \qquad k \geqslant 1.$$

Hence $\langle f(x), f(x^k) \rangle_{\mathbf{W}} = 0$ for $k \geqslant 1$. But since f is linear and one-to-one, $f(x) = 0 \Leftrightarrow x = 0$. Thus $f(x) \neq 0$, which implies $\{f(x^k)\}$ is not complete in **W**. Next, suppose $\{f(x^k)\}$ is not complete in **W**. Then there exists a $y \in \mathbf{W}$ with $y \neq 0$ such that

$$\langle y, f(x^k) \rangle_{\mathbf{W}} = 0, \qquad k \geqslant 1.$$

Since f is one-to-one and onto, it is invertible, in which case we have $\langle f^{-1}(y), x^k \rangle_{\mathbf{V}} = 0$ for $k \geqslant 1$. But since f is linear and invertible, $y \neq 0 \Leftrightarrow f^{-1}(y) \neq 0$. Hence $\{x^k\}$ is not complete in **V**. $\blacksquare$

Recall that the importance of the space $\mathbf{F}^n$ among finite-dimensional spaces lies in the fact that $\mathbf{F}^n$ is isomorphic to every n-dimensional vector space **V**. The space ℓ_2 is, in a sense, the infinite-dimensional counterpart of $\mathbf{F}^n$, as can be seen from the following result.

Proposition 9.2. The Hilbert space ℓ_2 is isomorphic to the Hilbert space $\mathbf{L_2(T, w)}$.

Proof. Let $\{x^k\}$ be a complete orthonormal set for $\mathbf{L_2(T, w)}$, and let $f: \ell_2 \to \mathbf{L_2(T, w)}$ be defined as follows:

$$f(a) = \sum_{k=1}^{\infty} a_k x^k.$$

We will show that f is linear, one-to-one, onto, and inner product preserving. Let $(\alpha, a, b) \in \mathbf{F} \times \ell_2 \times \ell_2$. Then,

$$\begin{aligned} f(\alpha a + b) &= \sum_{k=1}^{\infty} (\alpha a + b)_k x^k \\ &= \sum_{k=1}^{\infty} (\alpha a_k + b_k) x^k \\ &= \sum_{k=1}^{\infty} \alpha a_k x^k + \sum_{k=1}^{\infty} b_k x^k \\ &= \alpha \sum_{k=1}^{\infty} a_k x^k + \sum_{k=1}^{\infty} b_k x^k \\ &= \alpha f(a) + f(b). \end{aligned}$$

Hence f is linear. Suppose there exist $(a, b) \in \ell_2 \times \ell_2$ with $a \neq b$ such that $f(a) = f(b)$. Then since f is linear and $\{x^k\}$ is orthonormal:

$$\begin{aligned} 0 &= \|0\|^2 \\ &= \|f(a) - f(b)\|^2 \\ &= \|f(a - b)\|^2 \\ &= \left\| \sum_{k=1}^{\infty} (a - b)_k x^k \right\|^2 \\ &= \sum_{k=1}^{\infty} \|(a - b)_k x^k\|^2 \\ &= \sum_{k=1}^{\infty} |(a - b)_k|^2 \|x^k\|^2 \\ &= \sum_{k=1}^{\infty} |(a - b)_k|^2. \end{aligned}$$

Thus $a = b$, which is a contradiction. Hence f is one-to-one. Next, let $x \in \mathbf{L_2(T, w)}$. Then since $\{x^k\}$ is a complete orthonormal set for $\mathbf{L_2(T, w)}$, from Prop. 7.1 we have

$$x = \sum_{k=1}^{\infty} \langle x, x^k \rangle x^k.$$

Let $a_k = \langle x, x^k \rangle$ for $k \geqslant 1$. Then because $\{x^k\}$ is orthonormal,

$$\|x\|^2 = \left\| \sum_{k=1}^{\infty} a_k x^k \right\|^2,$$

$$
\begin{aligned}
&= \sum_{k=1}^{\infty} \|a_k x^k\|^2, \\
&= \sum_{k=1}^{\infty} |a_k|^2 \|x^k\|^2, \\
&= \sum_{k=1}^{\infty} |a_k|^2, \\
&= \|a\|.
\end{aligned}
$$

Thus for each $x \in \mathbf{L_2(T, w)}$ there exists an $a \in \ell_2$ such that $f(a) = x$. Hence f is an onto mapping. Finally, let $(a, b) \in \ell_2 \times \ell_2$. Then because $\{x^k\}$ is orthonormal,

$$
\begin{aligned}
\langle f(a), f(b) \rangle &= \left\langle \sum_{k=1}^{\infty} a_k x^k, \sum_{j=1}^{\infty} b_j x^j \right\rangle, \\
&= \sum_{k=1}^{\infty} a_k \left\langle x^k, \sum_{j=1}^{\infty} b_j x^j \right\rangle, \\
&= \sum_{k=1}^{\infty} \sum_{j=1}^{\infty} a_k \bar{b}_j \langle x^k, x^j \rangle, \\
&= \sum_{k=1}^{\infty} a_k \bar{b}_k \langle x^k, x^k \rangle, \\
&= \sum_{k=1}^{\infty} a_k \bar{b}_k \|x^k\|^2, \\
&= \sum_{k=1}^{\infty} a_k \bar{b}_k, \\
&= \langle a, b \rangle.
\end{aligned}
$$

Hence f is inner product preserving. Thus the Hilbert spaces ℓ_2 and $\mathbf{L_2(T, w)}$ are isomorphic. ■

The significance of Prop. 9.2 lies in the fact that instead of working with functions in the space $\mathbf{L_2(T, w)}$ we can choose a complete orthonormal set $\{x^k\}$ and then work with the Fourier coefficients of the functions. That is, we can work with sequences in the space ℓ_2, as shown in Figure 3.7. The Fourier coefficients can be thought of as the *coordinates* of a function with respect to the

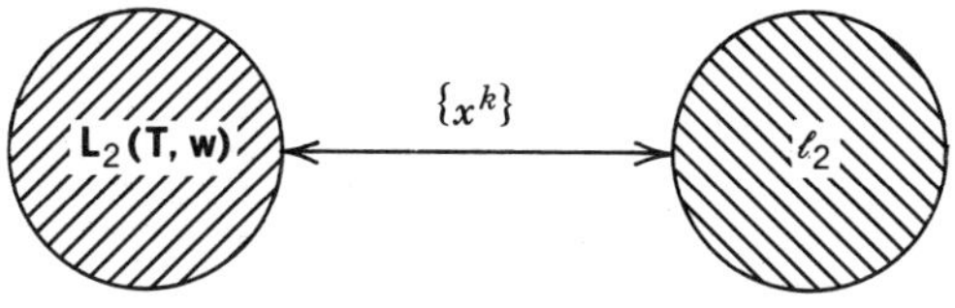

Figure 3.7. Isomorphism between $\mathbf{l_2}$ and $\mathbf{L_2(T, w)}$.

complete orthonormal set. That is, a complete orthonormal set is the infinite-dimensional equivalent of an orthonormal basis. Prop. 9.2 can be generalized (Aubin, 1979) in the sense that the space $\mathbf{L_2(T, w)}$ can be replaced by *any* separable Hilbert space.

3.10. ORTHOGONAL POLYNOMIALS

The purpose of this section is to develop and summarize some well-known examples of complete orthogonal sets. Consider the Hilbert space $\mathbf{L_2(T, w)}$ where $T = [\alpha, \beta]$ with $\alpha < \beta$. For each interval T in $\mathbf{R}$ one can find a weighting function w such that the elementary polynomials $\{t^k : k \geqslant 0\}$ belong to the space $\mathbf{L_2(T, w)}$. We can then apply the Gram–Schmidt procedure of Alg. 5.1 to construct a complete orthogonal set $\{f^k(t)\}$ for $\mathbf{L_2(T, w)}$ of the general form

$$f^k(t) = \sum_{j=0}^{k} a_j t^j, \qquad k \geqslant 0.$$

By scaling $f^k(t)$ with an appropriate scalar, we can standardize the norm such that $\|f^k\|^2 = g(k)$ for $k \geqslant 0$. In this manner we generate examples of the classic orthogonal polynomials. The classic orthogonal polynomials have many additional properties in common. Among these are the following. Each classic polynomial satisfies a three-term recurrence relation of the form

$$a_1(k) f^{k+1}(t) = a_2(k, t) f^k(t) + a_3(k) f^{k-1}(t).$$

In addition, each of the classic orthogonal polynomials is the solution of a linear second-order ordinary differential equation of the general form

$$b_1(t) \frac{d^2}{dt^2} f^k(t) + b_2(t) \frac{d}{dt} f^k(t) + c_k f^k(t) = 0.$$

Finally, each of the classic orthogonal polynomials can be expressed by its *Rodrigues formula*

$$f^k(t) = \frac{1}{\alpha_k w(t)} \frac{d^k}{dt^k} \{w(t)[h(t)]^k\}, \qquad k \geqslant 0.$$

The following are some examples of classic orthogonal polynomials that can be generated in this manner.

Example 10.1 (Legendre Polynomials: $T = [-1, 1]$). For the space $\mathbf{L_2([-1, 1])}$ where $w(t) = 1$, the *Legendre polynomials* $\{P^k(t)\}$ form a complete orthogonal set

$$P^k(t) = \frac{1}{2^k} \sum_{j=0}^{[k/2]} (-1)^j \binom{k}{j} \binom{2k-2j}{j} t^{k-2j}, \qquad k \geqslant 0.$$

Here $\binom{k}{j}$ denotes the binomial coefficient and $[k/2]$ denotes the largest integer less than or equal to $k/2$. The associated differential equation is

$$(1 - t^2) \frac{d^2}{dt^2} P^k(t) - 2t \frac{d}{dt} P^k(t) + k(k + 1)P^k(t) = 0.$$

In this case $\|P^k\|^2 = 2/(2k + 1)$. The first two Legendre polynomials are $P^0(t) = 1$ and $P^1(t) = t$, while the recurrence relation for generating the remaining ones is

$$(k + 1)P^{k+1}(t) = (2k + 1)tP^k(t) - kP^{k-1}(t).$$

Finally, the Rodrigues formula is

$$P^k(t) = \frac{(-1)^k}{k!2^k} \frac{d^k}{dt^k} \{(1 - t^2)^k\}.$$

Refer to Figure 3.8 for a graphical display of the first few Legendre polynomials. ■

Example 10.2 (Laguerre Polynomials: $T = [0, \infty)$). For the space $\mathbf{L_2([0, \infty), w)}$ with $w(t) = \exp(-t)$, the *Laguerre polynomials* $\{L^k(t)\}$ form a complete orthogonal set

$$L^k(t) = \sum_{j=1}^{k} \frac{(-1)^j}{j!} \binom{k}{k-j} t^j, \qquad k \geq 0.$$

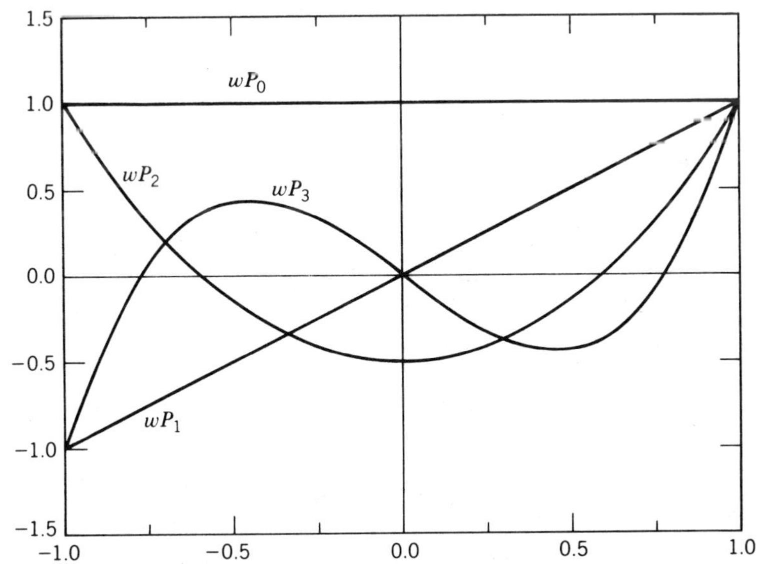

Figure 3.8. The Legendre polynomials.

The associated differential equation is

$$t \frac{d^2}{dt^2} L^k(t) + (1 - t) \frac{d}{dt} L^k(t) + kL^k(t) = 0.$$

In this case $\|L^k\|^2 = k!$. The first two Laguerre polynomials are $L^0(t) = 1$ and $L^1(t) = 1 - t$, while the recurrence relation for generating the remaining ones is

$$(k + 1)L^{k+1}(t) = (2k + 1 - t)L^k(t) - kL^{k-1}(t).$$

Finally, the Rodrigues formula is

$$L^k(t) = \frac{1}{k! \exp(-t)} \frac{d^k}{dt^k} \{t^k \exp(-t)\}.$$

Refer to Figure 3.9 for a graphical display of the first few Laguerre polynomials. ■

Example 10.3 (Hermite Polynomials: $T = (-\infty, \infty)$). For the space $\mathbf{L_2((-\infty, \infty), w)}$ with $w(t) = \exp(-t^2)$, the *Hermite polynomials* $\{H^k(t)\}$ form a complete orthogonal set

$$H_k(t) = k! \sum_{j=1}^{[k/2]} (-1)^j \frac{1}{j!(k - 2j)!} (2t)^{k-2j}, \qquad k \geqslant 0.$$

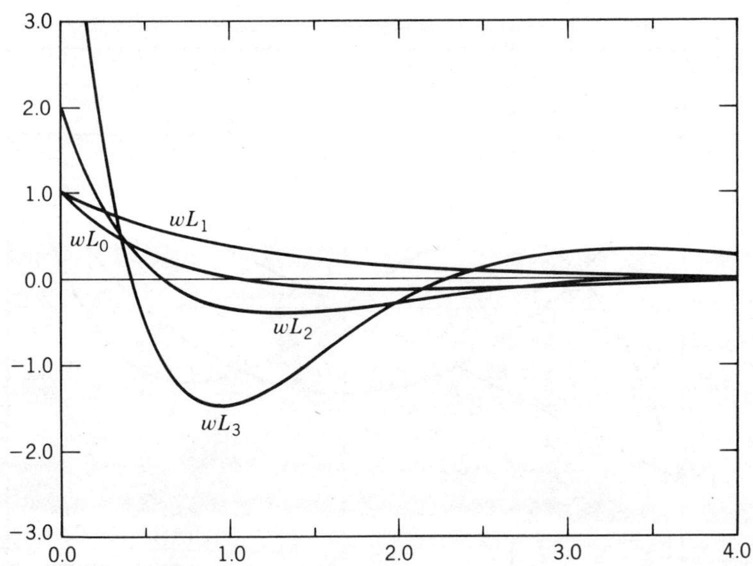

Figure 3.9. The Laguerre polynomials.

The associated differential equation is

$$\frac{d^2}{dt^2} H^k(t) - 2t \frac{d}{dt} H^k(t) + 2kH^k(t) = 0.$$

In this case $\|H^k\|^2 = k!2^k\pi^{1/2}$. The first two Hermite polynomials are $H^0(t) = 1$ and $H^1(t) = 2t$, while the recurrence relation for generating the remaining ones is

$$H^{k+1}(t) = 2tH^k(t) - 2kH^{k-1}(t).$$

Finally, the Rodrigues formula is

$$H^k(t) = (-1)^k \exp(t^2) \frac{d^k}{dt^k} \exp(-t^2).$$

Refer to Figure 3.10 for a graphical display of the first few Hermite polynomials. ■

Example 10.4 (Chebyshev Polynomials: $T = [-1, 1]$). For the space $\mathbf{L_2([-1, 1], w)}$ with $w = 1/\sqrt{1 - t^2}$, the *Chebyshev polynomials* $\{T^k(t)\}$ form a complete orthogonal set

$$T_k(t) = \frac{k}{2} \sum_{j=0}^{[k/2]} (-1)^j \frac{(k-j-1)!}{j!(k-2j)!} (2t)^{k-2j}, \qquad k \geq 0.$$

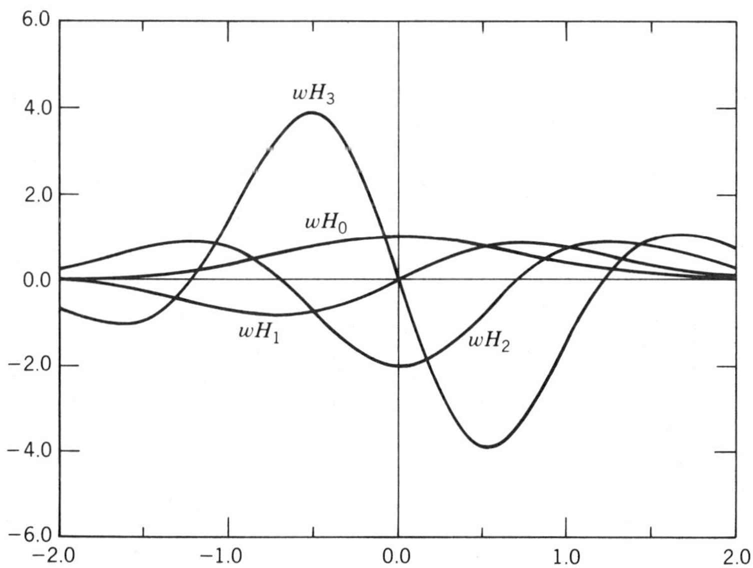

Figure 3.10. The Hermite polynomials.

The associated differential equation is

$$(1 - t^2)\frac{d^2}{dt^2}T^k(t) - t\frac{d}{dt}T^k(t) + k^2T^k(t) = 0.$$

In this case $\|T^0\|^2 = \pi$ and $\|T^k\|^2 = \pi/2$ for $k \geqslant 1$. The first two Chebyshev polynomials are $T^0(t) = 1$ and $T^1(t) = t$, while the recurrence relation for generating the remaining ones is

$$T^{k+1}(t) = 2tT^k(t) - T^{k-1}(t).$$

Finally, the Rodrigues formula is

$$T^k(t) = \frac{(-1)^k((1 - t^2)\pi)^{1/2}}{2^{k+1}\gamma(k + 1/2)}\frac{d^k}{dt^k}\{(1 - t^2)^{(k-1/2)}\}.$$

Here γ denotes the gamma function. Refer to Figure 3.11 for a graphical display of the first few Chebyshev polynomials. These are known as the Chebyshev polynomials of the first kind. ■

A tabular summary of the four families of classic polynomials presented in Exams. 10.1–10.4 is provided in Table 3.1. These examples of complete orthogonal sets are by no means exhaustive (Spiegel, 1968).

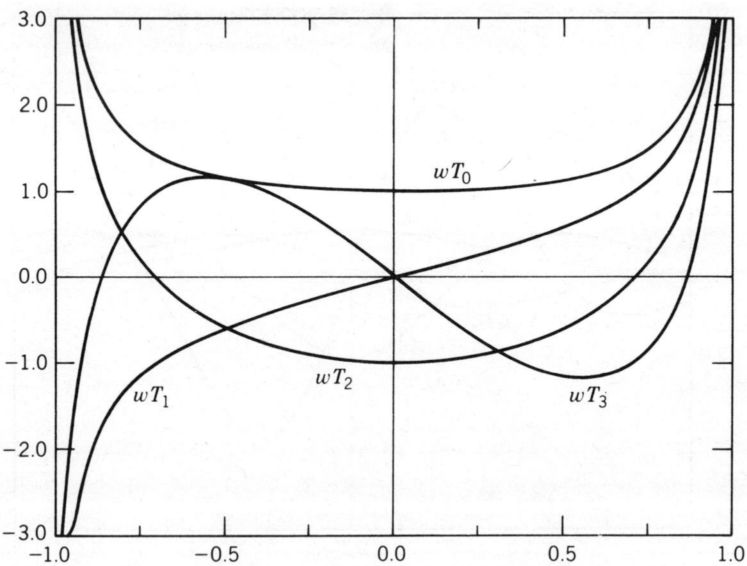

Figure 3.11. The Chebyshev polynomials.

TABLE 3.1. Classic Orthogonal Polynomials

Name	T	$w(t)$	$g(k)$
Legendre	$[-1, 1]$	1	$\frac{2}{2k+1}$
Laguerre	$[0, \infty)$	$\exp(-t)$	$k!$
Hermite	$(-\infty, \infty)$	$\exp(-t^2)$	$k!2\pi$
Chebyshev	$[-1, 1]$	$\sqrt{(1-t^2)}$	$\pi, \frac{\pi}{2}$

PROBLEMS

1. Show that the inner product proposed in Exam. 1.1 is a valid inner product for $\mathbf{F}^n$.

2. Show that the inner product proposed in Exam. 1.2 is a valid inner product for $\mathbf{F}^{n \times n}$.

3. Show that the inner product proposed in Exam. 1.3 is a valid inner product for $\mathbf{L}_2(\mathbf{T}, \mathbf{w})$.

4. As a basis for $\mathbf{R}^3$ consider the columns of X where

$$X = \begin{bmatrix} 1 & 0 & -1 \\ 0 & 1 & 1 \\ 1 & 2 & 0 \end{bmatrix}$$

   a. Let $x = [1, -2, 0]^T$. Find $[x]^{\mathbf{X}}$.
   b. Apply Alg. 5.1 to the columns of X to construct an orthogonal basis $\mathbf{Y} = \{y^1, y^2, y^3\}$.
   c. Use Prop. 7.1 to find $[x]^{\mathbf{Y}}$.

5. Consider the following complete orthogonal set for $\mathbf{L}_2([0, 1])$:

$$x^k(t) = \cos(\pi k t), \qquad k \geqslant 0.$$

   a. Let $\alpha \in (0, 1)$, and define $x(t) = 1/\alpha$ for $0 \leqslant t \leqslant \alpha$ and $x(t) = 0$ for $\alpha < t \leqslant 1$. Show that $x \in \mathbf{L}_2([0, 1])$.
   b. Find the orthogonal expansion of x relative to the complete orthogonal set $\{x^k\}$.

6. Show that an inner product on $\mathbf{V}$ is a continuous function mapping $\mathbf{V} \times \mathbf{V}$ into $\mathbf{F}$.

7. Consider the columns of the following matrix $X \in \mathbf{R}^{3\times 3}$:

$$X = \begin{bmatrix} 0 & 1 & 2 \\ 1 & -1 & -3 \\ 1 & 0 & -1 \end{bmatrix}.$$

a. Find an orthogonal basis for the subspace $\mathbf{U}$ spanned by x^1, x^2, and x^3.
b. Let $x = [1, 3, -2]^T$. Find the orthogonal projection of x onto $\mathbf{U}$.
c. Find the orthogonal projection of x onto $\mathbf{U}^\perp$.

8. Let $\mathbf{R}^{2\times 2}$ be the space of 2×2 real matrices, and let $\mathbf{R}_{sy}^{2\times 2}$ be the subspace of 2×2 symmetric matrices.

a. Find an orthogonal basis for $\mathbf{R}_{sy}^{2\times 2}$ using the inner product defined in Example 1.2.
b. Let A be the following matrix. Find the orthogonal projection of A onto $\mathbf{R}_{sy}^{2\times 2}$

$$A = \begin{bmatrix} 1 & 2 \\ 3 & 4 \end{bmatrix}.$$

c. Show that $(\mathbf{R}_{sy}^{2\times 2})^\perp = \mathbf{R}_{sk}^{2\times 2}$. That is, show that the orthogonal complement of the space of symmetric matrices is the space of skew-symmetric matrices.
d. Find the orthogonal projection of A onto $\mathbf{R}_{sk}^{2\times 2}$.

9. Let $\mathbf{V}$ and $\mathbf{W}$ be isomorphic inner product spaces. Show that $\mathbf{V}$ is a Hilbert space $\Leftrightarrow$ $\mathbf{W}$ is a Hilbert space.

10. Let $(\alpha, \beta) \in \mathbf{R}^2$ with $\alpha < \beta$. Find a complete orthogonal set for $\mathbf{L}_2([\alpha, \beta])$.

11. Let $\mathbf{V}$ be a complex inner product space, and let $j = \sqrt{-1}$. Verify the *polarization identity* for the space $\mathbf{V}$. That is, show that for each $(x, y) \in \mathbf{V}^2$

$$\langle x, y \rangle = \sum_{k=1}^{4} j^k \|x + j^k y\|^2.$$

REFERENCES

Aubin, Jean-Pierre, *Applied Functional Analysis*, Wiley, New York, 1979.

Bachman, G. B. and L. Narici, *Functional Analysis*, Academic, New York, 1966.

Luenberger, D. G., *Optimization by Vector Space Methods*, Wiley, New York, 1969.

Spiegel, M. R., *Mathematical Handbook of Formulas and Tables*, Schaum's Outline Series, McGraw-Hill, New York, 1968.

4 LINEAR TRANSFORMATIONS

In the previous chapters we examined the properties of vector spaces in general. At this point, we begin to focus on the properties of vectors belonging to one particular space, the space of linear transformations.

4.1. LINEAR TRANSFORMATIONS

Let **V** and **W** be two vector spaces defined over the same field **F**. Recall from Chapter 1 that the set of all functions mapping **V** into **W** is itself a vector space denoted **F**(**V**, **W**). Functions of this type are given a special name: they are called transformations.

Definition 1.1 (Transformation). Let **V** and **W** be vector spaces over the same field **F**. Then A is a *transformation* from **V** into **W** $\Leftrightarrow A: \mathbf{V} \to \mathbf{W}$. ■

Thus a transformation is simply a function mapping one vector space into another. Refer to Fig. 4.1 for a pictorial representation of a transformation between the spaces **V** and **W**.

Because the vectors in **V** and **W** can themselves be functions, a special notation is used for transformations between spaces. For each vector $x \in \mathbf{V}$, the corresponding vector in **W** is denoted Ax rather than $A(x)$. In particular, we refer to Ax as the image of the vector x under the transformation A. More generally, we define the image of a set of vectors under a transformation in the following manner.

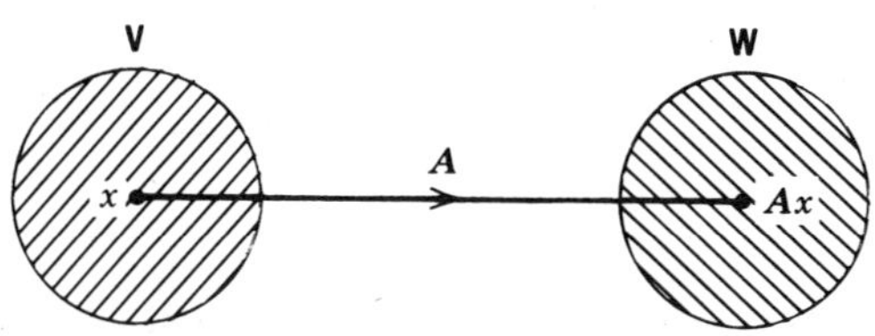

Figure 4.1. A transformation A from **V** into **W**.

Definition 1.2 (Image). Let $A: \mathbf{V} \to \mathbf{W}$, and let U be a subset of $\mathbf{V}$. Then the *image* of U under A is denoted AU and is defined

$$AU = \{Ax \in \mathbf{W} : x \in U\}. \qquad \blacksquare$$

The set of all transformations can be decomposed into a number of broad categories. To that end, the following two properties are fundamental.

Definition 1.3 (One-to-One). A transformation $A: \mathbf{V} \to \mathbf{W}$ is *one-to-one* or *injective* $\Leftrightarrow$ no two distinct vectors in $\mathbf{V}$ have the same image in $\mathbf{W}$. That is, for each $(x, y) \in \mathbf{V}^2$,

$$x \neq y \Rightarrow Ax \neq Ay. \qquad \blacksquare$$

Definition 1.4 (Onto). A transformation $A: \mathbf{V} \to \mathbf{W}$ is *onto* or *surjective* $\Leftrightarrow$ every vector in $\mathbf{W}$ is the image of at least one vector in $\mathbf{V}$. That is, for each $y \in \mathbf{W}$ there exists an $x \in \mathbf{V}$ such that

$$Ax = y. \qquad \blacksquare$$

Note that another way of defining A as onto or surjective is to say that the image of $\mathbf{V}$ under A is $\mathbf{W}$. When a transformation $A: \mathbf{V} \to \mathbf{W}$ is onto or surjective, we sometimes say that A maps $\mathbf{V}$ onto $\mathbf{W}$ rather than into $\mathbf{W}$. Transformations which are both one-to-one and onto are sometimes called *bijective* transformations.

Exercise 1.1. Show that the notions of one-to-one and onto are independent and meaningful by constructing examples of transformations of the form $A: \mathbf{R}^2 \to \mathbf{R}^2$ which are

1. Neither one-to-one nor onto,
2. One-to-one but not onto,
3. Onto but not one-to-one,
4. Both one-to-one and onto. ■

There is one more category of transformations that is of fundamental significance, the set of linear transformations.

Definition 1.5 (Linear). A transformation $A: \mathbf{V} \to \mathbf{W}$ is *linear* $\Leftrightarrow$ for each $(\alpha, \beta, x, y) \in \mathbf{F}^2 \times \mathbf{V}^2$:

$$A(\alpha x + \beta y) = \alpha Ax + \beta Ay \qquad \blacksquare$$

Thus a transformation A is linear if and only if the image of every linear combination of vectors is equal to the linear combination of the images. There are several important properties that set linear transformations apart from transformations in general. The following exercise summarizes some of the more useful ones.

Exercise 1.2 (Linear Transformations). Let $A:\mathbf{V} \to \mathbf{W}$ be a linear transformation. Show that for each $(\alpha, x, y) \in \mathbf{F} \times \mathbf{V}^2$,

1. $A(\alpha x) = \alpha A x$,
2. $A(x + y) = Ax + Ay$,
3. $A0 = 0$. ■

Properties 1 and 2 of Exer. 1.2 are sometimes referred to as the *homogeneity* and *additivity* properties of linear transformations. The set of all linear transformations between **V** and **W** forms a vector space. To see this, recall that a linear transformation is a special case of a function mapping a vector space **V** into a vector space **W**. Thus the set of all linear transformations is a subset of the vectors space **F(V, W)** defined in Chapter 1. This subset is nonempty since the transformation $Ax \triangleq 0$ for each $x \in \mathbf{V}$ is linear. Moreover, for each $(\alpha, \beta) \in \mathbf{F}^2$ if A and B are linear transformations from **V** into **W**, then $\alpha A + \beta B$ is also a linear transformation from **V** into **W**. Thus from Prop. 1.2.1, the set of linear transformations forms a subspace of **F(V, W)**. It is therefore a vector space in its own right. We denote this vector space as **L(V, W)**. That is,

$$\mathbf{L}(\mathbf{V}, \mathbf{W}) \triangleq \{A \in \mathbf{F}(\mathbf{V}, \mathbf{W}) : A \text{ is linear}\}.$$

Throughout the remainder of this chapter we focus our attention primarily on the properties of vectors in **L(V, W)**. There are many important examples of linear transformations. We will begin by considering a few of the more common ones.

Example 1.1 (Matrix Transformation). Let $\mathbf{V} = \mathbf{F}^{\mathbf{n}}$ and $\mathbf{W} = \mathbf{F}^{\mathbf{m}}$. Next, let A be any $m \times n$ *matrix*, with components in **F**. That is, $A \in \mathbf{F}^{\mathbf{m} \times \mathbf{n}}$. Then $A:\mathbf{F}^{\mathbf{n}} \to \mathbf{F}^{\mathbf{m}}$ is a linear transformation. As we shall see, this particular linear transformation is of special significance. ■

Example 1.2 (Derivative Transformation). Let $T = [\alpha, \beta]$ be an interval in **R**, let $\mathbf{V} = \mathbf{P}^{\mathbf{n}}(\mathbf{T})$, and let $\mathbf{W} = \mathbf{P}(\mathbf{T})$. Next, let A denote the *derivative* transformation. That is, for each $p \in \mathbf{V}$

$$(Ap)(t) = \frac{d}{dt}\, p(t), \qquad t \in T.$$

Then $A:\mathbf{P}^{\mathbf{n}}(\mathbf{T}) \to \mathbf{P}(\mathbf{T})$ is a linear transformation. ■

Example 1.3 (Integral Transformation). Let $T = [\alpha, \beta]$ be an interval in **R**, let $\mathbf{V} = \mathbf{C(T)}$, and let $\mathbf{W} = \mathbf{F(T)}$. Here **F(T)** is the space of functions mapping T into **R**, and **C(T)** is the space of continuous functions mapping T into **R**. Next, let A denote the *integral* transformation. That is, for each $f \in \mathbf{V}$

$$(Af)(t) = \int_\alpha^t f(\tau)\, d\tau, \qquad t \in T.$$

Then $A: \mathbf{C(T)} \to \mathbf{F(T)}$ is a linear transformation. ■

Example 1.4 (Inner Product Transformation). Let **V** be an inner product space over **F** and let $y \in \mathbf{V}$. Next, let $\mathbf{W} = \mathbf{F}$, and let A be defined as the *inner product* with y. That is, for each $x \in \mathbf{V}$

$$Ax = \langle x, y \rangle.$$

Then $A: \mathbf{V} \to \mathbf{F}$ is a linear transformation. Since $\mathbf{W} = \mathbf{F}$, we call this an example of a linear functional on **V**. ■

4.2. THE NULL SPACE

Associated with each linear transformation $A \in \mathbf{L(V, W)}$ is a special subspace of **V**. We call this subspace the null space of **A**.

Definition 2.1 (Null Space, Nullity). Let $A \in \mathbf{L(V, W)}$. Then the *null space* of A is denoted $\mathbf{N}(A)$ and is defined as

$$\mathbf{N}(A) = \{x \in \mathbf{V} : Ax = 0\}.$$

If $\mathbf{N}(A)$ is finite-dimensional, then the dimension of $\mathbf{N}(A)$ is called the *nullity* of A. ■

Thus the null space of A is the set of all vectors in **V** whose image under A is the zero vector in **W**. Refer to Figure 4.2 for a pictorial description of the null space of a linear transformation.

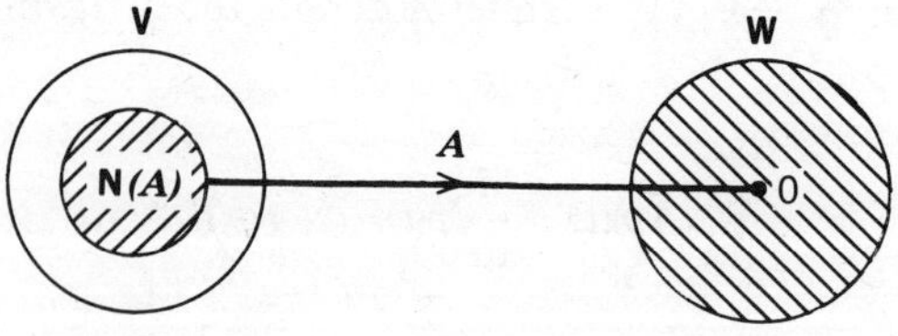

Figure 4.2. The null space of a linear transformation A.

Exercise 2.1 (Null Space). Let $A \in \mathbf{L}(\mathbf{V}, \mathbf{W})$. Show that $\mathbf{N}(A)$ is a subspace of $\mathbf{V}$. ■

The content of the null space determines whether or not a linear transformation is one-to-one or injective. Consider, in particular, the following result.

Proposition 2.1 (One-to-One). Let $A \in \mathbf{L}(\mathbf{V}, \mathbf{W})$. Then A is one-to-one $\Leftrightarrow \mathbf{N}(A) = \{\mathbf{0}\}$.

Proof. Suppose $\mathbf{N}(A) \neq \{\mathbf{0}\}$. Then there exists an $x \in \mathbf{N}(A)$ with $x \neq 0$. Because A is linear, from Exer. 1.2 and Def. 2.1 we have

$$\begin{aligned} Ax &= 0, \\ &= A0. \end{aligned}$$

Hence A is not one-to-one. Next, suppose A is not one-to-one. Then there exists $(x, y) \in \mathbf{V}^2$ with $x \neq y$ such that $Ax = Ay$. Let $z = x - y$. Then since A is linear, we have

$$\begin{aligned} Az &= A(x - y), \\ &= Ax - Ay, \\ &= 0. \end{aligned}$$

But $z \neq 0$. Thus $\mathbf{N}(A) \neq \{\mathbf{0}\}$. ■

Note that the zero vector *always* belongs to the null space. If this is the only vector in the null space, then and only then is the linear transformation one-to-one.

Example 2.1 (Null Space). Consider the derivative transformation discussed in Exam. 1.2. In this case,

$$\mathbf{N}(A) = \mathbf{P}^0(\mathbf{T}).$$

That is, all polynomials of degree zero get mapped into the zero polynomial by the derivative transformation. It follows from Prop. 2.1 that the derivative transformation defined in Exam. 1.2 is not one-to-one. ■

4.3 THE RANGE SPACE

Associated with each linear transformation $A \in \mathbf{L}(\mathbf{V}, \mathbf{W})$ is a special subspace of $\mathbf{W}$. We call this subspace the range space of A.

Definition 3.1 (Range Space, Rank). Let $A \in \mathbf{L}(\mathbf{V}, \mathbf{W})$. Then the *range space* of A is denoted $\mathbf{R}(A)$ and is defined

$$\mathbf{R}(A) = \{Ax \in \mathbf{W} : x \in \mathbf{V}\}.$$

If $\mathbf{R}(A)$ is finite-dimensional, then the dimension of $\mathbf{R}(A)$ is called the *rank* of A. ■

Thus the range space of A is simply the image of $\mathbf{V}$ under A. Refer to Figure 4.3 for a pictorial description of the range space of a linear transformation.

Exercise 3.1 (Range Space). Let $A \in \mathbf{L}(\mathbf{V}, \mathbf{W})$. Show that $\mathbf{R}(A)$ is a subspace of $\mathbf{W}$. ■

The content of the range space determines whether or not a linear transformation is onto or surjective. Consider, in particular, the following result.

Proposition 3.1 (Onto). Let $A \in \mathbf{L}(\mathbf{V}, \mathbf{W})$. Then A is onto $\Leftrightarrow \mathbf{R}(A) = \mathbf{W}$. ■

The proof of Prop. 3.1 follows directly from Def. 3.1 and Def. 1.4. As an example of an illustration of Prop. 3.1 we again examine the derivative transformation.

Example 3.1 (Range Space). Consider the derivative transformation $A : \mathbf{P}^n(\mathbf{T}) \rightarrow \mathbf{P}(\mathbf{T})$ discussed in Exam. 1.2. In this case we have

$$\mathbf{R}(A) = \mathbf{P}^{n-1}(\mathbf{T}).$$

That is, polynomials of degree n get mapped into polynomials of degree $n - 1$ by the derivative transformation. It follows from Prop. 3.1 that the derivative transformation of Exam. 1.2 is not onto. ■

The null space and range space of a linear transformation are by no means independent. Indeed, if $A \in \mathbf{L}(\mathbf{V}, \mathbf{W})$, and if $\mathbf{V}$ is finite-dimensional, then knowing either the nullity or the rank of A is equivalent to knowing both. In particular, we have the following result.

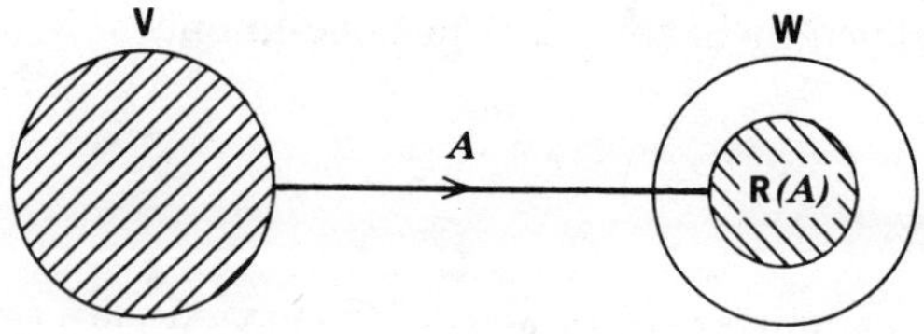

Figure 4.3. The range space of a linear transformation A.

Proposition 3.2 (Nullity and Rank). Let $A \in \mathbf{L}(\mathbf{V}, \mathbf{W})$ where $\mathbf{V}$ is finite-dimensional. Then

$$\text{nullity}(A) + \text{rank}(A) = \dim(\mathbf{V}).$$

Proof. Let p and q denote the dimensions of $\mathbf{V}$ and $\mathbf{N}(A)$, respectively. Next, let $\{x^1, x^2, \ldots, x^q\}$ be a basis for $\mathbf{N}(A)$. Augment this basis with vectors $\{x^{q+1}, x^{q+2}, \ldots, x^p\}$ such that $\mathbf{B} = \{x^1, x^2, \ldots, x^p\}$ forms a basis for $\mathbf{V}$. Then for each $x \in \mathbf{V}$,

$$x = \sum_{k=1}^{p} a_k x^k.$$

Here $a = [x]^{\mathbf{B}}$. Since A is linear, we have

$$\begin{aligned} Ax &= A\left(\sum_{k=1}^{p} a_k x^k\right) \\ &= \sum_{k=1}^{p} a_k A x^k \\ &= \sum_{k=q+1}^{p} a_k A x^k. \end{aligned}$$

Hence $\{Ax^{q+1}, Ax^{q+2}, \ldots, Ax^p\}$ span $\mathbf{R}(A)$. It remains to demonstrate that they are linearly independent. Suppose there exist $\{b_{q+1}, b_{q+2}, \ldots, b_p\}$ not all zero such that

$$\sum_{k=q+1}^{p} b_k A x^k = 0.$$

Then since A is linear, it follows that

$$\sum_{k=q+1}^{p} b_k x^k \in \mathbf{N}(A).$$

Now, $\{x^1, x^2, \ldots, x^q\}$ form a basis for $\mathbf{N}(A)$. Thus there exist $\{b_1, b_2, \ldots, b_q\}$ such that

$$\sum_{k=1}^{q} b_k x^k = -\sum_{k=q+1}^{p} b_k x^k.$$

It follows that there exists a vector $b \in \mathbf{F^n}$ with $b \neq 0$ such that

$$\sum_{k=1}^{p} b_k x^k = 0.$$

Thus the vectors $\{x^1, x^2, \ldots, x^p\}$ are linearly dependent. But this is a contradiction since **B** is a basis for **V**. Hence $\{Ax^{q+1}, Ax^{q+2}, \ldots, Ax^p\}$ must be linearly independent. It follows that they form a basis for **R**(A). Thus $\text{rank}(A) = p - q$ in which case,

$$\text{nullity}(A) + \text{rank}(A) = \dim(\mathbf{V}). \qquad \blacksquare$$

Refer to Figure 4.4 for a pictorial representation of the relationship between the nullity and the rank of a linear transformation defined on a finite-dimensional vector space.

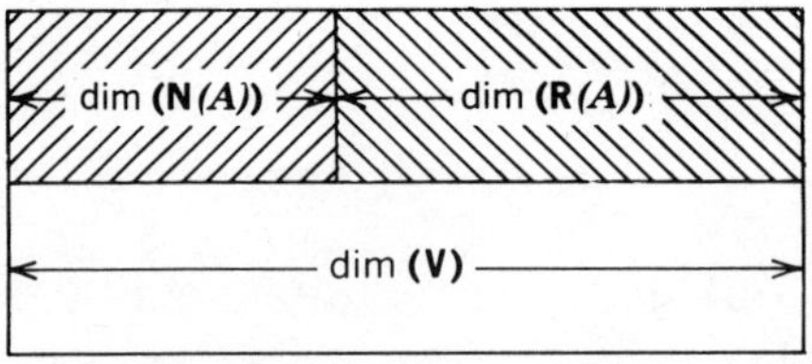

Figure 4.4. Nullity and rank of a linear transformation A.

4.4. INVERSE TRANSFORMATIONS

One of the questions of practical importance that arises in the study of transformations in general is the following. Given the image of a vector under a transformation, under what conditions can we recover the identity of that vector? To concisely state these conditions we make use of a special transformation which maps a vector space onto itself called the identity transformation.

Definition 4.1 (Identity). Let **V** be a vector space. Then the *identity* transformation on **V** is denoted $I_{\mathbf{V}}$ or I and is defined

$$I_{\mathbf{V}} x \triangleq x, \qquad x \in \mathbf{V}. \qquad \blacksquare$$

Thus the image of a vector under the identity transformation I is the vector itself. That is, every vector remains *invariant* under the identity transformation. Given the notion of the identity transformation, we can now define the inverse of a transformation in the following manner.

Definition 4.2 (Inverse). Let $A \in \mathbf{F}(\mathbf{V}, \mathbf{W})$. Then A is *invertible* $\Leftrightarrow$ there exists a transformation A^{-1} mapping **W** into **V** such that

1. $A^{-1}A = I_{\mathbf{V}}$,
2. $AA^{-1} = I_{\mathbf{W}}$.

In this case A^{-1} is unique, and we call A^{-1} the *inverse of A*. $\blacksquare$

Thus A^{-1} is the inverse of A if and only if the *composition* $\mathbf{A}^{-1}\mathbf{A}$ is the identity on $\mathbf{V}$, and the composition AA^{-1} is the identity on $\mathbf{W}$. We sometimes refer to property 1 by saying that A^{-1} is a *left-inverse* of A. Similarly, we refer to property 2 by saying that A^{-1} is a *right-inverse* of A. Thus the inverse A^{-1} is both a left inverse and a right inverse of A. Refer to Figure 4.5 for a pictorial description of the relationship between a transformation and its inverse.

Exercise 4.1 (Uniqueness of Inverse). Let $A \in \mathbf{F}(\mathbf{V}, \mathbf{W})$. Show that if the inverse of A exists, then it is unique. ■

The notions of invertible transformations, one-to-one transformations, and onto transformations are interrelated. Consider, in particular, the following result.

Proposition 4.2 (Inverses). Let $A \in \mathbf{F}(\mathbf{V}, \mathbf{W})$. Then A is invertible $\Leftrightarrow$

1. A is one-to-one,
2. A is onto.

Proof. Suppose A is invertible. Let $(x, y) \in \mathbf{V}^2$. Then

$$\begin{aligned} Ax = Ay &\Rightarrow A^{-1}Ax = A^{-1}Ay \\ &\Rightarrow I_{\mathbf{V}}x = I_{\mathbf{V}}y \\ &\Rightarrow x = y. \end{aligned}$$

Hence A is one-to-one. Next, let $y \in \mathbf{W}$, and define $x = A^{-1}y$. Then A is onto because

$$\begin{aligned} Ax &= AA^{-1}y \\ &= I_{\mathbf{W}}y \\ &= y. \end{aligned}$$

Next, suppose A is one-to-one and onto. Then for each $y \in \mathbf{W}$ there exists an $x \in \mathbf{V}$ such that $Ax = y$. But because A is one-to-one, $x \in \mathbf{V}$ is unique. Let A^{-1} be

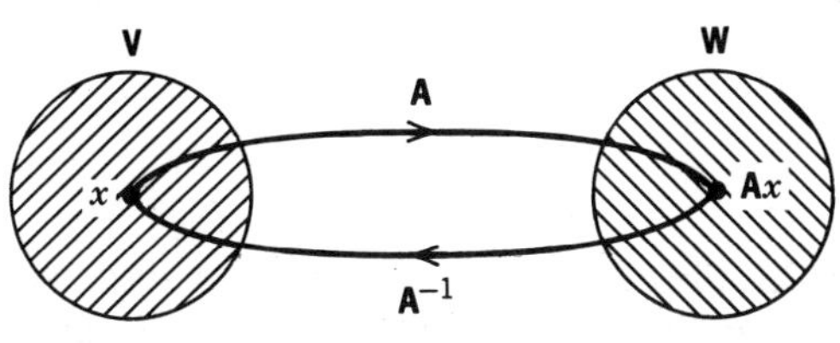

Figure 4.5. A transformation A and its inverse A^{-1}.

defined $A^{-1}y = x$ for each $y \in \mathbf{W}$. Then for each $y \in \mathbf{W}$,

$$\begin{aligned} AA^{-1}y &= Ax \\ &= y. \end{aligned}$$

Thus AA^{-1} is the identity on $\mathbf{W}$. Moreover, for each $x \in \mathbf{V}$ we have

$$\begin{aligned} A^{-1}Ax &= A^{-1}y \\ &= x. \end{aligned}$$

Hence $A^{-1}A$ is the identity on $\mathbf{V}$. Thus A is invertible and A^{-1} is the inverse. ■

Exercise 4.2. Show the notions of invertibility and linearity are independent and meaningful by constructing examples of transformations of the form $A: \mathbf{R}^2 \to \mathbf{R}^2$ that are

1. Neither linear nor invertible,
2. Linear but not invertible,
3. Invertible but not linear,
4. Both linear and invertible. ■

Note that Prop. 4.1 is valid whether or not the transformation A is linear. In the case of linear transformations, the criterion for invertibility becomes more specific as can be seen from the following result.

Proposition 4.2 (Invertibility). Let $A \in \mathbf{L}(\mathbf{V}, \mathbf{W})$. Then A is invertible $\Leftrightarrow$

1. $\mathbf{N}(A) = \{\mathbf{0}\}$,
2. $\mathbf{R}(A) = \mathbf{W}$. ■

The proof of Prop. 4.2 follows directly from Prop. 4.1, Prop. 2.1, and Prop. 3.1. Refer to Figure 4.6 for a pictorial summary of the criteria for invertibility of both transformations in general and linear transformations in particular.

Exercise 4.3. Let $A \in \mathbf{L}(\mathbf{V}, \mathbf{W})$ where $\mathbf{V}$ is finite-dimensional. Show that if A is invertible, then

$$\dim(\mathbf{W}) = \dim(\mathbf{V}).$$ ■

In view of Prop. 4.2, the derivative transformation discussed in Exam. 2.1 and Exam. 3.1 is not invertible. As an example of a transformation which is invertible, consider the following.

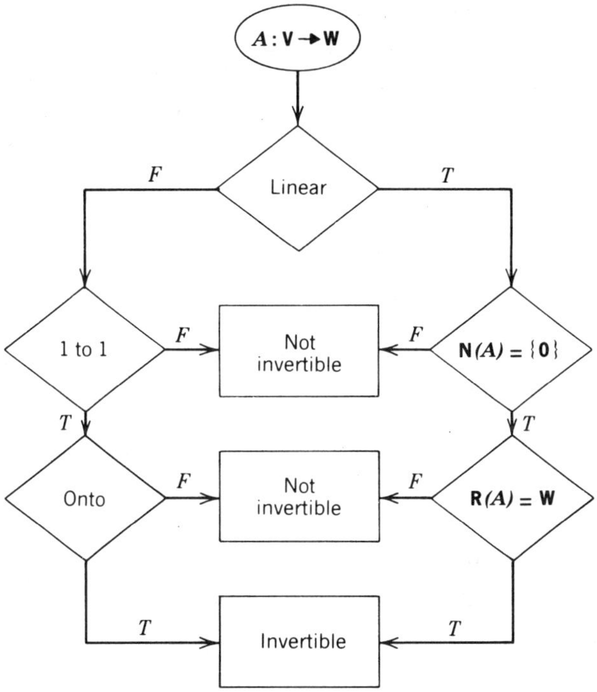

Figure 4.6. Invertibility criteria.

Example 4.1 (Weighting Transformation). Let $T = [\alpha, \beta]$ be a bounded interval in $\mathbf{R}$, and let $\mathbf{V} = \mathbf{C(T)}$. Next, let $w: T \to (0, \infty)$ be a continuous positive weighting function. Define the *weighting transformation* A as follows. For each $f \in \mathbf{V}$,

$$(Af)(t) = w(t)f(t), \qquad t \in T.$$

Then $A: \mathbf{C(T)} \to \mathbf{C(T)}$ is an invertible linear transformation. Indeed, in this case the inverse of A is

$$(A^{-1}f)(t) = \frac{f(t)}{w(t)}. \qquad \blacksquare$$

As one might expect, the composition of two invertible transformations is again an invertible transformation. In computing the inverse, the order of the constituent transformations is interchanged as can be seen from the following result.

Proposition 4.3 (Compositions). Let $B \in \mathbf{F(V, W)}$, and let $A \in \mathbf{F(W, U)}$. Then the composition $AB \in \mathbf{F(V, U)}$ is invertible $\Leftrightarrow$ A and B are invertible, in which case

$$(AB)^{-1} = B^{-1}A^{-1}.$$

Proof. Let $\boldsymbol{C} = \boldsymbol{B}^{-1}\boldsymbol{A}^{-1}$. Then $\boldsymbol{C}$ is a left-inverse of $\boldsymbol{AB}$ because

$$\begin{aligned} \boldsymbol{CAB} &= \boldsymbol{B}^{-1}\boldsymbol{A}^{-1}\boldsymbol{AB} \\ &= \boldsymbol{B}^{-1}\boldsymbol{I}_{\mathbf{W}}\boldsymbol{B} \\ &= \boldsymbol{B}^{-1}\boldsymbol{B} \\ &= \boldsymbol{I}_{\mathbf{V}}. \end{aligned}$$

Furthermore, $\boldsymbol{C}$ is also a right-inverse of $\boldsymbol{AB}$ as

$$\begin{aligned} \boldsymbol{ABC} &= \boldsymbol{ABB}^{-1}\boldsymbol{A}^{-1} \\ &= \boldsymbol{AI}_{\mathbf{W}}\boldsymbol{A}^{-1} \\ &= \boldsymbol{AA}^{-1} \\ &= \boldsymbol{I}_{\mathbf{U}}. \end{aligned}$$

From Exer. 4.1, the inverse is unique. Thus the composition $\boldsymbol{AB}$ is invertible $\Leftrightarrow$ $\boldsymbol{A}$ and $\boldsymbol{B}$ are invertible and in this case the inverse is given by $\boldsymbol{C} = \boldsymbol{B}^{-1}\boldsymbol{A}^{-1}$. ■

Exercise 4.4 (Restriction of W to R(A)). Let $A \in \mathbf{L}(\mathbf{V}, \mathbf{R}(A))$. Show that the transformation $\boldsymbol{A}$ is invertible $\Leftrightarrow$ $\mathbf{N}(A) = \{\mathbf{0}\}$. ■

4.5. BOUNDED TRANSFORMATIONS

Thus far our discussion of transformations between spaces has made no reference to the topological structure of those spaces. Suppose now that **V** and **W** are normed vector spaces. We can then identify an important subspace of **L(V, W)** called the space of bounded linear transformations.

Definition 5.1 (Bounded). Let $A \in \mathbf{F}(\mathbf{V}, \mathbf{W})$ where **V** and **W** are normed vector spaces. Then $\boldsymbol{A}$ is *bounded* $\Leftrightarrow$ there exists an $\alpha > 0$ called a *bound* such that for each $x \in \mathbf{V}$

$$\|Ax\|_W \leqslant \alpha \|x\|_V.$$ ■

Note that transformations do not have to be linear to be bounded. The set of transformations mapping **V** into **W** that are both *bounded* and *linear* is denoted **B(V, W)**. That is,

$$\mathbf{B}(\mathbf{V}, \mathbf{W}) \triangleq \{\mathbf{A} \in \mathbf{L}(\mathbf{V}, \mathbf{W}) : A \text{ is bounded}\}.$$

Linear combinations of bounded linear transformations are again bounded linear transformations. Thus the set **B(V, W)** forms a vector space, as is indicated in the following exercise.

Exercise 5.1. Show that **B(V, W)** is a subspace of **L(V, W)**. ■

Interestingly enough, within the space **L(V, W)** the notion of boundedness is equivalent to the notion of continuity. In particular, we have the following result.

Proposition 5.1 (Boundedness and Continuity). Let $A \in \mathbf{L}(\mathbf{V}, \mathbf{W})$. Then A is continuous $\Leftrightarrow A$ is bounded.

Proof. Suppose A is continuous but not bounded. Then for each $k \geqslant 1$ there exists an $x^k \in \mathbf{V}$ such that

$$\|Ax^k\|_{\mathbf{W}} > k\|x^k\|_{\mathbf{V}}.$$

Let $y^k = x^k/(k\|x^k\|_{\mathbf{V}})$, for $k \geqslant 1$. Then, clearly, $y^k \to 0$ as $k \to \infty$. Since A is linear and continuous, it then follows that $Ay^k \to 0$ as $k \to \infty$. But for each $k \geqslant 1$

$$\begin{aligned} \|Ay^k\|_{\mathbf{W}} &= \left\| \frac{Ax^k}{(k\|x^k\|_{\mathbf{V}})} \right\|_{\mathbf{W}} \\ &= \frac{\|A^k x\|_W}{k\|x^k\|_V} \\ &> 1. \end{aligned}$$

This contradicts the fact that $Ay^k \to 0$ as $k \to \infty$. Thus if A is continuous, then A is bounded. Next, suppose A is bounded. Then since A is linear, from Def. 5.1 it follows that A is continuous at $x = 0$. Let $x \in \mathbf{V}$, and suppose

$$x^k \to x \quad \text{as} \quad k \to \infty.$$

Then since A is linear and continuous at $x = 0$, as $k \to \infty$ we have

$$\begin{aligned} Ax^k &= A(x^k - x + x) \\ &= A(x^k - x) + Ax \\ &\to A0 + Ax \\ &= Ax. \end{aligned}$$

Hence A is continuous at x. Thus A is continuous. ■

Thus linear transformations are bounded if and only if they are continuous. Of course for a nonlinear transformation mapping a normed vector space into a normed vector space, this equivalence between continuity and boundedness no longer applies. If we restrict our attention to linear transformations defined on finite-dimensional normed vector spaces, then they are all bounded and therefore all continuous, as can be seen from the following exercise.

Exercise 5.2 (Boundedness). Let $A \in \mathbf{L}(\mathbf{V}, \mathbf{W})$. Show that if $\mathbf{V}$ is finite-dimensional, then $A \in \mathbf{B}(\mathbf{V}, \mathbf{W})$. ■

The space $\mathbf{B}(\mathbf{V}, \mathbf{W})$ of bounded linear transformations from $\mathbf{V}$ into $\mathbf{W}$ can be made into a normed vector space. In particular, the norms of $\mathbf{V}$ and $\mathbf{W}$ generate or induce a norm in $\mathbf{B}(\mathbf{V}, \mathbf{W})$ in the following manner.

Definition 5.2 (Induced Norm). Let $A \in \mathbf{B}(\mathbf{V}, \mathbf{W})$. Then the *induced norm* of A is defined

$$\|A\| = \sup_{x \neq 0} \left\{ \frac{\|Ax\|_{\mathbf{W}}}{\|x\|_{\mathbf{V}}} \right\}.$$ ■

Exercise 5.3 (Induced Norm). Show that the induced norm in $\mathbf{B}(\mathbf{V}, \mathbf{W})$ exists and satisfies the axioms of a norm. ■

For the induced norm of Def. 5.2, it is actually not necessary to find the supremum of $\|Ax\|$ for all nonzero vectors x in the space $\mathbf{V}$. Due to the linearity of A, it is sufficient to examine the values of $\|Ax\|$ along the surface of the unit ball in $\mathbf{V}$, as seen in the following exercise.

Exercise 5.4. Let $A \in \mathbf{B}(\mathbf{V}, \mathbf{W})$. Show that the induced norm of A as defined in Def. 5.2 can also be computed as follows:

$$\|A\| = \sup_{\|x\|=1} \{ \|Ax\|_{\mathbf{W}} \}.$$ ■

When we restrict our attention to the normed vector space $\mathbf{B}(\mathbf{V}, \mathbf{W})$, the important practical issue of the existence of an inverse can be resolved by the following result.

Proposition 5.2 (Bounded Inverse). Let $A \in \mathbf{B}(\mathbf{V}, \mathbf{W})$. Then A is invertible on $\mathbf{R}(A)$ and $A^{-1} \in \mathbf{B}(\mathbf{R}(A), \mathbf{V}) \Leftrightarrow$ there exists a lower bound $\beta > 0$ such that for each $x \in \mathbf{V}$,

$$\|Ax\|_{\mathbf{W}} \geqslant \beta \|x\|_{\mathbf{V}}.$$

Proof. To show that A^{-1} exists on $\mathbf{R}(A)$ it suffices from Exer. 4.4 to show that $\mathbf{N}(A) = \{\mathbf{0}\}$. Let $x \in \mathbf{V}$, and suppose $Ax = 0$. If $\|Ax\|$ is bounded as in Prop. 5.2, then

$$\begin{aligned} 0 &= \|0\|_{\mathbf{W}} \\ &= \|Ax\|_{\mathbf{W}} \\ &\geqslant \beta\|x\|_{\mathbf{V}}. \end{aligned}$$

Since $\beta > 0$, it follows that $x = 0$. Hence $\mathbf{N}(A) = \{\mathbf{0}\}$. Thus from Exer. 4.4, A^{-1} exists on $\mathbf{R}(A)$. Next, we show that $A^{-1} \in \mathbf{B}(\mathbf{R}(A), \mathbf{V})$. Let $(x, y) \in \mathbf{R}(A)^2$. Then there exist $(u, v) \in \mathbf{V}^2$ such that $Au = x$ and $Av = y$. Let $(\alpha, \beta) \in \mathbf{F}^2$. Then

$$\begin{aligned} A^{-1}(\alpha x + \beta y) &= A^{-1}(\alpha Au + \beta Av) \\ &= A^{-1}A(\alpha u + \beta v) \\ &= I_{\mathbf{V}}(\alpha u + \beta v) \\ &= \alpha u + \beta v \\ &= \alpha A^{-1}x + \beta A^{-1}y. \end{aligned}$$

Hence A^{-1} is linear. Let $y \in \mathbf{R}(A)$. Then there exists an $x \in \mathbf{V}$ such that $Ax = y$. Since $\|Ax\|_{\mathbf{W}}$ is bounded as in Prop. 5.2, we have

$$\begin{aligned} \|y\|_{\mathbf{W}} &= \|Ax\|_{\mathbf{W}} \\ &\geqslant \beta\|x\|_{\mathbf{V}} \\ &= \beta\|A^{-1}y\|_{\mathbf{V}}. \end{aligned}$$

It then follows from Def. 5.1 that $A^{-1} \in \mathbf{B}(\mathbf{R}(A), \mathbf{V})$ with bound $\alpha = 1/\beta$. To establish the other half of the proof, suppose that A^{-1} exists on $\mathbf{R}(A)$, and suppose that $A^{-1} \in \mathbf{B}(\mathbf{R}(A), \mathbf{V})$. Then there exists an $\alpha > 0$ such that for each $x \in \mathbf{V}$

$$\begin{aligned} \|x\|_{\mathbf{V}} &= \|I_{\mathbf{V}}x\|_{\mathbf{V}} \\ &= \|A^{-1}Ax\|_{\mathbf{V}} \\ &\leqslant \alpha\|Ax\|_{\mathbf{W}}. \end{aligned}$$

If we select $\beta = 1/\alpha$, then A is bounded from below as in Prop. 5.2. ■

One of the important questions which arises in the study of normed vector spaces is the question of completeness. The following result provides us with a simple sufficient condition for the completeness of the space $\mathbf{B}(\mathbf{V}, \mathbf{W})$.

Proposition 5.3 (Completeness of B(V, W)). Let $\mathbf{V}$ and $\mathbf{W}$ be normed vector spaces. Then $\mathbf{B}(\mathbf{V}, \mathbf{W})$ is a Banach space if $\mathbf{W}$ is a Banach space.

Proof. We must show that every Cauchy sequence in $\mathbf{B(V, W)}$ converges to a vector in $\mathbf{B(V, W)}$. Let $\{A_{\mathbf{k}}\}$ be a Cauchy sequence in $\mathbf{B(V, W)}$. Then for each $\alpha > 0$ there exists an $N \geqslant 1$ such that

$$(k, j) \geqslant (N, N) \Rightarrow \|A_{\mathbf{k}} - A_{\mathbf{j}}\| < \alpha.$$

Let $x \in \mathbf{V}$. Then for $(k, j) \geqslant (N, N)$ we have

$$\begin{aligned}\|A_{\mathbf{k}}x - A_{\mathbf{j}}x\|_{\mathbf{W}} &= \|(A_{\mathbf{k}} - A_{\mathbf{j}})x\|_{\mathbf{W}} \\ &\leqslant \|A_{\mathbf{k}} - A_{\mathbf{j}}\|(\|x\|_{\mathbf{V}}) \\ &< \alpha\|x\|_{\mathbf{V}}.\end{aligned}$$

It follows that $\{A_{\mathbf{k}}x\}$ is a Cauchy sequence in $\mathbf{W}$ for each $x \in \mathbf{V}$. Since $\mathbf{W}$ is complete, for each $x \in \mathbf{V}$ the sequence $\{A_{\mathbf{k}}x\}$ converges to a vector $Ax \in \mathbf{W}$. Now for each $x \in \mathbf{V}$ define

$$Ax = \lim_{k \to \infty} \{A_{\mathbf{k}}x\}.$$

Since the norm is continuous, for each $x \in \mathbf{V}$ and each $j \geqslant N$ we have

$$\begin{aligned}\|(A - A_{\mathbf{j}})x\|_{\mathbf{W}} &= \left\|\left(\lim_{k \to \infty} \{A_{\mathbf{k}}\} - A_{\mathbf{j}}\right)x\right\|_{\mathbf{W}}, \\ &= \left\|\lim_{k \to \infty} \{A_{\mathbf{k}} - A_{\mathbf{j}}\}x\right\|_{\mathbf{W}}, \\ &= \lim_{k \to \infty} \{\|(A_{\mathbf{k}} - A_{\mathbf{j}})x\|_{\mathbf{W}}\}, \\ &< \alpha\|x\|_{\mathbf{V}}.\end{aligned}$$

It therefore follows that for each $j \geqslant N$:

$$\begin{aligned}\|A - A_{\mathbf{j}}\| &= \sup_{x \neq 0}\left\{\frac{\|(A - A_{\mathbf{j}})x\|_{\mathbf{W}}}{\|x\|_{\mathbf{V}}}\right\}, \\ &\leqslant \alpha.\end{aligned}$$

Hence $A_{\mathbf{k}} \to A$ as $k \to \infty$. It remains to demonstrate that $A \in \mathbf{B(V, W)}$. Clearly, since $A_{\mathbf{k}} \in \mathbf{L(V, W)}$ for $k \geqslant N$, we have $A \in \mathbf{L(V, W)}$. Next, because $\{A_{\mathbf{k}}\}$ is a Cauchy sequence, it is uniformly bounded. That is, there exists a $\beta > 0$ such that for each $k \geqslant N$

$$\|A_{\mathbf{k}}\| < \beta.$$

But because the norm is continuous, for each $x \in \mathbf{V}$ we then have

$$\|Ax\|_{\mathbf{W}} = \lim_{k \to \infty} \{\|A_k x\|_{\mathbf{W}}\},$$
$$< \beta \|x\|_{\mathbf{V}}.$$

Hence $A \in \mathbf{B(V, W)}$. Thus if $\mathbf{W}$ is complete, then $\mathbf{B(V, W)}$ is complete. ■

In the event that both $\mathbf{V}$ and $\mathbf{W}$ are Banach spaces, it is possible to develop a simple sufficient condition that ensures that a linear transformation is bounded. To see this it is useful to first consider the notion of a graph of a transformation.

Definition 5.3 (Graph). Let $A \in \mathbf{F(V, W)}$. Then the *graph* of A is the following subset of $\mathbf{V} \times \mathbf{W}$:

$$\mathbf{G}(A) = \{(x, Ax) \in \mathbf{V} \times \mathbf{W} : x \in \mathbf{V}\}.$$ ■

Thus the graph of a transformation mapping $\mathbf{V}$ into $\mathbf{W}$ is a subset of the product space $\mathbf{V} \times \mathbf{W}$. In the event that the transformation A is linear, the graph of A is in fact a subspace of $\mathbf{V} \times \mathbf{W}$ as can be seen from the following exercise.

Exercise 5.5 (Graph). Show that $\mathbf{G}(A)$ is a subspace of $\mathbf{V} \times \mathbf{W} \Leftrightarrow A \in \mathbf{L(V, W)}$. ■

Refer to Figure 4.7 for a simple example of a graph of a transformation. Note that if $A : \mathbf{R} \to \mathbf{R}$, then the $\mathbf{G}(A)$ is a curve, while if $A : \mathbf{R}^2 \to \mathbf{R}$, then $\mathbf{G}(A)$ is a surface. For linear transformations, the curve is a straight line through the origin, while the surface is a plane through the origin.

When a linear transformation A is bounded, the graph of A has an additional property: it contains its boundary, as can be seen from the following exercise.

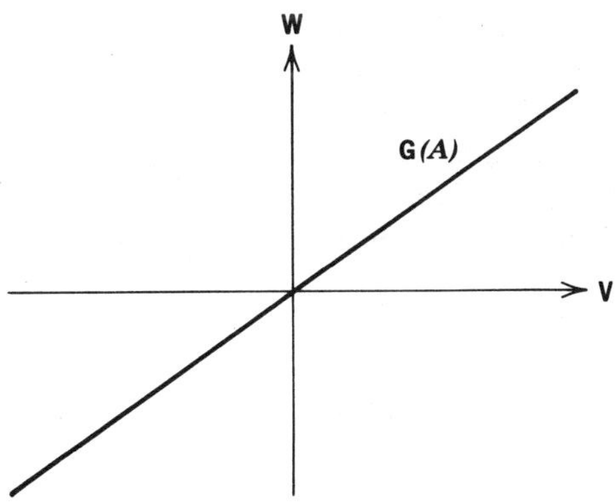

Figure 4.7. Graph of a linear transformation $A : \mathbf{R} \to \mathbf{R}$.

Exercise 5.6 (Closed Graph). Show that if $A \in \mathbf{B(V, W)}$, then $\mathbf{G}(A)$ is closed. ■

The converse of Exer. 5.6 does not hold. However, if both **V** and **W** are Banach spaces, then it is valid. Thus we have the following result.

Proposition 5.4 (Closed Graph Theorem). Let $A \in \mathbf{L(V, W)}$ where **V** and **W** are Banach spaces. Then $\mathbf{G}(A)$ is closed $\Leftrightarrow A \in \mathbf{B(V, W)}$.

The reader can refer to Bachman and Narici (1966) for a proof of the closed graph theorem. The following exercise provides an example of a linear transformation that is not bounded. Therefore the graph of this transformation is not closed.

Exercise 5.7 (Unbounded Linear Transformation). Let $T = [\alpha, \beta]$ be an interval in **R**, and let $\mathbf{W} = \mathbf{C}_\infty(\mathbf{T})$ where $\mathbf{C}_\infty(\mathbf{T})$ is the space of continuous functions mapping T into **R** with the ∞-norm. Next, let A be the derivative transformation:

$$(Ax)(t) = \frac{d}{dt} x(t), \qquad t \in T.$$

Finally, let **V** be the space of all functions in **W** whose derivatives are also in **W**. That is, $\mathbf{V} = \{x \in \mathbf{W} : Ax \in \mathbf{W}\}$. Show that $A \in \mathbf{L(V, W)}$ but $A \notin \mathbf{B(V, W)}$. ■

4.6. FUNCTIONALS

At this point we turn our attention to linear transformations of a special kind. We restrict the space **W** to the field **F** to generate a special class of transformations called functionals.

Definition 6.1 (Functional). Let **V** be a vector space over the field **F**. Then f is a *functional* on $\mathbf{V} \Leftrightarrow f : \mathbf{V} \to \mathbf{F}$. ■

Thus a functional is simply a function which maps a space into the field over which that space is defined. Since a field can be viewed as a one-dimensional vector space, it follows that a functional is a special case of a transformation. The notion of a functional allows us to associate with each normed vector space another normed vector space called its dual.

Definition 6.2 (Dual Space). Let **V** be a normed vector space over the field **F**. Then the *dual* of **V** is denoted **V*** and is defined as the vector space of bounded linear functionals on **V**. That is,

$$\mathbf{V}^* \triangleq \mathbf{B(V, F)}. \qquad ■$$

Dual spaces have a number of interesting properties. One of these is that the dual space $\mathbf{V}^*$ is always complete, as can be seen from the following result.

Proposition 6.1 (Completeness of Dual). The dual of every normed vector space is a Banach space.

Proof. The dual of $\mathbf{V}$ is $\mathbf{B}(\mathbf{V}, \mathbf{F})$. But every field $\mathbf{F}$ is a complete vector space. Hence from Prop. 5.3, $\mathbf{V}^*$ is a Banach space. ■

If the space $\mathbf{V}$ is a Hilbert space, then it is isomorphic to its dual $\mathbf{V}^*$. This is a consequence of the following fundamental result called the Riesz representation theorem for a bounded linear functional on a Hilbert space.

Proposition 6.2 (Riesz). Let $\mathbf{V}$ be a Hilbert space, and let $\mathbf{V}^*$ be its dual. Then for each $f \in \mathbf{V}^*$ there exists a unique $y \in \mathbf{V}$ such that

$$f(x) = \langle x, y \rangle, \qquad x \in \mathbf{V}.$$

Proof. Let $f \in \mathbf{V}^*$ and let $\mathbf{N}(\mathbf{f})$ be the null space of f. Since f is bounded and linear, from Prop. 5.1 it is continuous. Thus $\mathbf{N}(\mathbf{f})$ is a closed subspace of $\mathbf{V}$. Suppose $\mathbf{N}(\mathbf{f}) = \mathbf{V}$. Pick $y = 0$. Then for each $x \in \mathbf{V}$ we have $f(x) = \langle x, 0 \rangle$. Next, suppose $\mathbf{N}(\mathbf{f}) \neq \mathbf{V}$. Then there exists a $z \in \mathbf{N}(\mathbf{f})^\perp$ with $z \neq 0$. Here $\mathbf{N}(\mathbf{f})^\perp$ represents the orthogonal complement of $\mathbf{N}(\mathbf{f})$. Let $y = \alpha z$ where $\alpha = \overline{f(z)}/\|z\|^2$. Then for each $\beta \in \mathbf{F}$ we have

$$\begin{aligned} f(\beta z) &= \beta f(z) \\ &= \beta \bar{\alpha} \|z\|^2 \\ &= \beta \alpha \langle z, z \rangle \\ &= \beta \langle z, \alpha z \rangle \\ &= \beta \langle z, y \rangle \\ &= \langle \beta z, y \rangle. \end{aligned}$$

Next, let $x \in \mathbf{N}(\mathbf{f})$. Then

$$\begin{aligned} f(x) &= 0 \\ &= \bar{\alpha} \langle x, z \rangle \\ &= \langle x, \alpha z \rangle \\ &= \langle x, y \rangle. \end{aligned}$$

More generally, let $x \in \mathbf{V}$, and let $\beta = f(x)/f(z)$. Then

$$\begin{aligned} f(x - \beta z) &= f(x) - \beta f(z) \\ &= 0. \end{aligned}$$

Thus $x - \beta z \in \mathbf{N(f)}$. It follows that $f(x - \beta z) = \langle x - \beta z, y\rangle$. Thus for an arbitrary $x \in \mathbf{V}$ we have

$$\begin{aligned} f(x) &= f(x - \beta z + \beta z) \\ &= f(x - \beta z) + f(\beta z) \\ &= \langle x - \beta z, y\rangle + \langle \beta z, y\rangle \\ &= \langle x, y\rangle. \end{aligned}$$

To show that y is unique, suppose there exists a $w \in \mathbf{V}$ such that $f(x) = \langle x, w\rangle$ for each $x \in \mathbf{V}$. Then for each $x \in \mathbf{V}$

$$\begin{aligned} \langle x, y - w\rangle &= \langle x, y\rangle - \langle x, w\rangle \\ &= f(x) - f(x) \\ &= 0. \end{aligned}$$

Now, let $x = y - w$. It then follows that $x = 0$. Hence y is unique. ■

In view of Prop. 6.2 we see that there is an isomorphism between a Hilbert space $\mathbf{V}$ and its dual $\mathbf{V}^*$. This isomorphic relationship between a vector space and its dual can be extended to certain Banach spaces, as can be seen from the following two results. The reader is referred to Dunford and Schwartz (1958) for proofs of these results.

Proposition 6.3 (Dual of ℓ_p). Let $1 \leqslant p \leqslant \infty$. Then ℓ_p^* is isomorphic to ℓ_q whenever $1/p + 1/q = 1$.

Proposition 6.4 (Dual of $\mathbf{L_p(T, w)}$). Let $T = [\alpha, \beta]$ be an interval in $\mathbf{R}$, let w be a positive weighting function on T, and let $1 \leqslant p \leqslant \infty$. Then $\mathbf{L_p^*(T, w)}$ is isomorphic to $\mathbf{L_q(T, w)}$ whenever $1/p + 1/q = 1$. ■

PROBLEMS

1. Let $\mathbf{A}: \mathbf{R} \to \mathbf{R}^2$ be defined as the transformation $\mathbf{A}x = [x^2, e^{-x}]^T$.

a. Is A linear?

b. Is A one-to-one?

c. Is A onto?

d. Is A invertible?

In each case support your answer with a proof or a counterexample.

2. Let $T = [\alpha, \beta]$ be a bounded interval in **R**, let $\mathbf{V} = \mathbf{C(T)}$, and let $w \in \mathbf{C(T)}$. Consider the following transformation mapping **V** into **V**:

$$(Ax)(t) = w(t)x(t).$$

Suppose Q is the subset of T on which the function w is zero. That is, $Q = \{t \in T : w(t) = 0\}$.

a. Show that A is linear.

b. Find an expression for $\mathbf{N}(A)$ in terms of Q.

c. Find an expression for $\mathbf{R}(A)$ in terms of Q.

d. Under what conditions on Q is A invertible? Find A^{-1}.

3. Let $A \in \mathbf{B}(\mathbf{V}, \mathbf{W})$. Show that $\mathbf{N}(A)$ is a closed subspace of **V**.

4. Let $A \in \mathbf{L}(\mathbf{V}, \mathbf{W})$. Show that $A \in \mathbf{B}(\mathbf{V}, \mathbf{W}) \Leftrightarrow A$ maps bounded sets into bounded sets.

5. Let $A \in \mathbf{L}(\mathbf{V}, \mathbf{W})$ be onto. Show that $A \in \mathbf{B}(\mathbf{V}, \mathbf{W})$, A^{-1} exists, and $\mathbf{A}^{-1} \in \mathbf{B}(\mathbf{W}, \mathbf{V}) \Leftrightarrow$ there exist $\beta \geqslant \alpha > 0$ such that for each $x \in \mathbf{V}$:

$$\alpha \|x\|_{\mathbf{V}} \leqslant \|Ax\|_{\mathbf{W}} \leqslant \beta \|x\|_{\mathbf{V}}.$$

6. Let $1 \leqslant p \leqslant \infty$, and let $y \in \mathbf{L_q}([\alpha, \beta])$ where $1/p + 1/q = 1$. Show that $f \subset \mathbf{L}_{\mathbf{p}}^{*}([\alpha, \beta])$ if

$$f(x) - \int_{\alpha}^{\beta} x(t)y(t)\, dt.$$

7. Let $\mathbf{V} = \mathbf{C}_{\infty}([\alpha, \beta])$, and let $y \in \mathbf{V}$. For each $x \in \mathbf{V}$ let

$$f(x) = \int_{\alpha}^{\beta} x(t)y(t)\, dt.$$

a. Show that $f \in \mathbf{C}_{\infty}^{*}([\alpha, \beta])$.

b. Find $\|f\|$.

8. Let $\mathbf{V} = \mathbf{F^n}$, let $\mathbf{W} = \mathbf{F^m}$, and let $A : \mathbf{V} \to \mathbf{W}$ be an $m \times n$ matrix with components in **F**. Let $\|A\|_p$ be the norm induced by the pth Holder norm. Show that $\|A\|_1$ is the maximum column-sum norm. That is,

$$\|A\|_1 = \max_{j=1}^{n} \left\{ \sum_{k=1}^{m} |A_{kj}| \right\}.$$

9. Let $\mathbf{V} = \mathbf{F^n}$, let $\mathbf{W} = \mathbf{F^m}$, and let $A:\mathbf{V} \to \mathbf{W}$ be an $m \times n$ matrix with components in $\mathbf{F}$. Let $\|A\|_p$ be the norm induced by the pth Holder norm. Show that $\|A\|_\infty$ is the maximum row-sum norm. That is,

$$\|A\|_\infty = \max_{k=1}^{m} \left\{ \sum_{j=1}^{n} |A_{kj}| \right\}.$$

REFERENCES

Bachman, G. and L. Narici, *Functional Analysis*, Academic, New York, 1966.

Dunford, N. and J. Schwartz, *Linear Operators, Part I. General Theory*, Wiley-Interscience, New York, 1958.

5 FINITE-DIMENSIONAL LINEAR TRANSFORMATIONS

The theory of linear transformations from a vector space **V** to a vector space **W** can be expanded upon considerably if we place restrictions on the dimensions of **V** and **W**.

5.1. FINITE-DIMENSIONAL LINEAR TRANSFORMATIONS

Let **V** and **W** be vector spaces defined over the same field **F**. Recall that if **V** is finite-dimensional, then **L(V, W)** = **B(V, W)**. That is, every linear transformation defined on a finite-dimensional space **V** is bounded and therefore continuous. Of particular interest is the case when **W** is also finite-dimensional. We then say that the transformation itself is finite-dimensional.

Definition 1.1 (Dimension). Let **V** and **W** be vector spaces over the same field **F**, and let $\mathbf{A}: \mathbf{V} \to \mathbf{W}$. Then **A** is a *finite-dimensional* transformation $\Leftrightarrow$ both **V** and **W** are finite-dimensional in which case the *dimension* of A is denoted $\dim(A)$ and is defined

$$\dim(A) \triangleq \dim(\mathbf{V}) \dim(\mathbf{W}). \qquad \blacksquare$$

Note that in Def. 1.1 no reference is made to the transformation A being linear. If the transformation A is linear and if the space **V** on which it is defined is finite-dimensional, then we can always restrict **W** to produce a finite-dimensional transformation, as can be seen from the following exercise.

Exercise 1.1. Let $A \in \mathbf{L}(\mathbf{V}, \mathbf{R}(A))$. Show that A is finite-dimensional $\Leftrightarrow$ **V** is finite-dimensional in which case $\dim(A) \leqslant (\dim(\mathbf{V}))^2$. $\blacksquare$

Although there are many examples of finite-dimensional linear transformations, the most important example is the $m \times n$ matrix (Cullen, 1966). Indeed, let **V** and **W** be vector spaces of dimension n and m, respectively. Recall from Chapter 1 that the space **V** is isomorphic to the space $\mathbf{F}^n$. Similarly, the space $\mathbf{W}$ is isomorphic to the space $\mathbf{F^m}$. In view of these observations, we can show that the space $\mathbf{L(V, W)}$ is isomorphic to the space $\mathbf{F^{m \times n}}$. That is, each vector in the space $\mathbf{L(V, W)}$ corresponds to a vector in the space $\mathbf{F^{m \times n}}$ and vice versa. Hence every finite-dimensional linear transformation can be represented by a matrix. The importance of this observation lies in the fact that it allows us to investigate the properties of finite-dimensional linear transformations, in general, by focusing our attention on the properties of $m \times n$ matrices in particular (Bellman, 1960).

5.2. MATRIX REPRESENTATIONS

Let $\mathbf{V}$ and $\mathbf{W}$ be vector spaces of dimension n and m, respectively, and suppose $A \in \mathbf{L(V, W)}$. If $\mathbf{B}$ is a basis for $\mathbf{V}$, then each vector $x \in \mathbf{V}$ can be represented by its coordinates, a vector $[x]^{\mathbf{B}} \in \mathbf{F^n}$. Similarly, if $\mathbf{C}$ is a basis for $\mathbf{W}$, then the vector $Ax \in \mathbf{W}$ can be represented by its coordinates, a vector $[Ax]^{\mathbf{C}} \in \mathbf{F^m}$. The linear transformation from the coordinates of x to the coordinates of Ax is an $m \times n$ matrix. This matrix, $A \in \mathbf{F^{m \times n}}$, can be thought of as the coordinates of $A \in \mathbf{L(V, W)}$ relative to the bases $\{\mathbf{B}, \mathbf{C}\}$. In particular, we have the following result.

Proposition 2.1 (Matrix Representation). Let $\mathbf{B} = \{b^1, b^2, \ldots, b^n\}$ and $\mathbf{C} = \{c^1, c^2, \ldots, c^m\}$ be bases for $\mathbf{V}$ and $\mathbf{W}$, respectively. Next, let A be the $m \times n$ matrix whose kth column is defined $a^k = [Ab^k]^{\mathbf{C}}$ for $1 \leqslant k \leqslant n$. Then $\mathbf{L(V, W)}$ is isomorphic to $\mathbf{F^{m \times n}}$ and

$$[Ax]^{\mathbf{C}} = A[x]^{\mathbf{B}}, \qquad x \in \mathbf{V}.$$

Proof. Let $x \in \mathbf{V}$, and let $b = [x]^{\mathbf{B}}$. Then since $\mathbf{A}$ is linear,

$$\begin{aligned}
[Ax]^{\mathbf{C}} &= \left[A \sum_{k=1}^{n} b_k b^k \right]^{\mathbf{C}} \\
&= \left[\sum_{k=1}^{n} b_k A b^k \right]^{\mathbf{C}} \\
&= \sum_{k=1}^{n} b_k [Ab^k]^{\mathbf{C}} \\
&= \sum_{k=1}^{n} b_k a^k \\
&= Ab \\
&= A[x]^{\mathbf{B}}.
\end{aligned}$$

Next, to show that the spaces $\mathbf{L(V, W)}$ and $\mathbf{F^{m \times n}}$ are isomorphic, let $T: \mathbf{L(V, W)} \to \mathbf{F^{m \times n}}$ be the transformation defined in Prop. 2.1. That is, for each

$\boldsymbol{A} \in \mathbf{L(V, W)}$ let $A \in \mathbf{F^{m \times n}}$ be the matrix whose kth column is

$$a^k = [\boldsymbol{A}b^k]^{\mathbf{C}}, \qquad 1 \leqslant k \leqslant n.$$

We must show that $\boldsymbol{T}$ is linear, one-to-one, and onto. To that end let $(\alpha, \boldsymbol{A}, \boldsymbol{B}) \in \mathbf{F} \times \mathbf{L(V, W)}^2$. Then for each $x \in \mathbf{V}$

$$\begin{aligned}[(\alpha\boldsymbol{A} + \boldsymbol{B})x]^{\mathbf{C}} &= \alpha[\boldsymbol{A}x]^{\mathbf{C}} + [\boldsymbol{B}x]^{\mathbf{C}}, \\ &= \alpha A[x]^{\mathbf{B}} + B[x]^{\mathbf{B}}, \\ &= (\alpha A + B)[x]^{\mathbf{B}}.\end{aligned}$$

Thus $\boldsymbol{T}(\alpha\boldsymbol{A} + \boldsymbol{B}) = \alpha\boldsymbol{TA} + \boldsymbol{TB}$. Hence $\boldsymbol{T}$ is linear. Next, it is clear that the only linear transformation $\boldsymbol{A}$ that yields the zero matrix, $a^k = 0$ for $1 \leqslant k \leqslant n$, is the zero transformation. Thus $\mathbf{N}(\boldsymbol{T}) = \{\mathbf{0}\}$. Hence the linear transformation $\boldsymbol{T}$ is one-to-one. Finally, let $A \in \mathbf{F^{m \times n}}$, and define $\boldsymbol{A}$ as follows:

$$\boldsymbol{A}x = \sum_{j=1}^{n} (A[x]^{\mathbf{B}})_j c^j.$$

Then for each $1 \leqslant k \leqslant n$

$$\begin{aligned}[\boldsymbol{A}b^k]^{\mathbf{C}} &= \left[\sum_{j=1}^{n} (A[b^k]^{\mathbf{B}})_j c^j\right]^C \\ &= A[b^k]^{\mathbf{B}} \\ &= Ai^k \\ &= a^k.\end{aligned}$$

Hence $\boldsymbol{TA} = A$. Thus $\boldsymbol{T}$ is onto. It follows that $\boldsymbol{T}$ is an isomorphism between $\mathbf{L(V, W)}$ and $\mathbf{F^{m \times n}}$. ∎

In view of Prop. 2.1, *every* finite-dimensional linear transformation can be represented by an $m \times n$ matrix. Refer to Figure 5.1 for a pictorial description of the isomorphic relationship between the spaces $\mathbf{L(V, W)}$ and $\mathbf{F^{m \times n}}$.

We call $A \in \mathbf{F^{m \times n}}$ the *matrix representation* of $\boldsymbol{A}$ induced by the bases $\mathbf{B}$ and $\mathbf{C}$. Normally, the bases are understood and left implicit. However, in those instances where it is useful to denote the bases explicitly, we use the following notation:

$$A \triangleq [\boldsymbol{A}]^{\mathbf{B,C}}.$$

Note from the statement of Prop. 2.1 that **boldfaced** type is used to distinguish linear transformations from matrices. In particular, bold italic

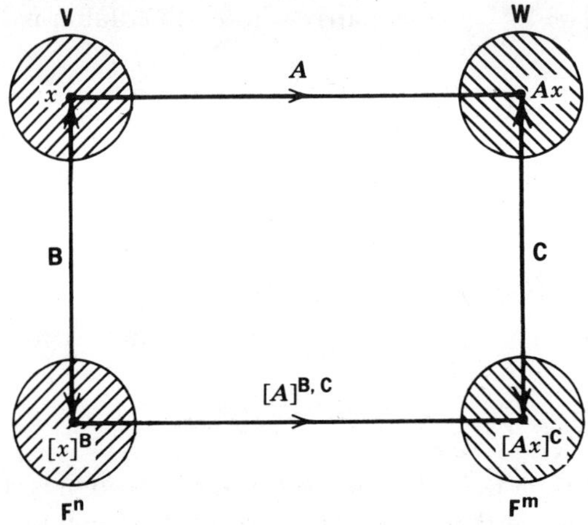

Figure 5.1. Isomorphism between **L(V, W)** and $\mathbf{F}^{m \times n}$.

uppercase letters of the form A are used to represent linear transformations, while corresponding regular italic upper case letters of the form A are used to denote matrix representations. The following are some examples of linear transformations and their matrix representations.

Example 2.1 (Integral). Let $\mathbf{V} = \mathbf{P}^n([0, \infty))$. Consider the integral transformation on **V**. That is, for each $x \in \mathbf{V}$

$$(Ax)(t) \triangleq \int_0^t x(\tau)\, d\tau, \qquad 0 \leqslant t < \infty.$$

Then $A \in \mathbf{L}(\mathbf{V}, \mathbf{W})$ where $\mathbf{W} = \mathbf{P}^{n+1}([0, \infty))$. Next, let **B** be the standard basis of elementary polynomials. That is,

$$\mathbf{B} = \{1, t, t^2, \ldots\}, \qquad 0 \leqslant t < \infty.$$

The matrix representation of A induced by the bases of elementary polynomials is then computed as follows:

$$\begin{aligned} a^k &= [At^{k-1}]^{\mathbf{B}} \\ &= \left[\int_0^t \tau^{k-1}\, d\tau\right]^{\mathbf{B}} \\ &= [t^k/k]^{\mathbf{B}} \\ &= (1/k)i^{k+1}. \end{aligned}$$

Recall that i^k denotes the kth column of the identity matrix I. Thus for the case $n = 2$ we have

$$A = \begin{bmatrix} 0 & 0 & 0 \\ 1 & 0 & 0 \\ 0 & \frac{1}{2} & 0 \\ 0 & 0 & \frac{1}{3} \end{bmatrix}.$$

Note that this matrix maps the coordinates of a polynomial into the coordinates of its integral. For example, suppose

$$x(t) = t^2 - 3t + 4.$$

We then have $[x]^{\mathbf{B}} = [4, -3, 1]^{\mathrm{T}}$. To obtain the coordinates of Ax we use

$$\begin{aligned} [Ax]^{\mathbf{B}} &= A[x]^{\mathbf{B}}, \\ &= \begin{bmatrix} 0 & 0 & 0 \\ 1 & 0 & 0 \\ 0 & \frac{1}{2} & 0 \\ 0 & 0 & \frac{1}{3} \end{bmatrix} \begin{bmatrix} 4 \\ -3 \\ 1 \end{bmatrix}, \\ &= [0, 4, -\tfrac{3}{2}, \tfrac{1}{3}]^{\mathrm{T}}. \end{aligned}$$

Thus the integral of $x(t)$ is $\frac{1}{3}t^3 - \frac{3}{2}t^2 + 4t$. ■

Example 2.2 (Derivative). Let $\mathbf{V} = \mathbf{P}^n([0, \infty))$. Consider the derivative transformation on $\mathbf{V}$. That is, for each $x \in \mathbf{V}$

$$(Ax)(t) \triangleq \frac{d}{dt} x(t), \qquad 0 \leqslant t < \infty.$$

Then $A \in \mathbf{L}(\mathbf{V}, \mathbf{W})$ where $\mathbf{W} = \mathbf{P}^{n-1}([0, \infty))$. Next, let $\mathbf{B}$ be the standard basis of elementary polynomials. That is,

$$\mathbf{B} = \{1, t, t^2, \ldots\}, \qquad 0 \leqslant t < \infty.$$

The matrix representation of A induced by the bases of elementary polynomials is then computed as follows:

$$\begin{aligned} a^k &= [At^{k-1}]^{\mathbf{B}} \\ &= \left[\frac{d}{dt} t^{k-1}\right]^{\mathbf{B}} \\ &= [(k-1)t^{k-2}]^{\mathbf{B}} \\ &= (k-1)i^{k-1}. \end{aligned}$$

Here i^k denotes the kth column of the identity matrix I. Thus for the case $n = 3$ we have

$$A = \begin{bmatrix} 0 & 1 & 0 & 0 \\ 0 & 0 & 2 & 0 \\ 0 & 0 & 0 & 3 \end{bmatrix}.$$

Note that this matrix maps the coordinates of a polynomial into the coordinates of its derivative. For example, suppose

$$x(t) = t^3 - 2t^2 + 5t - 4.$$

We then have $[x]^{\mathbf{B}} = [-4, 5, -2, 1]^{\mathrm{T}}$. To obtain the coordinates of Ax we use

$$\begin{aligned} [Ax]^{\mathbf{B}} &= A[x]^{\mathbf{B}}, \\ &= \begin{bmatrix} 0 & 1 & 0 & 0 \\ 0 & 0 & 2 & 0 \\ 0 & 0 & 0 & 3 \end{bmatrix} \begin{bmatrix} -4 \\ 5 \\ -2 \\ 1 \end{bmatrix}, \\ &= [5, -4, 3]^{\mathrm{T}}. \end{aligned}$$

Thus the derivative of $x(t)$ is $3t^2 - 4t + 5$. ■

Exercise 2.1 (Orthonormal Basis). Let $\mathbf{B} = \{b^1, b^2, \ldots, b^n\}$ and $\mathbf{C} = \{c^1, c^2, \ldots, c^m\}$ be bases for $\mathbf{V}$ and $\mathbf{W}$, respectively. Suppose $\mathbf{W}$ is an inner product space, and suppose $\mathbf{C}$ is an orthonormal basis. Let $A \in \mathbf{L}(\mathbf{V}, \mathbf{W})$. Show that the matrix representation of A induced by $\{\mathbf{B}, \mathbf{C}\}$ can be expressed as follows:

$$A_{kj} = \langle Ab^j, c^k \rangle, \qquad 1 \leqslant k \leqslant m, \quad 1 \leqslant j \leqslant n. \qquad ■$$

5.3. COORDINATE TRANSFORMATIONS

The matrix representation theorem tells us that by choosing bases for $\mathbf{V}$ and $\mathbf{W}$, we can represent each vector in $\mathbf{L}(\mathbf{V}, \mathbf{W})$ by its coordinates, a vector in $\mathbf{F}^{\mathbf{m} \times \mathbf{n}}$. To develop a strategy for picking bases for $\mathbf{V}$ and $\mathbf{W}$, consider the problem of transforming from one matrix representation to another. That is, given the representation of $A \in \mathbf{L}(\mathbf{V}, \mathbf{W})$ relative to one pair of bases, how do we find its representation relative to a second pair of bases? Here each matrix represents the same underlying linear transformation, but in a different coordinate system. Consequently, going from one matrix representation to another is a coordinate transformation problem.

Proposition 3.1 (Coordinate Transformation). Let A be a linear transformation from $\mathbf{V}$ to $\mathbf{W}$. Let $\mathbf{B} = \{b^1, b^2, \ldots, b^n\}$ and $\mathbf{C} = \{c^1, c^2, \ldots, c^n\}$ be bases for $\mathbf{V}$, and let $\mathbf{D} = \{d^1, d^2, \ldots, d^m\}$ and $\mathbf{E} = \{e^1, e^2, \ldots, e^m\}$ be bases for $\mathbf{W}$. Finally, let P be the $n \times n$ matrix whose kth column is $p^k = [c^k]^{\mathbf{B}}$, and let Q be the $m \times n$ matrix whose kth column is $q^k = [d^k]^{\mathbf{E}}$. Then

$$[A]^{\mathbf{C},\mathbf{E}} = Q[A]^{\mathbf{B},\mathbf{D}}P.$$

Proof. First we show that the matrix P maps $\mathbf{C}$-coordinates into $\mathbf{B}$-coordinates. For each $x \in \mathbf{V}$ we have

$$\begin{aligned}
[x]^{\mathbf{B}} &= \left[\sum_{k=1}^{n} [x]_k^{\mathbf{C}} c^k\right]^{\mathbf{B}} \\
&= \sum_{k=1}^{n} [x]_k^{\mathbf{C}} [c^k]^{\mathbf{B}} \\
&= \sum_{k=1}^{n} [x]_k^{\mathbf{C}} p^k \\
&= P[x]^{\mathbf{C}}.
\end{aligned}$$

Using a similar argument we can show that the matrix Q maps $\mathbf{D}$-coordinates into $\mathbf{E}$-coordinates. That is, for each $y \in \mathbf{W}$ we have $[y]^{\mathbf{E}} = Q[y]^{\mathbf{D}}$. Thus for each $x \in \mathbf{V}$

$$\begin{aligned}
Q[A]^{\mathbf{B},\mathbf{D}}P[x]^{\mathbf{C}} &= Q[A]^{\mathbf{B},\mathbf{D}}[x]^{\mathbf{B}}, \\
&= Q[Ax]^{\mathbf{D}} \\
&= [Ax]^{\mathbf{E}}.
\end{aligned}$$

Because $x \in \mathbf{V}$ is arbitrary, it follows from Prop. 2.1 that

$$[A]^{\mathbf{C},\mathbf{E}} = Q[A]^{\mathbf{B},\mathbf{D}}P. \qquad \blacksquare$$

Thus, to obtain one matrix representation from another, premultiplication and postmultiplication by coordinate transformation matrices is required. Refer to Figure 5.2 for a pictorial description which summarizes the relationship between a linear transformation and two of its matrix representations.

A matrix is itself an example of a linear transformation. As such, it should be possible to find a matrix representation of a matrix. The following exercise shows that the components of a matrix can be thought of as the coordinates induced by the standard bases for $\mathbf{F}^{\mathbf{n}}$ and $\mathbf{F}^{\mathbf{m}}$.

Exercise 3.1 (Natural Representation). Let $\mathbf{I} = \{i^1, i^2, \ldots\}$ denote the standard bases for $\mathbf{F}^{\mathbf{n}}$ and $\mathbf{F}^{\mathbf{m}}$. Show that for each $A \in \mathbf{F}^{\mathbf{m} \times \mathbf{n}}$ if B is the matrix representation of A induced by the standard bases, then $B = A$. $\blacksquare$

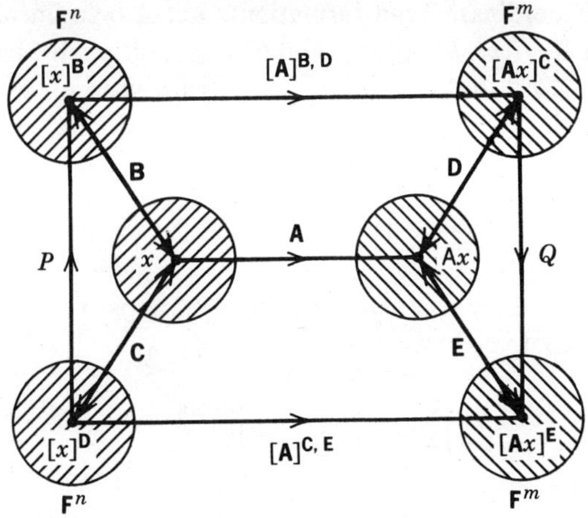

Figure 5.2. Coordinate transformations in $\mathbf{F}^{m \times n}$.

As an example of a linear transformation and two of its matrix representations, consider the derivative transformation defined on the space of nth degree polynomials.

Example 3.1 (Coordinate Transformation). Let $\mathbf{V} = \mathbf{P}^n([0, \infty))$, and let A be the derivative transformation on $\mathbf{V}$. That is, for each $x \in \mathbf{V}$,

$$(Ax)(t) \triangleq \frac{d}{dt} x(t), \qquad 0 \leqslant t < \infty.$$

Then $A \in \mathbf{L}(\mathbf{V}, \mathbf{W})$ where $\mathbf{W} = \mathbf{P}^{n-1}([0, \infty))$. Next, let $\mathbf{B}$ be the standard basis of elementary polynomials. That is,

$$\mathbf{B} = \{1, t, t^2, \ldots\}, \qquad 0 \leqslant t < \infty.$$

If $B \in \mathbf{F}^{n \times (n+1)}$ is the matrix representation of A induced by the bases of elementary polynomials, then from Exam. 2.2 for the case $n = 3$ we have

$$A = \begin{bmatrix} 0 & 1 & 0 & 0 \\ 0 & 0 & 2 & 0 \\ 0 & 0 & 0 & 3 \end{bmatrix}.$$

Next, consider the following alternative basis for the space of polynomials:

$$\mathbf{D} = \{1, 2t, 3t^2, \ldots\}, \qquad 0 \leqslant t < \infty.$$

Let $D \in \mathbf{F}^{n \times (n+1)}$ be the matrix representation of A induced by the bases associated with $\mathbf{D}$. Then according to Prop. 3.1 we have

$$D = QBP.$$

The first coordinate transformation matrix P is computed as $p^k = [kt^{k-1}]^{\mathbf{B}} = ki^k$ for $1 \leqslant k \leqslant n+1$. Similarly, the second coordinate transformation matrix Q is computed as $q^k = [t^{k-1}]^{\mathbf{D}} = (1/k)i^k$ for $1 \leqslant k \leqslant n$. Thus for the case $n = 3$ we have

$$\begin{aligned} D &= QBP \\ &= \begin{bmatrix} 1 & 0 & 0 \\ 0 & \frac{1}{2} & 0 \\ 0 & 0 & \frac{1}{3} \end{bmatrix} \begin{bmatrix} 0 & 1 & 0 & 0 \\ 0 & 0 & 2 & 0 \\ 0 & 0 & 0 & 3 \end{bmatrix} \begin{bmatrix} 1 & 0 & 0 & 0 \\ 0 & 2 & 0 & 0 \\ 0 & 0 & 3 & 0 \\ 0 & 0 & 0 & 4 \end{bmatrix} \\ &= \begin{bmatrix} 1 & 0 & 0 \\ 0 & \frac{1}{2} & 0 \\ 0 & 0 & \frac{1}{3} \end{bmatrix} \begin{bmatrix} 0 & 2 & 0 & 0 \\ 0 & 0 & 6 & 0 \\ 0 & 0 & 0 & 12 \end{bmatrix} \\ &= \begin{bmatrix} 0 & 2 & 0 & 0 \\ 0 & 0 & 3 & 0 \\ 0 & 0 & 0 & 4 \end{bmatrix}. \end{aligned}$$

As a check we can use Prop. 2.1 to compute the matrix D directly. In this case for $1 \leqslant k \leqslant n+1$ we have

$$\begin{aligned} d^k &= [A(kt^{k-1})]^{\mathbf{D}} \\ &= [(d/dt)(kt^{k-1})]^{\mathbf{D}} \\ &= [k(k-1)t^{k-2}]^{\mathbf{D}} \\ &= ki^{k-1}, \qquad i^0 = 0. \end{aligned}$$

■

5.4. ECHELON MATRICES

At this point we turn our attention to matrices exclusively. Since a matrix is a special case of a finite-dimensional linear transformation, every matrix has an infinite number of representations. Certain of these representations are more convenient to use than others. As a prelude to developing an algorithm for finding useful representations, consider the following simple operations one might perform on the rows of a matrix (Noble and Deniel, 1977).

Definition 4.1 (Elementary Row Operations). The following operations are called *elementary row operations* of the first, second, and third type, respectively:

1. Interchange rows k and j.
2. Multiply row k by a nonzero scalar α.
3. Add α times row j to row k. ■

To determine the effect of a sequence of elementary row operations on a general matrix, let us begin by examining the effects each of these operations have on the identity matrix. The resulting matrices are called elementary row matrices.

Definition 4.2 (Elementary Row Matrices). An *elementary row matrix* is a matrix obtained by applying an elementary row operation to the identity matrix. Here,

$E(k, j)$: interchange rows k and j of I,
$E(k, \alpha)$: multiply row k of I by $\alpha \neq 0$,
$E(k, j, \alpha)$: add α times row j to row k of I. ■

We refer to the matrices $E(k, j)$, $E(k, \alpha)$, and $E(k, j, \alpha)$ as elementary row matrices of the first, second, and third type, respectively. Since elementary row operations do not alter the rank of a matrix, the elementary row matrices are all nonsingular. Indeed, the following exercise shows that the inverse of an elementary row matrix is an elementary row matrix of the same type.

Exercise 4.1 (Elementary Row Matrices). Show that the elementary row matrices are invertible and that their inverses are elementary row matrices of the same type. ■

Elementary row matrices are useful for representing the effects of elementary row operations on general matrices. In particular, performing an elementary row operation is equivalent to premultiplying by the corresponding elementary row matrix, as can be seen from the following result.

Proposition 4.1 (Elementary Row Operations). Let B be the matrix obtained from A by applying a sequence of p elementary row operations. If $E_1, E_2, \ldots, E_p$ denote the associated elementary row matrices, then

$$B = E_p E_{p-1} \cdots E_1 A. \qquad ■$$

The proof of Prop. 4.1 follows from Def. 4.2 and the definition of matrix multiplication. In view of Prop. 4.1 and Exer. 4.1, if B is obtained from A by elementary row operations, then A can be obtained from B by elementary row operations. Thus it is meaningful to speak of two matrices being equivalent to one another in the following sense.

Definition 4.3 (Row Equivalence). Two matrices are *row equivalent* $\Leftrightarrow$ one can be obtained from the other by a sequence of elementary row operations. ■

All matrices are row equivalent to a matrix of a special form called the row echelon form. Rather than attempt to characterize the row echelon form explicitly, we instead define it implicitly by means of the following algorithm. Here we assume that A is an $m \times n$ matrix.

Algorithm 4.1 (Row Echelon Form)

1. Set $i = 1$ and $j = 1$.
2. Compute the *pivot* index p such that

$$|A_{pj}| = \max_{k=i}^{m} \{|A_{kj}|\}.$$

3. If $A_{pj} = 0$, go to 8.
4. Interchange row i with row p.
5. Divide row i by A_{ij}.
6. Subtract A_{kj} times row i from row k for each $k \neq i$.
7. If $i = m$, stop. Otherwise set $i = i + 1$.
8. If $j = n$, stop. Otherwise set $j = j + 1$.
9. Go to 2. ■

The matrix B obtained by applying Alg. 4.1 to a matrix A is called the *row echelon form* of A. Algorithm 4.1 is known as the *Gauss–Jordan procedure* for row reducing a matrix. Note that steps 4, 5 and 6 correspond to elementary row operations of the first, second, and third type, respectively. Row echelon matrices have several properties which set them apart from matrices in general. The following exercise summarizes these properties.

Exercise 4.2 (Row Echelon Matrices). Let $A \in \mathbf{F}^{m \times n}$, and let B be the row echelon form of A. Show that B has the following properties:

1. Zero rows of B occur below nonzero rows.
2. B is an upper triangular matrix.
3. The first nonzero component of each nonzero row is 1.
4. If B has r nonzero rows, $i^1, i^2, \ldots, i^r$ are columns of B.
5. Column i^k occurs to the left of column i^{k+1}. ■

The following example illustrates an application of Alg. 4.1 to a 3×4 matrix.

Example 4.1 (Row Reduction). Apply Alg. 4.1 to row reduce the following matrix A to its row echelon form B,

$$
A = \begin{bmatrix} 1 & 0 & -2 & 1 \\ 0 & 3 & 1 & 2 \\ -2 & 3 & 5 & 0 \end{bmatrix} \rightarrow \begin{bmatrix} -2 & 3 & 5 & 0 \\ 0 & 3 & 1 & 2 \\ 1 & 0 & -2 & 1 \end{bmatrix}
$$

$$
\rightarrow \begin{bmatrix} 1 & -\frac{3}{2} & -\frac{5}{2} & 0 \\ 0 & 3 & 1 & 2 \\ 1 & 0 & -2 & 1 \end{bmatrix} \rightarrow \begin{bmatrix} 1 & -\frac{3}{2} & -\frac{5}{2} & 0 \\ 0 & 3 & 1 & 2 \\ 0 & \frac{3}{2} & \frac{1}{2} & 1 \end{bmatrix}
$$

$$
\rightarrow \begin{bmatrix} 1 & -\frac{3}{2} & -\frac{5}{2} & 0 \\ 0 & 1 & \frac{1}{3} & \frac{2}{3} \\ 0 & \frac{3}{2} & \frac{1}{2} & 1 \end{bmatrix} \rightarrow \begin{bmatrix} 1 & 0 & -2 & 1 \\ 0 & 1 & \frac{1}{3} & \frac{2}{3} \\ \hdashline 0 & 0 & 0 & 0 \end{bmatrix}
$$

$$
= B. \qquad \blacksquare
$$

Algorithm 4.1 can be described as an implementation of Gauss–Jordan elimination with partial pivoting. The pivot is obtained in steps 2–4 and then normalized to unity in step 5. The normalization step facilitates the elimination operation, which is then implemented in step 6. Here all entries in column j except the pivot, $A_{ij} = 1$, are eliminated or set to zero. Step 6 is sometimes modified such that only the entries below the pivot are eliminated. This variation is more efficient in certain instances and is referred to as *Gaussian elimination.*

In step 2 of Alg. 4.1 the pivot is selected to be the component of A in column j at or below row i having the largest magnitude. This maximization step is employed to enhance the numerical stability of Alg. 4.1 when it is implemented on a computer. This is the *partial pivoting* feature. Alternatively, *full pivoting* can be used, in which the search for the best component of A to pivot on extends to the right of column j as well as below row i.

For well-conditioned low-dimensional pencil and paper computations such as those illustrated in this text, finding the maximum component to pivot on is not really necessary. Instead, the pivot can be *any* nonzero component of A in column j at or below row i. Thus if $A_{ij} \neq 0$, it will usually work fine. Alternatively, if one of the components of A below A_{ij} is equal to unity, it would also make an effective pivot because the normalization in step 5 is then unnecessary.

Recall that every matrix is a linear transformation and therefore has a null space associated with it. The following exercise shows that matrices that are row equivalent to one another have the same null space.

Exercise 4.3 (Null Space Invariance). Let $A \in \mathbf{F}^{m \times n}$ and $B \in \mathbf{F}^{m \times n}$. Show that if B is row equivalent to A, then $\mathbf{N(B)} = \mathbf{N(A)}$. $\blacksquare$

Given elementary row operations and elementary row matrices, we could now develop the dual notions of elementary column operations and elementary column matrices. However, rather than go through a parallel development for the column echelon form of a matrix, let us consider instead the use of the matrix transpose.

Definition 4.4 (Transpose). Let $A \in \mathbf{F}^{m \times n}$. Then the *transpose* of A is the matrix $A^T \in \mathbf{F}^{n \times m}$ defined as follows. For $1 \leqslant k \leqslant n$, $1 \leqslant j \leqslant m$

$$A^T_{kj} = A_{jk}.$$

More generally, the *Hermitian transpose* of A is denoted A^H and is defined $A^H = \bar{A}^T$. ■

Thus transposing a matrix involves interchanging its rows with its columns. In particular, the transpose of a column vector is a row vector and vice versa. For matrices with complex elements, the Hermitian transpose of A is just the transpose of the complex conjugate of A. With the help of preprocessing and postprocessing using the transpose operation, Alg. 4.1 can be modified to form its *column echelon form* counterpart. In this case performing an elementary column operation is equivalent to postmultiplying by the associated elementary column matrix. The following exercises summarize some additional fundamental properties of the matrix transpose operation.

Exercise 4.4 (Transpose of Inverse). Show that if A is an invertible matrix, then the inverse and transpose operations commute. That is,

$$(A^{-1})^T = (A^T)^{-1}.$$ ■

Exercise 4.5 (Transpose of Composition). Show that if $A \in \mathbf{F}^{m \times n}$ and $B \in \mathbf{F}^{n \times p}$, then the transpose of the composition AB is the composition of the transposes in reverse order,

$$(AB)^T = B^T A^T.$$ ■

The row echelon form of a matrix provides us with a considerable amount of information regarding the matrix. For example, we can obtain the nullity and rank by inspection. To see this, it is useful to first consider the spaces generated or spanned by the rows and the columns of a matrix.

Definition 4.5 (Row and Column Spaces). Let $A \in \mathbf{F}^{m \times n}$. Then the *row space* of A is the space spanned by the rows of A, and the *column space* of A is the space spanned by the columns of A. ■

Recall that the vector Ax can be regarded as a linear combination of the columns of A where the weighting coefficients of the linear combination are the components of x. Given this observation, we conclude that the column space of

A is synonymous with the range space of A, as can be seen from the following exercise.

Exercise 4.6 (Column Space). Let $A \in \mathbf{F}^{\mathbf{m} \times \mathbf{n}}$, and let $\mathbf{W}$ be the column space of A. Show that $\mathbf{W} = \mathbf{R(A)}$. ∎

Algorithm 4.1 provides us with a ready means of finding a basis for the row space of a matrix. In doing so we also determine the nullity and rank of the matrix, as can be seen from the following result.

Proposition 4.2 (Nullity and Rank). Let $A \in \mathbf{F}^{\mathbf{m} \times \mathbf{n}}$, and let B be the row echelon form of A. Then the r nonzero rows of B form a basis for the row space of A, and

1. $\text{rank}(A) = r$,
2. $\text{nullity}(A) = n - r$.

Proof. Let $\mathbf{U}$ be the row space of A. Then since A and B are row equivalent, the r nonzero rows of B span $\mathbf{U}$. Since B is a row echelon matrix with r nonzero rows, from step 6 of Alg. 4.1, $\{i^1, i^2, \ldots, i^r\}$ are columns of B. Thus the r nonzero rows of B are linearly independent. Hence they form a basis for the row space $\mathbf{U}$. Next, note from Exer. 4.3 that $\text{nullity}(B) = \text{nullity}(A)$ because row equivalent matrices have the same null space. It then follows from Prop. 4.3.2 that $\text{rank}(B) = \text{rank}(A)$. Let $\mathbf{W}$ be the column space of B. Then from Exer. 4.6, $\text{rank}(B) = \dim(\mathbf{W})$. But $\dim(\mathbf{W}) = r$ since B has r nonzero rows and $\{i^1, i^2, \ldots, i^r\}$ are columns of B. Hence from Prop. 4.3.2

$$\text{rank}(A) = r,$$

$$\text{nullity}(A) = n - r. \qquad ∎$$

Thus the number of nonzero rows in the row echelon form of a matrix is the rank of the matrix. If there are no zero rows in the row echelon form and if the matrix is square, then the matrix is invertible. That is, the row echelon form of a nonsingular matrix is the identity matrix as can be seen from the following exercise.

Exercise 4.7 (Invertibility). Let $A \in \mathbf{F}^{\mathbf{n} \times \mathbf{n}}$, and let B be the row echelon form of A. Show that A is invertible $\Leftrightarrow B = I$. ∎

Thus invertible matrices are row equivalent to the identity matrix. More generally, if B is the row echelon form of A, then the nonzero rows of B form a basis for the row space of A. As an example of an application of Prop. 4.2 and Exer. 4.7, consider the following.

Example 4.2 (Row Space). Suppose we apply Alg. 4.1 (without partial pivoting) to the following matrix A to find its row echelon form B:

$$A = \begin{bmatrix} 1 & 0 & 2 \\ 3 & -1 & 0 \\ 1 & -1 & -4 \end{bmatrix} \to \begin{bmatrix} 1 & 0 & 2 \\ 0 & -1 & -6 \\ 0 & -1 & -6 \end{bmatrix}$$

$$\to \begin{bmatrix} 1 & 0 & 2 \\ 0 & 1 & 6 \\ 0 & -1 & -6 \end{bmatrix} \to \begin{bmatrix} 1 & 0 & 2 \\ 0 & 1 & 6 \\ \hline 0 & 0 & 0 \end{bmatrix}$$

$$= B.$$

Thus from Exer. 4.7, A is not invertible. From Prop. 4.2 we have $\text{rank}(A) = 2$, $\text{nullity}(A) = 1$ and $\{c^1, c^2\}$ is a basis for the row space of A where

$$c^1 = [1, 0, 2],$$

$$c^2 = [0, 1, 6]. \qquad \blacksquare$$

Note that Prop. 4.2 can be used to find the dimension of, and a basis for, the space spanned by *any* finite set of vectors. All we have to do is to first arrange those vectors as the *rows* of a matrix and then find the row echelon form.

Exercise 4.8 (Range Space). Let B be the row echelon form of A^T. Show that the nonzero rows of B form a basis for **R(A)**. ■

5.5. ROW REDUCTION ALGORITHMS

The discussion of null spaces, range spaces, and inverses in Chapter 4 applied to linear transformations in general. If we focus our attention on matrices in particular, these concepts can be expanded upon somewhat. We begin by examining the following fundamental relationship between the rows and columns of a matrix.

Proposition 5.1 (Linear Independence). Let $A \in \mathbf{F}^{m \times n}$, and let $r = \text{rank}(A)$. Then A has r linearly independent rows and r linearly independent columns.

Proof. Since every matrix A is row equivalent to its row echelon form B, the number of linearly independent rows of A is equal to the number of nonzero rows of B. From Prop. 4.2, the number of nonzero rows of B is $r = \text{rank}(A)$. Thus the number of linearly independent rows of A is r. Similarly, from Exer. 4.6 the column space of A is the range space of A. Thus the number of linearly independent columns of A is also r. ■

In view of Prop. 5.1, the rank of a matrix is bounded from above by the number of rows and the number of columns. In particular, we have the following exercise.

Exercise 5.1 (Rank Bound). Let $A \in \mathbf{F}^{m \times n}$. Show that the rank of A is constrained by the following inequality:

$$\operatorname{rank}(A) \leqslant \min\{m, n\}. \qquad \blacksquare$$

If the rank of a matrix A is equal to the number of rows or the number of columns, then the rank is as large as possible. In this case we say that the matrix is of full rank.

Definition 5.1 (Full Rank). An $m \times n$ matrix A is of *full rank* $\Leftrightarrow \operatorname{rank}(A) = \min\{m, n\}$. ■

Being of full rank is one of two necessary conditions for matrix invertibility. The other is that the matrix be square. The following simple adaptation of Alg. 4.1 provides us with a relatively efficient means of finding the inverse of a square matrix or determining that an inverse does not exist.

Algorithm 5.1 (Matrix Inversion)

1. Given an $n \times n$ matrix A, form $B = [A, I]$.
2. Use Alg. 4.1 to reduce B to its row echelon form C.
3. Partition C conformably with B as $C = [C_1, C_2]$. ■

Proposition 5.2 (Matrix Inversion). Let C be the matrix constructed by Alg. 5.1. Then A is invertible $\Leftrightarrow C_1 = I$, in which case $A^{-1} = C_2$.

Proof. Since C is the row echelon form of B, from Prop. 4.1 there exists an $n \times n$ matrix E such that,

$$C = EB.$$

Here E is a product of elementary row matrices. If we now partition C conformably with B, then from Alg. 5.1 we have

$$\begin{aligned} [C_1, C_2] &= E[B_1, B_2] \\ &= E[A, I] \\ &= [EA, EI] \\ &= [EA, E]. \end{aligned}$$

Now C_1 is the row echelon form of A. Thus from Exer. 4.7, A is invertible $\Leftrightarrow C_1 = I$. But

$$\begin{aligned} C_1 = I &\Rightarrow EA = I \\ &\Rightarrow E = A^{-1} \\ &\Rightarrow C_2 = A^{-1}. \end{aligned}$$

■

As an example of an application of Alg. 5.1, consider the following 3×3 matrix.

Example 5.1 (Matrix Inverse). Determine whether or not the 3×3 matrix A is invertible and, if so, find its inverse. Here

$$A = \begin{bmatrix} 1 & 0 & 2 \\ 2 & -1 & 3 \\ 4 & 1 & 8 \end{bmatrix}.$$

Applying Alg. 5.1 we have

$$\begin{aligned} B &= [A, I] \\ &= \begin{bmatrix} 1 & 0 & 2 & 1 & 0 & 0 \\ 2 & -1 & 3 & 0 & 1 & 0 \\ 4 & 1 & 8 & 0 & 0 & 1 \end{bmatrix} \rightarrow \begin{bmatrix} 1 & 0 & 2 & 1 & 0 & 0 \\ 0 & -1 & -1 & -2 & 1 & 0 \\ 0 & 1 & 0 & -4 & 0 & 1 \end{bmatrix} \\ &\rightarrow \begin{bmatrix} 1 & 0 & 2 & 1 & 0 & 0 \\ 0 & 1 & 1 & 2 & -1 & 0 \\ 0 & 1 & 0 & -4 & 0 & 1 \end{bmatrix} \rightarrow \begin{bmatrix} 1 & 0 & 2 & 1 & 0 & 0 \\ 0 & 1 & 1 & 2 & -1 & 0 \\ 0 & 0 & -1 & -6 & 1 & 1 \end{bmatrix} \\ &\rightarrow \begin{bmatrix} 1 & 0 & 2 & 1 & 0 & 0 \\ 0 & 1 & 1 & 2 & -1 & 0 \\ 0 & 0 & 1 & 6 & -1 & -1 \end{bmatrix} \rightarrow \left[\begin{array}{ccc:ccc} 1 & 0 & 0 & -11 & 2 & 2 \\ 0 & 1 & 0 & -4 & 0 & 1 \\ 0 & 0 & 1 & 6 & -1 & -1 \end{array}\right] \\ &= [I, A^{-1}]. \end{aligned}$$

Thus A is invertible and

$$A^{-1} = \begin{bmatrix} -11 & 2 & 2 \\ -4 & 0 & 1 \\ 6 & -1 & -1 \end{bmatrix}.$$

■

Next, we consider the problem of finding bases for the null space and range space of a matrix. Since the range space is simply the column space, finding a basis for the range space is a simple matter. Indeed from Exer. 4.8, the nonzero rows of the row echelon form of A^T form a basis for **R(A)**. The problem of finding a basis for the null space of a matrix is also important because many problems can be cast in the form of finding a basis for the null space of a suitably defined matrix. For example, finding all of the eigenvectors of A associated with an eigenvalue λ reduces to the problem of finding a basis for the null space of the matrix $B(\lambda) = \lambda I - A$.

As it turns out, a slight modification of Alg. 5.1 provides us with a means of obtaining a basis for the null space of the matrix A. As an added bonus, we also get a basis for the range space and, if A is invertible, the transpose of the inverse. Consequently, the following algorithm is quite useful.

Algorithm 5.2 (Null and Range Spaces)

1. Given an $m \times n$ matrix A, form the matrix $B = [A^{\mathrm{T}}, I]$.
2. Use Alg. 4.1 to find C, the row echelon form of B.
3. Partition C as follows where $r = \operatorname{rank}(A)$:

$$C = \begin{array}{c} \begin{array}{cc} m & n \end{array} \\ \left[\begin{array}{c|c} C_1 & C_2 \\ \hline 0 & C_3 \end{array}\right] \end{array} \begin{array}{l} \\ r \\ n - r \end{array}$$ ■

Proposition 5.3 (Null and Range Space). Let C be the matrix constructed by Alg. 5.2. Then C has the following properties:

1. The rows of C_1 form a basis for **R(A)**.
2. The rows of C_3 form a basis for **N(A)**.
3. A is invertible $\Leftrightarrow C_1 = I$, in which case $A^{-1} = C_2^{\mathrm{T}}$.

Proof. Let D be the $n \times m$ matrix consisting of the first m columns of C. That is,

$$D = \begin{bmatrix} C_1 \\ 0 \end{bmatrix}.$$

Then from Alg. 5.2, D is the row echelon form of A^{T}. Hence from Exer. 4.8, the rows of C_1 form a basis for **R(A)**. Next, as C is row equivalent to B, there exists an invertible matrix E such that

$$C = EB.$$

Partitioning E conformably with C we get

$$\begin{bmatrix} C_1 & C_2 \\ 0 & C_3 \end{bmatrix} = \begin{bmatrix} E_1 \\ E_2 \end{bmatrix} [A^{\mathrm{T}} \quad I]$$

$$= \begin{bmatrix} E_1 A^{\mathrm{T}} & E_1 \\ E_2 A^{\mathrm{T}} & E_2 \end{bmatrix}.$$

Thus $E_2 A^T = 0$ or, from Exer. 4.5, $AE_2^T = 0^T$. Hence the row of E_2 belong to **N(A)**. But E is a product of elementary matrices and is therefore invertible. Hence the rows of E_2 are linearly independent. Finally, E_2 has $n - r$ rows where $r = \text{rank}(A)$. Thus the rows of $E_2 = C_3$ form a basis for **N(A)**. Next, since D is the row echelon form of A^T, it follows from Prop. 5.2 that A^T is invertible $\Leftrightarrow C_1 = I$, in which case $(A^T)^{-1} = C_2$. Thus from Exer. 4.4,

$$\begin{aligned} C_2^T &= ((A^T)^{-1})^T \\ &= ((A^T)^T)^{-1} \\ &= A^{-1}. \end{aligned}$$

■

As an example of an application of Alg. 5.2 to a 3×4 matrix consider the following.

Example 5.2 (Null and Range Spaces). Find bases for the null space and the range space of the 3×4 matrix A where

$$A = \begin{bmatrix} 1 & 0 & 3 & -2 \\ -2 & 1 & 0 & 3 \\ 0 & 1 & 6 & -1 \end{bmatrix}.$$

Applying Alg. 5.2 we have

$$\begin{aligned} B &= [A^T, I] \\ &= \begin{bmatrix} 1 & -2 & 0 & 1 & 0 & 0 & 0 \\ 0 & 1 & 1 & 0 & 1 & 0 & 0 \\ 3 & 0 & 6 & 0 & 0 & 1 & 0 \\ -2 & 3 & -1 & 0 & 0 & 0 & 1 \end{bmatrix} \\ &\to \begin{bmatrix} 1 & -2 & 0 & 1 & 0 & 0 & 0 \\ 0 & 1 & 1 & 0 & 1 & 0 & 0 \\ 0 & 6 & 6 & -3 & 0 & 1 & 0 \\ 0 & -1 & -1 & 2 & 0 & 0 & 1 \end{bmatrix} \\ &\to \left[\begin{array}{ccc:cccc} 1 & 0 & 2 & 1 & 2 & 0 & 0 \\ 0 & 1 & 1 & 0 & 1 & 0 & 0 \\ \hdashline 0 & 0 & 0 & -3 & -6 & 1 & 0 \\ 0 & 0 & 0 & 2 & 1 & 0 & 1 \end{array}\right]. \end{aligned}$$

Thus we have $\text{rank}(A) = 2$ and $\{r^1, r^2\}$ is a basis for **R(A)** where

$$\begin{aligned} r^1 &= [1, 0, 2]^T, \\ r^2 &= [0, 1, 1]^T. \end{aligned}$$

Similarly, nullity$(A) = 2$ and $\{n^1, n^2\}$ is a basis for **N(A)** where

$$n^1 = [-3, -6, 1, 0]^T,$$
$$n^2 = [2, 1, 0, 1]^T.$$ ■

Exercise 5.2 (Rank Invariance). Show that the rank of a matrix is invariant under premultiplication or postmultiplication by an invertible matrix. ■

5.6. LINEAR ALGEBRAIC SYSTEMS

In engineering applications, a considerable number of problems can be formulated in such a manner that they come down to a question of solving a linear system of algebraic equations of the following generic form:

$$Ax = b.$$

Here $A \in \mathbf{F}^{m \times n}$ and $b \in \mathbf{F}^m$ vary from application to application. The problem is to find a solution vector $x \in \mathbf{F}^n$ that is consistent with this *linear algebraic system.* That is, we must find an inverse image of the vector b under the linear transformation A. Depending upon the nature of A and b, this system of equations may have no solution, one solution, or an infinite number of solutions. In particular, we have the following result.

Proposition 6.1 (Existence and Uniqueness). Let $A \in \mathbf{F}^{m \times n}$ and $b \in \mathbf{F}^m$. Then the linear algebraic system $Ax = b$ has the following properties:

1. If $b \notin \mathbf{R(A)}$, no solution exists.
2. If $b \in \mathbf{R(A)}$ and $\mathbf{N(A)} = \{\mathbf{0}\}$, a unique solution exists.
3. If $b \in \mathbf{R(A)}$ and $\mathbf{N(A)} \neq \{\mathbf{0}\}$, infinitely many solutions exist.

Proof. Suppose $b \notin \mathbf{R(A)}$. Then for each $x \in \mathbf{F}^n$, $Ax \neq b$. Hence there is no solution. Next, suppose $b \in \mathbf{R(A)}$. Then there exists an $x \in \mathbf{F}^n$ such that $Ax = b$. If there exists a $y \in \mathbf{F}^n$ with $y \neq x$ such that $Ay = b$, then as A is linear,

$$\begin{aligned} A(x - y) &= Ax - Ay \\ &= b - b \\ &= 0. \end{aligned}$$

Hence $x - y \in \mathbf{N(A)}$. Thus if $b \in \mathbf{R(A)}$, then $Ax = b$ has an infinite number of solutions $\Leftrightarrow \mathbf{N(A)} \neq \{\mathbf{0}\}$. Otherwise it has exactly one solution. ■

Refer to Figure 5.3 for a pictorial description which summarizes the conditions for the existence and uniqueness of solutions of linear algebraic systems.

Although a linear algebraic system of equations can have an infinite number of solutions, it has at most one solution that is orthogonal to the null space of A. In particular we have the following result which decomposes the general solution into two parts.

Proposition 6.2 (General Solution). Let $b \in \mathbf{R(A)}$. Then there exists a unique $x^0 \in \mathbf{N(A)}^\perp$ called the *minimum norm solution* such that, $Ax^0 = b$. Furthermore, all solutions of $Ax = b$ are of the form

$$x = x^0 + x^1, \qquad x^1 \in \mathbf{N(A)}.$$

Proof. Let $b \in \mathbf{R(A)}$. Then from Prop. 6.1, $Ax = b$ has at least one solution x. Using the projection theorem from Chapter 3, we can decompose the space $\mathbf{F^n}$ into the following direct sum:

$$\mathbf{F^n} = \mathbf{N(A)}^\perp \oplus \mathbf{N(A)}.$$

Then for each $x \in \mathbf{F^n}$ there exist unique $(x^0, x^1) \in \mathbf{N(A)}^\perp \times \mathbf{N(A)}$ such that

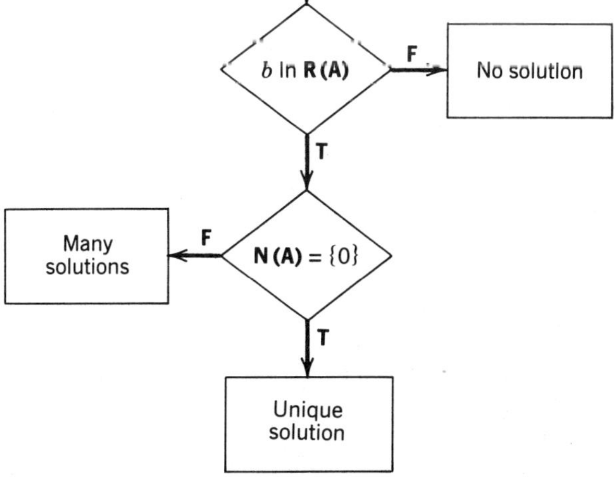

Figure 5.3. Existence and uniqueness of solutions of $Ax = b$.

$x = x^0 + x^1$. Suppose x is a solution of $Ax = b$. Then

$$\begin{aligned} b &= Ax \\ &= A(x^0 + x^1) \\ &= Ax^0 + Ax^1 \\ &= Ax^0 + 0 \\ &= Ax^0. \end{aligned}$$

Hence x^0, the orthogonal projection of x onto $\mathbf{N(A)}^\perp$, is also a solution. To show that the vector x^0 is unique, let y be another solution. Then there exist unique $(y^0, y^1) \in \mathbf{N(A)}^\perp \times \mathbf{N(A)}$ with $y = y^0 + y^1$ such that $Ay = b$. Hence

$$\begin{aligned} 0 &= b - b \\ &= Ax - Ay \\ &= A(x^0 + x^1) - A(y^0 + y^1) \\ &= Ax^0 - Ay^0 \\ &= A(x^0 - y^0). \end{aligned}$$

Thus $x^0 - y^0 \in \mathbf{N(A)}$. But it is also true that $x^0 - y^0 \in \mathbf{N(A)}^\perp$. Hence $x^0 - y^0 = 0$. Consequently, the solution of $Ax = b$ belonging to $\mathbf{N(A)}^\perp$ is unique. ■

Thus the general solution of $Ax = b$ consists of the minimum norm solution $x^0 \in \mathbf{N(A)}^\perp$ plus any vector belonging to $\mathbf{N(A)}$. Refer to Figure 5.4 for a pictorial representation of the decomposition of the general solution into its two parts.

As the name implies, the feature that distinguishes the minimum norm solution x^0 from all other solutions is that it is optimal in the sense of minimizing $\|x\|$, as can be seen from the following exercise.

Exercise 6.1 (Minimum Norm Solution). Use Prop. 6.2 and the Pythagorean theorem from Chapter 3 to show that if x is any solution of $Ax = b$ and if x^0 is

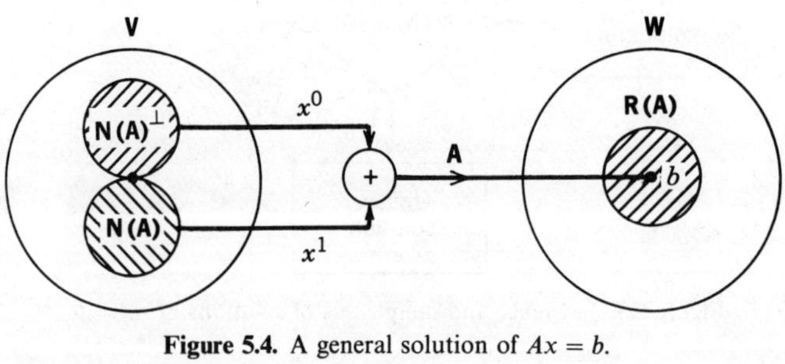

Figure 5.4. A general solution of $Ax = b$.

the solution belonging to $\mathbf{N(A)}^{\perp}$,

$$\|x^0\| \leqslant \|x\|. \qquad \blacksquare$$

To characterize a complete set of solutions of $Ax = b$, let $r = \text{rank}(A)$, and let $b \in \mathbf{R(A)}$, in which case the linear algebraic system has at least one solution. If $b \neq 0$, then the minimum norm solution does not belong to $\mathbf{N(A)}$. Hence there are a total of $n - r + 1$ linearly independent solutions, which we refer to as a fundamental system of solutions.

Definition 6.1 (Fundamental Solutions). Let $Ax = b$ with $r = \text{rank}(A)$, and let $b \in \mathbf{R(A)} - \{\mathbf{0}\}$. Then any set of $n - r + 1$ linearly independent solutions of $Ax = b$ is called a *fundamental system* of solutions. ■

Note that if $b \notin \mathbf{R(A)}$, then there are no solutions, whereas if $b \in \mathbf{R(A)}$, but $b = 0$, then there are $n - r$ linearly independent solutions. If A is of full column rank and $b = 0$, then the only solution is $x = 0$, and $\{0\}$ is not a linearly independent set.

Exercise 6.2 (Fundamental Solutions). Let $Ax = b$, and let $r = \text{rank}(A)$. Suppose $\{x^1, x^2, \ldots, x^{n-r}\}$ is a basis for $\mathbf{N(A)}$ and $b \in \mathbf{R(A)} - \{\mathbf{0}\}$. Show that if x^0 is the minimum norm solution, then the following is a fundamental system of solutions of $Ax = b$:

$$S = \{x^0,\ x^0 + x^1,\ x^0 + x^2, \ldots, x^0 + x^{n-r}\}. \qquad \blacksquare$$

A simple adaptation of Alg. 5.2 provides us with the minimum norm solution as well as a basis for $\mathbf{N(A)}$. In view of Exer. 6.2, it therefore furnishes us with a means of obtaining a fundamental system of solutions of $Ax = b$.

Algorithm 6.1 (Linear Algebraic Systems)

1. Given the equation $Ax = b$, form the following matrix B:

$$B = \begin{array}{c} \begin{array}{ccc} m & 1 & n \end{array} \\ \left[\begin{array}{c|c|c} -b^{\mathrm{T}} & 1 & 0 \\ \hline A^{\mathrm{T}} & 0 & I \end{array}\right] \end{array} \begin{array}{l} \\ 1 \\ n \end{array}.$$

2. Use Alg. 4.1 to compute C, the row echelon form of B.

3. Use Prop. 4.2 to compute $r = \text{rank}(A)$.

4. Partition C as follows:

$$C = \begin{array}{c} \begin{array}{ccc} m & 1 & n \end{array} \\ \left[\begin{array}{c|c|c} C_1 & C_2 & C_3 \\ \hline C_4 & C_5 & C_6 \\ \hline 0 & C_7 & C_8 \end{array}\right] \end{array} \begin{array}{l} \\ r \\ 1 \\ n - r \end{array}. \qquad \blacksquare$$

Proposition 6.3 (Fundamental Solutions). Let C be the matrix constructed by Alg. 6.1. Then the system $Ax = b$ has a solution $\Leftrightarrow C_4 = 0$, in which case C_6^T is the minimum norm solution, and the rows of C_8 form a basis for **N(A)**.

Proof. Let $D = [-b, A]$. From Prop. 6.1, at least one solution exists $\Leftrightarrow b \in \mathbf{R(A)}$. But

$$\begin{aligned} b \in \mathbf{R(A)} &\Leftrightarrow \operatorname{rank}(D) = r \\ &\Leftrightarrow C_4 = 0. \end{aligned}$$

Thus the system of equations has a solution $\Leftrightarrow C_4 = 0$. Suppose $C_4 = 0$. Then as C is a row echelon matrix, we must have $C_7 = 0$. Let E be the resulting $(n - r + 1) \times (n + 1)$ matrix in the lower right corner of C. That is,

$$E = \begin{bmatrix} C_5 & C_6 \\ 0 & C_8 \end{bmatrix}.$$

Now from Alg. 5.2, the rows of E form a basis for **N(D)**. Hence we have

$$\begin{aligned} 0 &= DE^T \\ &= [-b \quad A]\begin{bmatrix} C_5 & C_6 \\ 0 & C_8 \end{bmatrix}^T \\ &= [-b \quad A]\begin{bmatrix} C_5^T & 0 \\ C_6^T & C_8^T \end{bmatrix} \\ &= [-bC_5^T + AC_6^T, AC_8^T]. \end{aligned}$$

If we now express this as two separate equations this yields

$$\begin{aligned} AC_6^T &= bC_5^T, \\ AC_8^T &= 0. \end{aligned}$$

Hence the rows of C_8 belong to **N(A)**. Moreover, since the rows of E form a basis for **N(D)**, the rows of C_8 are linearly independent. Finally since there are $n - r$ linearly independent rows in C_8, they form a basis for **N(A)**. It remains to demonstrate that C_6^T is the minimum norm solution. Because the rows of E form a basis for **N(D)**, we must have $C_5 \neq 0$. But as C is a row echelon matrix with $C_5 \neq 0$, we then have $C_5^T = 1$. Thus $AC_6^T = b$ in which case C_6^T is a solution. To show that it is the minimum norm solution, it is sufficient to show that $C_8 C_6^T = 0$, which is left as an exercise. ■

As an example of an application of Alg. 6.1, consider the following 3×3 system.

Example 6.1 (Linear Algebraic System). Find a fundamental system of solutions for the following linear algebraic system:

$$\begin{bmatrix} 1 & 0 & -2 \\ 0 & -1 & -1 \\ 2 & 3 & -1 \end{bmatrix} \begin{bmatrix} x_1 \\ x_2 \\ x_3 \end{bmatrix} = \begin{bmatrix} -1 \\ -1 \\ 1 \end{bmatrix}.$$

Applying Alg. 6.1, we have

$$B = \begin{bmatrix} -b^{\mathrm{T}} & 1 & 0 \\ A^{\mathrm{T}} & 0 & I \end{bmatrix}$$

$$= \begin{bmatrix} 1 & 1 & -1 & 1 & 0 & 0 & 0 \\ 1 & 0 & 2 & 0 & 1 & 0 & 0 \\ 0 & -1 & 3 & 0 & 0 & 1 & 0 \\ -2 & -1 & -1 & 0 & 0 & 0 & 1 \end{bmatrix}$$

$$\rightarrow \begin{bmatrix} 1 & 1 & -1 & 1 & 0 & 0 & 0 \\ 0 & -1 & 3 & -1 & 1 & 0 & 0 \\ 0 & -1 & 3 & 0 & 0 & 1 & 0 \\ 0 & 1 & -3 & 2 & 0 & 0 & 1 \end{bmatrix}$$

$$\rightarrow \begin{bmatrix} 1 & 1 & -1 & 1 & 0 & 0 & 0 \\ 0 & 1 & -3 & 1 & -1 & 0 & 0 \\ 0 & -1 & 3 & 0 & 0 & 1 & 0 \\ 0 & 1 & -3 & 2 & 0 & 0 & 1 \end{bmatrix}$$

$$\rightarrow \begin{bmatrix} 1 & 0 & 2 & 0 & 1 & 0 & 0 \\ 0 & 1 & -3 & 1 & -1 & 0 & 0 \\ 0 & 0 & 0 & 1 & -1 & 1 & 0 \\ 0 & 0 & 0 & 1 & 1 & 0 & 1 \end{bmatrix}$$

$$\rightarrow \begin{bmatrix} 1 & 0 & 2 & 0 & 1 & 0 & 0 \\ 0 & 1 & 3 & 0 & 0 & -1 & 0 \\ 0 & 0 & 0 & 1 & -1 & 1 & 0 \\ 0 & 0 & 0 & 0 & 2 & -1 & 1 \end{bmatrix}$$

$$\rightarrow \begin{bmatrix} 1 & 0 & 2 & 0 & 1 & 0 & 0 \\ 0 & 1 & -3 & 0 & 0 & -1 & 0 \\ 0 & 0 & 0 & 1 & -1 & 1 & 0 \\ 0 & 0 & 0 & 0 & 1 & -\frac{1}{2} & \frac{1}{2} \end{bmatrix}$$

$$\rightarrow \begin{bmatrix} 1 & 0 & 2 & 0 & 0 & \frac{1}{2} & -\frac{1}{2} \\ 0 & 1 & -3 & 0 & 0 & -1 & 0 \\ 0 & 0 & 0 & 1 & 0 & \frac{1}{2} & \frac{1}{2} \\ 0 & 0 & 0 & 0 & 1 & -\frac{1}{2} & \frac{1}{2} \end{bmatrix}$$

$$= C.$$

Next, we must determine the rank of A. Finding the row echelon form of A we get

$$A = \begin{bmatrix} 1 & 0 & -2 \\ 0 & -1 & -1 \\ 2 & 3 & -1 \end{bmatrix} \rightarrow \begin{bmatrix} 1 & 0 & -2 \\ 0 & -1 & -1 \\ 0 & 3 & 3 \end{bmatrix} \rightarrow \begin{bmatrix} 1 & 0 & -2 \\ 0 & 1 & 1 \\ 0 & 3 & 3 \end{bmatrix} \rightarrow \left[\begin{array}{ccc} 1 & 0 & -2 \\ 0 & 1 & 1 \\ \hdashline 0 & 0 & 0 \end{array}\right].$$

Thus $r = \text{rank}(A) = 2$. Hence we partition C as follows:

$$C = \left[\begin{array}{ccc:c:ccc} 1 & 0 & 2 & 0 & 0 & \frac{1}{2} & -\frac{1}{2} \\ 0 & 1 & -3 & 0 & 0 & -1 & 0 \\ \hdashline 0 & 0 & 0 & 1 & 0 & \frac{1}{2} & \frac{1}{2} \\ \hdashline 0 & 0 & 0 & 0 & 1 & -\frac{1}{2} & \frac{1}{2} \end{array}\right].$$

Since $C_4 = 0$, it follows that $C_6^T = [0, \frac{1}{2}, \frac{1}{2}]^T$ is a particular solution and $C_8^T = [1, -\frac{1}{2}, \frac{1}{2}]^T$ is a basis for **N(A)**. Thus $\{x^0, x^0 + x^1\}$ is a fundamental system of solutions where

$$x^0 = [0, \tfrac{1}{2}, \tfrac{1}{2}]^T,$$

$$x^1 = [1, -\tfrac{1}{2}, \tfrac{1}{2}]^T.$$

■

5.7. GENERALIZED INVERSES

The discussion of the linear algebraic system $Ax = b$ has thus far focused on the case where one or more solutions exist. In engineering applications, a problem that is at least as important corresponds to the case where no solution exists. This case typically arises when there are more constraints than degrees of freedom. In these instances we are interested in finding vectors which best approximate the solution in some sense. To concisely formulate this problem, it is useful to consider something called a generalized inverse of a matrix.

Definition 7.1 (Generalized Inverse). Let $A \in \mathbf{F}^{m \times n}$. Then $B \in \mathbf{F}^{n \times m}$ is a *generalized inverse* of $A \Leftrightarrow B$ satisfies at least property 1 or 2 where

1. $ABA = A$,
2. $BAB = B$,
3. $(AB)^H = AB$,
4. $(BA)^H = BA$.

■

Note that the four properties listed in Def. 7.1 are all satisfied by the matrix A^{-1} when it exists. Hence B is a generalization of the matrix inverse A^{-1}. If B satisfies all four properties then it is called a *Moore–Penrose inverse* or *pseudoinverse* of A and is denoted A^+. The utility of the Moore–Penrose inverse lies in the fact that it always exists and is unique. Moreover, it is the key to finding an approximate solution to $Ax = b$ when $b \notin \mathbf{R(A)}$. To see this, let $r(x)$ denote the *residual vector* of the system $Ax = b$. That is,

$$r(x) \triangleq Ax - b.$$

Then x is a solution of $Ax = b$ if and only if $r(x) = 0$. More generally, if a solution does not exist, we can attempt to find a vector x which minimizes $\|r(x)\|$. If the Euclidean norm is used (the second Holder norm), a minimizing x is referred to as a *least squares* solution of $Ax = b$. Typically, there may be many vectors x which minimize the norm of the residual vector. Of all the vectors which minimize $\|r(x)\|$, the vector for which $\|x\|$ is smallest is called the *minimum least squares* solution of $Ax = b$. The pseudoinverse of A is the key to obtaining the minimum least squares solution, as can be seen from the following result.

Proposition 7.1 (Moore–Penrose Inverse). Let $A \in \mathbf{F}^{\mathbf{m} \times \mathbf{n}}$, and let A^+ be the Moore–Penrose inverse of A. Then the minimum least squares solution of the linear algebraic system $Ax = b$ is

$$x^0 = A^+ b. \qquad \blacksquare$$

The reader can refer to Ben-Israel and Greville (1974) for a proof of Prop. 7.1. Note that if $b \in \mathbf{R(A)}$, then at least one solution of $Ax = b$ exists. Hence $\|r(x^0)\| = 0$ in this case. Of all the vectors x for which $r(x) = 0$, x^0 is the vector with minimum norm. In view of Exer. 6.1, it follows that A^+b is the unique solution of $Ax = b$ that is orthogonal to $\mathbf{N(A)}$. Thus Prop. 7.1 provides us with another way to compute the unique solution in $\mathbf{N(A)}^\perp$ when multiple solutions of $Ax = b$ exist. More generally, if $b \notin \mathbf{R(A)}$, in which case no solution exists, then A^+b is the smallest vector which minimizes the Euclidean norm of the residual $r(x)$. Refer to Figure 5.5 for a pictorial representation of the essential characteristics of the Moore–Penrose inverse.

In general, the Moore–Penrose inverse can be quite difficult to compute. However, there are two important special cases when simple explicit expressions can be obtained for A^+. The first case arises when the matrix A has full row rank.

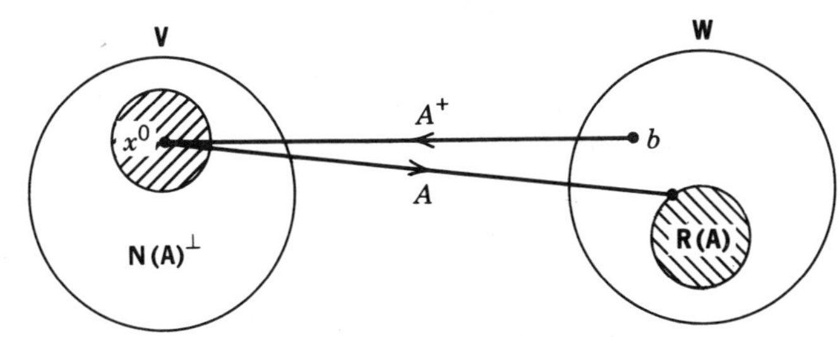

Figure 5.5. The minimum least squares solution of $Ax = b$.

Proposition 7.2 (Underdetermined System). Let $A \in \mathbf{F}^{m \times n}$, and suppose $\operatorname{rank}(A) = m$. Then the Moore–Penrose inverse of A is

$$A^+ = A^{\mathrm{H}}(AA^{\mathrm{H}})^{-1}.$$

Proof. If $\operatorname{rank}(A) = m$, then the square matrix $AA^{\mathrm{H}} \in \mathbf{F}^{m \times m}$ has m linearly independent rows and columns. Thus $(AA^{\mathrm{H}})^{-1}$ exists. By substituting the expression for A^+ into Def. 7.1, one can then verify that properties 1 through 4 are satisfied. ■

Note that if $\operatorname{rank}(A) = m$, then we must have $m \leqslant n$. If $m < n$, then there are fewer equations than unknown components of x. In this case we refer to $Ax = b$ as an *underdetermined system* of equations. Here at least one solution will exist for all vectors b. If $m < n$, then multiple solutions exist. In this case the vector $x^0 = A^+b$ is the unique solution in $\mathbf{N(A)}^\perp$.

Example 7.1 (Underdetermined System). Let $Ax = b$ be the following underdetermined system of two equations in three unknowns:

$$\begin{bmatrix} 1 & 2 & 1 \\ 2 & -1 & 0 \end{bmatrix} x = \begin{bmatrix} 0 \\ 3 \end{bmatrix}.$$

It is clear from inspection that $\operatorname{rank}(A) = 2$. Thus, applying Prop. 7.2 to compute the Moore–Penrose inverse of A, we have

$$AA^{\mathrm{H}} = \begin{bmatrix} 1 & 2 & 1 \\ 2 & -1 & 0 \end{bmatrix} \begin{bmatrix} 1 & 2 \\ 2 & -1 \\ 1 & 0 \end{bmatrix} = \begin{bmatrix} 6 & 0 \\ 0 & 5 \end{bmatrix}.$$

Since AA^{H} is diagonal, its inverse can be obtained by inspection. Thus the Moore–Penrose inverse of A is

$$\begin{aligned} A^+ &= A^{\mathrm{H}}(AA^{\mathrm{H}})^{-1} \\ &= \begin{bmatrix} 1 & 2 \\ 2 & -1 \\ 1 & 0 \end{bmatrix} \begin{bmatrix} \frac{1}{6} & 0 \\ 0 & \frac{1}{5} \end{bmatrix} = \begin{bmatrix} \frac{1}{6} & \frac{2}{5} \\ \frac{1}{3} & -\frac{1}{5} \\ \frac{1}{6} & 0 \end{bmatrix}. \end{aligned}$$

Finally, the minimum least squares solution of $Ax = b$ is

$$\begin{aligned} x^0 &= A^+b \\ &= \begin{bmatrix} \frac{1}{6} & \frac{2}{5} \\ \frac{1}{3} & -\frac{1}{5} \\ \frac{1}{6} & 0 \end{bmatrix} \begin{bmatrix} 0 \\ 3 \end{bmatrix} = \begin{bmatrix} \frac{6}{5} \\ -\frac{3}{5} \\ 0 \end{bmatrix}. \end{aligned}$$

Note that the residual vector in this case is

$$r(x^0) = Ax^0 - b$$

$$= \begin{bmatrix} 1 & 2 & 1 \\ 2 & -1 & 0 \end{bmatrix} \begin{bmatrix} \frac{6}{5} \\ -\frac{3}{5} \\ 0 \end{bmatrix} - \begin{bmatrix} 0 \\ 3 \end{bmatrix} = \begin{bmatrix} 0 \\ 0 \end{bmatrix}.$$

Thus $x^0 = [\frac{6}{5}, -\frac{3}{5}, 0]^T$ is a solution. Moreover, among all solutions it is the solution for which $\|x^0\| = 1.34$ is a minimum. ■

A simple explicit expression can also be obtained for the Moore–Penrose inverse when A has full column rank.

Proposition 7.3 (Overdetermined System). Let $A \in \mathbf{F}^{m \times n}$, and suppose $\operatorname{rank}(A) = n$. Then the Moore–Penrose inverse of A is

$$A^+ = (A^H A)^{-1} A^H.$$

Proof. If $\operatorname{rank}(A) = n$, then the square matrix $A^H A \in \mathbf{F}^{n \times n}$ has n linearly independent rows and columns. Thus $(A^H A)^{-1}$ exists. By substituting the expression for A^+ into Def. 7.1, one can then verify that properties 1 through 4 are satisfied. ■

Note that if $\operatorname{rank}(A) = n$, then we must have $m \geqslant n$. If $m > n$, then there are more equations than unknown components of x. In this case we refer to $Ax = b$ as an *overdetermined system* of equations. Here at most one solution will exist for each vector b. If $b \notin \mathbf{R}(\mathbf{A})$, then no solutions exist. In this case the vector $x = A^+ b$ minimizes $\|r(x)\|$, and among all vectors with this property it is the one with the smallest norm.

Example 7.2 (Overdetermined System). Let $Ax = b$ be the following overdetermined sytem of three equations in two unknowns:

$$\begin{bmatrix} 1 & 2 \\ 2 & -1 \\ 1 & 0 \end{bmatrix} x = \begin{bmatrix} 3 \\ 0 \\ 2 \end{bmatrix}.$$

It is clear from inspection that $\operatorname{rank}(A) = 2$. Thus, applying Prop. 7.3 to compute the Moore–Penrose inverse of A, we have

$$A^H A = \begin{bmatrix} 1 & 2 & 1 \\ 2 & -1 & 0 \end{bmatrix} \begin{bmatrix} 1 & 2 \\ 2 & -1 \\ 1 & 0 \end{bmatrix} = \begin{bmatrix} 6 & 0 \\ 0 & 5 \end{bmatrix}.$$

Since $A^{H}A$ is diagonal, its inverse can be obtained by inspection. Thus the Moore–Penrose inverse of A is

$$A^{+} = (A^{H}A)^{-1}A^{H},$$
$$= \begin{bmatrix} \frac{1}{6} & 0 \\ 0 & \frac{1}{5} \end{bmatrix} \begin{bmatrix} 1 & 2 & 1 \\ 2 & -1 & 0 \end{bmatrix} = \begin{bmatrix} \frac{1}{6} & \frac{1}{3} & \frac{1}{6} \\ \frac{2}{5} & -\frac{1}{5} & 0 \end{bmatrix}.$$

Finally, the minimum least squares solution of $Ax = b$ is

$$x^{0} = A^{+}b$$
$$= \begin{bmatrix} \frac{1}{6} & \frac{1}{3} & \frac{1}{6} \\ \frac{2}{5} & -\frac{1}{5} & 0 \end{bmatrix} \begin{bmatrix} 3 \\ 0 \\ 2 \end{bmatrix} = \begin{bmatrix} \frac{5}{6} \\ \frac{6}{5} \end{bmatrix}.$$

Note that the residual vector in this case is

$$r(x^{0}) = Ax^{0} - b$$
$$= \begin{bmatrix} 1 & 2 \\ 2 & -1 \\ 1 & 0 \end{bmatrix} \begin{bmatrix} \frac{5}{6} \\ \frac{6}{5} \end{bmatrix} - \begin{bmatrix} 3 \\ 0 \\ 2 \end{bmatrix} = \begin{bmatrix} \frac{7}{30} \\ \frac{7}{15} \\ -\frac{7}{6} \end{bmatrix}.$$

Thus there is no solution. However, x^{0} minimizes the norm of the residual vector with $\|r(x^{0})\| = 1.278$. Moreover, among all vectors for which $\|r(x)\| = 1.278$, the vector $x^{0} = [\frac{5}{6}, \frac{6}{5}]^{T}$ is the one whose norm is the smallest. ■

PROBLEMS

1. Let **V** and **W** be the spaces spanned by $\mathbf{B} = \{b^{1}, b^{2}, \ldots, b^{n}\}$ and $\mathbf{C} = \{c^{1}, c^{2}, \ldots, c^{m}\}$, respectively, where $b^{k}(t) = \cos(kt)$ and $c^{k}(t) = \sin(kt)$ for $0 \leqslant t \leqslant 2\pi$. Consider the following transformation on **V**:

$$(Ax)(t) = \frac{d^{3}}{dt^{3}}x(t) - \frac{d}{dt}x(t), \qquad 0 \leqslant t \leqslant 2\pi.$$

a. Show that $A: \mathbf{V} \to \mathbf{W}$ and that A is linear.

b. Show that **B** and **C** are bases for **V** and **W**, respectively.

c. Find the matrix representation of A in the coordinate system induced by $\{\mathbf{B}, \mathbf{C}\}$.

2. Consider the following matrix:

$$A = \begin{bmatrix} 1 & 2 & 0 & -1 \\ 3 & 0 & 1 & 2 \\ 2 & 0 & 1 & -1 \end{bmatrix}.$$

a. Let $\mathbf{B} = \{b^1, b^2, b^3\}$ be the following set of vectors in $\mathbf{R}^3$. Show that $\mathbf{B}$ is a basis for $\mathbf{R}^3$,

$$b^1 = [1, 0, 0]^T,$$
$$b^2 = [1, 1, 0]^T,$$
$$b^3 = [1, 1, 1]^T.$$

b. Let $\mathbf{C} = \{c^1, c^2, c^3, c^4\}$ be the following set of vectors in $\mathbf{R}^4$. Show that $\mathbf{C}$ is a basis for $\mathbf{R}^4$,

$$c^1 = [1, 0, 0, 0]^T,$$
$$c^2 = [0, 2, 0, 0]^T,$$
$$c^3 = [0, 0, 3, 0]^T,$$
$$c^4 = [0, 0, 0, 4]^T.$$

c. Find the representation of A in the coordinate system induced by $\{\mathbf{B}, \mathbf{C}\}$.

3. Find the row echelon form of the following 5×4 matrix:

$$A = \begin{bmatrix} 1 & 2 & -1 & 4 \\ 0 & 1 & 0 & 1 \\ 2 & 1 & 0 & 3 \\ -1 & 0 & -1 & 0 \\ 0 & 0 & 2 & -1 \end{bmatrix}.$$

4. Consider the following 3×4 matrix:

$$A = \begin{bmatrix} 1 & 2 & 0 & -2 \\ -1 & -1 & 2 & 1 \\ 1 & 3 & 2 & -3 \end{bmatrix}.$$

a. Find a basis for $\mathbf{N(A)}$.

b. What is nullity(A)?

c. Find a basis for $\mathbf{R(A)}$.

d. What is rank(A)?

5. Consider the following matrix:

$$A = \begin{bmatrix} 1 & 0 & 2 & -1 \\ 0 & 1 & 1 & 0 \\ -2 & 1 & -3 & 2 \end{bmatrix}.$$

a. Find a basis for **N(A)**.
b. What is nullity(A)?
c. What is rank(A)?
d. Let $x = [2, 0, -1, 1]^T$. Is $x \in \mathbf{N(A)}$?

6. Consider the following matrix:

$$A = \begin{bmatrix} 1 & 0 & 3 \\ 1 & 1 & 5 \\ 0 & 2 & 4 \\ 1 & -1 & 1 \end{bmatrix}.$$

a. Find a basis for **R(A)**.
b. What is rank(A)?
c. What is nullity(A)?
d. Let $y = [3, 0, 0, 0]^T$. Is $y \in \mathbf{R(A)}$?

7. Let **W** be the space spanned by the vectors $\{x^1, x^2, x^3\}$ where

$$x^1 = [1, 0, 2]^T,$$
$$x^2 = [1, -1, 0]^T,$$
$$x^3 = [3, -2, 2]^T.$$

a. Find a basis for **W**.
b. What is dim(**W**)?
c. Find an orthogonal basis for **W**.
d. Find the orthogonal projection of $x = [2, -1, 0]^T$ onto **W**.
e. Find the orthogonal projection of x onto $\mathbf{W}^\perp$.

8. Let $A \in \mathbf{F}^{m \times k}$ and $B \in \mathbf{F}^{k \times n}$. Show that

$$\text{rank}(AB) \leqslant \min\{\text{rank}(A), \text{rank}(B)\}.$$

9. Find A^{-1} where A is the following 4×4 matrix:

$$A = \begin{bmatrix} 1 & 0 & 2 & 1 \\ 0 & 1 & 1 & -3 \\ -2 & 1 & 0 & 3 \\ 1 & -1 & 2 & 0 \end{bmatrix}.$$

10. Find a fundamental system of solutions to the following linear algebraic system:

$$\begin{bmatrix} 1 & 2 & 0 \\ 0 & 1 & 3 \\ 2 & 0 & -1 \end{bmatrix} \begin{bmatrix} x_1 \\ x_2 \\ x_3 \end{bmatrix} = \begin{bmatrix} 0 \\ -1 \\ 4 \end{bmatrix}.$$

11. Consider the following linear algebraic system. Find conditions on b for which this system has at least one solution:

$$\begin{bmatrix} 1 & 3 & 0 \\ -1 & 2 & 1 \\ 3 & -1 & 2 \end{bmatrix} \begin{bmatrix} x_1 \\ x_2 \\ x_3 \end{bmatrix} = \begin{bmatrix} b_1 \\ b_2 \\ b_3 \end{bmatrix}.$$

12. Let $A \in \mathbf{F}^{m \times n}$. Show that $\text{rank}(A^T) = \text{rank}(A)$.

13. Consider the following 3×2 matrix A:

$$A = \begin{bmatrix} 3 & 1 \\ 1 & -2 \\ 2 & 0 \end{bmatrix}.$$

a. Find A^+.
b. Find the minimum least squares solution of $Ax = b$ if $b = [1, 0, -1]^T$.

14. Consider the following 2×3 matrix A:

$$A = \begin{bmatrix} 1 & -1 & 2 \\ 2 & 0 & 3 \end{bmatrix}.$$

a. Find A^+.
b. Find the minimum least squares solution of $Ax = b$ if $b = [1, -1]^T$.

REFERENCES

Bellman, R. E., *Introduction to Matrix Algebra*, McGraw-Hill, New York, 1960.

Ben-Israel, A. and T. N. E. Greville, *Generalized Inverses, Theory and Applications*, Wiley, New York, 1974.

Cullen, C. G., *Matrices and Linear Transformations*, Addison-Wesley, New York, 1966.

Noble, B. and J. W. Deniel, *Applied Linear Algebra*, McGraw-Hill, New York, 1977.

6 LINEAR OPERATORS

Recall that a linear transformation is function mapping a vector space **V** into a vector space **W**. That is, a linear transformation is a vector in the space **L(V, W)**. In the last chapter we examined the class of linear transformations which arises when both **V** and **W** are finite-dimensional vector spaces. Another important special case occurs when the vector space **W** is identical to the vector space **V**. We refer to the space **L(V, V)** as the space of linear operators on **V**.

6.1. LINEAR OPERATORS

Because a linear operator is an important special case of a linear transformation, all of the results developed thus far concerning linear transformations, in general, apply to linear operators in particular. However, there are certain additional properties that are peculiar to linear operators themselves (Bachman and Narici, 1966; Luenberger, 1969). It is with these properties that we will be concerned.

Definition 1.1 (Operator). Let $A : \mathbf{V} \to \mathbf{W}$ where **V** and **W** are vector spaces. Then A is an *operator* on $\mathbf{V} \Leftrightarrow \mathbf{W} = \mathbf{V}$. ■

Thus operators map vector spaces back into themselves. An operator is finite-dimensional if and only if the space on which it is defined is finite-dimensional. Refer to Figure 6.1 for a pictorial description of an operator A on a vector space **V**.

As with transformations, in general, operators may be one-to-one, onto, invertible, or linear. The following are some important examples of operators.

Example 1.1 (Square Matrix). Let $\mathbf{V} = \mathbf{F}^n$, and let $A \in \mathbf{F}^{n \times n}$. Then A is a finite-dimensional linear operator on **V**. In this case, $\dim(A) = n^2$. As we shall see in the next chapter, the square matrix is the most imprtant example of a finite-dimensional linear operator. ■

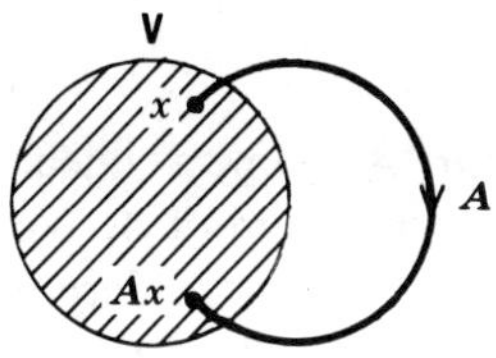

Figure 6.1. A linear operator A on a vector space **V**.

Example 1.2 (Convolution). Let $\mathbf{V} = \mathbf{L}_\infty(\mathbf{R})$, and consider the following integral transformation on **V**. For each $x \in \mathbf{V}$

$$(Ax)(t) = \int_R h(t-\tau)x(\tau)\, d\tau, \qquad t \in \mathbf{R}.$$

We refer to the function h as the *kernel.* Suppose the kernel is absolutely integrable. That is, $h \in \mathbf{L}_1(\mathbf{R})$. Then

$$\begin{aligned}
\|Ax\|_\infty &= \sup_{t\in\mathbf{R}} \left\{ \left| \int_{\mathbf{R}} h(t-\tau)x(\tau)\, d\tau \right| \right\} \\
&\leqslant \sup_{t\in\mathbf{R}} \left\{ \int_{\mathbf{R}} |h(t-\tau)x(\tau)|\, d\tau \right\} \\
&= \sup_{t\in\mathbf{R}} \left\{ \int_{\mathbf{R}} |h(t-\tau)|\,|x(\tau)|\, d\tau \right\} \\
&\leqslant \sup_{t\in\mathbf{R}} \left\{ \int_{\mathbf{R}} |h(t-\tau)|\, d\tau \right\} \|x\|_\infty \\
&= \sup_{t\in\mathbf{R}} \left\{ \int_{\mathbf{R}} |h(\tau)|\, d\tau \right\} \|x\|_\infty \\
&= \sup_{t\in\mathbf{R}} \{\|h\|_1\} \|x\|_\infty \\
&= \|h\|_1 \|x\|_\infty.
\end{aligned}$$

Hence $h \in \mathbf{L}_1(\mathbf{R}) \Rightarrow Ax \in \mathbf{V}$. Thus A is a linear operator on **V**. We call A the *convolution* operator. It is an important example of an infinite-dimensional linear operator. ∎

Although Def. 1.1 makes no mention of linearity, our principal concern will be with linear operators. Thus we will focus on the properties of vectors in the space **L(V, V)**. Exam. 1.1 and Exam. 1.2 are illustrations of two linear operators that appear repeatedly in engineering applications.

6.2. THE SPECTRUM

One of the features that distinguish a linear operator A from a linear transformation is a subset of the complex plane called the spectrum of A. We define the spectrum of A and its complement, the resolvent set, by considering properties of the inverse of $B(\lambda) = \lambda I - A$ as λ varies over the complex plane.

Definition 2.1 (Resolvent Set and Spectrum). Let $A \in \mathbf{L}(\mathbf{V}, \mathbf{V})$ and let $\lambda \in C$. Then λ belongs to the *resolvent set* of $A \Leftrightarrow$

1. $\lambda I - A$ is one-to-one,
2. $\mathbf{R}(\lambda I - A)$ is dense in $\mathbf{V}$,
3. $(\lambda I - A)^{-1}$ is bounded on $\mathbf{R}(\lambda I - A)$.

Otherwise λ belongs to the *spectrum* of A. We denote the resolvent set by $\rho(A)$ and the spectrum by $\sigma(A)$. ■

Thus each linear operator A induces a decomposition of the complex plane into the resolvent set $\rho(A)$ and the spectrum $\sigma(A)$. The spectrum of a linear operator can be further decomposed into three subsets as follows.

Definition 2.2 (Spectra). Let $A \in \mathbf{L}(\mathbf{V}, \mathbf{V})$ and $\lambda \in \sigma(A)$. Then the following three possibilities arise:

1. If $\lambda I - A$ is not one-to-one, then λ belongs to $\sigma_p(A)$, the *point spectrum* of A.
2. If $\lambda I - A$ is one-to-one, but $\mathbf{R}(\lambda I - A)$ is not dense in $\mathbf{V}$, then λ belongs to $\sigma_r(A)$, the *residual spectrum* of A.
3. If $\lambda I - A$ is one-to-one and $\mathbf{R}(\lambda I - A)$ is dense in $\mathbf{V}$, but $(\lambda I - A)^{-1}$ is not bounded on $\mathbf{R}(\lambda I - A)$, then λ belongs to $\sigma_c(A)$, the *continuous spectrum* of A. ■

Thus the spectrum of A is the disjoint union of the point, residual and continuous spectra of A. The spectra and the resolvent set of a linear operator are summarized in a flowchart in Figure 6.2.

Elements of the point spectrum of A are called *eigenvalues* of A. For a finite-dimensional linear operator A, the set of eigenvalues is the spectrum of A, as can be seen from the following exercise.

Exercise 2.1 (Eigenvalues). Let $A \in \mathbf{L}(\mathbf{V}, \mathbf{V})$. Show that if $\mathbf{V}$ is finite-dimensional, then $\sigma(A) = \sigma_p(A)$. ■

We illustrate the notion of the spectrum of a linear operator with several examples. The first example is a linear operator with a nonempty continuous

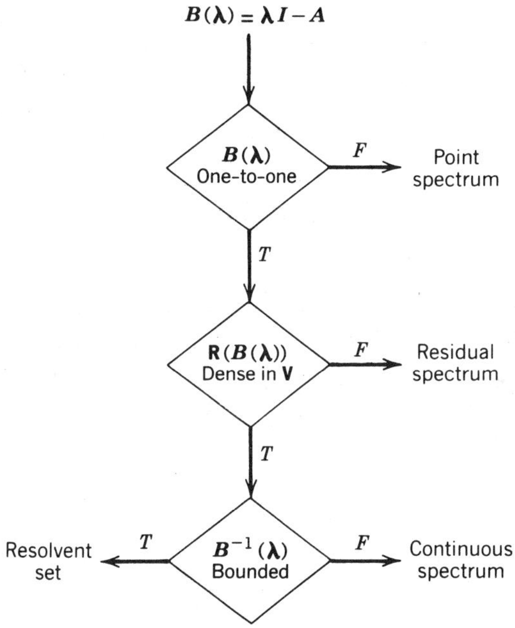

Figure 6.2. The spectrum and resolvent set of a linear operator A.

spectrum. The second is a linear operator with a nonempty residual spectrum. In view of Exer. 2.1, both operators are infinite-dimensional. Finally, a third example is included which illustrates the spectrum of a finite-dimensional linear operator.

Example 2.1 [Continuous Spectrum (Bachman and Narici, 1966)]. Let $\mathbf{V}$ be the Hilbert space $\mathbf{L}_2(\mathbf{R})$, and let $\{x^k\}$ be a complete orthonormal set in $\mathbf{V}$. Consider the following linear operator on $\mathbf{V}$. For each $x \in \mathbf{V}$

$$Ax = \sum_{k=1}^{\infty} \frac{a_k}{k} x^k.$$

Here $a_k = \langle x, x^k \rangle$ for $k \geqslant 1$. Thus $Ax^k = x^k/k$. The operator $\boldsymbol{\lambda I} - A$ can be expressed

$$(\boldsymbol{\lambda I} - A)x = \sum_{k=1}^{\infty} \left(\lambda - \frac{1}{k}\right) a_k x^k$$

$$\triangleq \sum_{k=1}^{\infty} (\lambda - \lambda_k) a_k x^k.$$

Thus for each $k \geqslant 1$ we have $x^k \in \mathbf{N}(\boldsymbol{\lambda_k I} - A)$, where $\lambda_k = 1/k$. It then follows from Def. 2.2 that $\lambda_k \in \sigma_p(A)$ for $k \geqslant 1$. Next, we will show that $0 \in \sigma_c(A)$. Suppose

$-Ax = 0$. Then

$$\begin{aligned} 0 &= \|-Ax\|^2 \\ &= \left\| -\sum_{k=1}^{\infty} \frac{a_k}{k} x^k \right\|^2 \\ &= \sum_{k=1}^{\infty} \frac{|a_k|^2}{k^2}. \end{aligned}$$

Hence $a_k = 0$ for $k \geqslant 1$. Thus $x = 0$ which implies that $-A$ is one-to-one. It follows that $(-A)^{-1}$ exists on $\mathbf{R}(-A)$. Next, we show that $\mathbf{R}(-A)$ is dense in $\mathbf{V}$. Note that $(-A)(-kx^k) = x^k$ for $k \geqslant 1$. Hence $\{x^k\} \in \mathbf{R}(-A)$. But $\{x^k\}$ is a complete orthonormal set for $\mathbf{V}$. Hence if $\mathbf{U}$ is the subspace of $\mathbf{V}$ spanned by $\{x^k\}$, then the closure of $\mathbf{U}$ is $\mathbf{V}$. Thus $\mathbf{R}(-A)$ is dense in $\mathbf{V}$. It remains to show that $(-A)^{-1}$ is unbounded. Now,

$$\begin{aligned} \|-Ax^k\| &= \|-\lambda_k x^k\| \\ &= |\lambda_k| \\ &= \frac{1}{k}. \end{aligned}$$

If $(-A)^{-1}$ is bounded, then from Prop. 4.5.2 there exists an $\alpha > 0$ such that for each $x \in \mathbf{V}$

$$\|-Ax\| \geqslant \alpha \|x\|.$$

Thus $(-A)^{-1}$ is unbounded. It then follows from Def. 2.2 that $0 \in \sigma_c(A)$. Note that 0 is a limit point of the set $\sigma_p(A)$. To complete the picture for this operator, it can be shown (Bachman and Narici, 1966) that the residual spectrum of A is empty. Thus to summarize we have

$$\begin{aligned} \sigma_p(A) &= \{1, \tfrac{1}{2}, \tfrac{1}{3}, \tfrac{1}{4}, \ldots\} \\ \sigma_r(A) &= \varnothing, \\ \sigma_c(A) &= \{0\}. \end{aligned}$$

■

Example 2.2 [Residual Spectrum (Bachman and Narici, 1966)]. Let $\mathbf{V} = \mathbf{L}_2(\mathbf{R})$, and let $\{x^k\}$ be a complete orthonormal set in $\mathbf{V}$. Consider the following *destruction* operator on $\mathbf{V}$. For each $x \in \mathbf{V}$,

$$Ax = \sum_{k=1}^{\infty} \frac{a_k}{k} x^{k+1}.$$

Here $a_k \triangleq \langle x, x^k \rangle$ for $k \geqslant 1$. Thus $Ax^k = x^{k+1}/k$. Suppose $Ax = 0$. Then

$$\begin{aligned} 0 &= \|Ax\|^2 \\ &= \left\| \sum_{k=1}^{\infty} \frac{a_k}{k} x^{k+1} \right\|^2 \\ &= \sum_{k=1}^{\infty} \frac{|a_k|^2}{k^2} \end{aligned}$$

Thus $a_k = 0$ for $k \geqslant 1$. It follows that $x = 0$. Consequently, A is one-to-one and A^{-1} exists on $\mathbf{R}(A)$. From the definition of A it is clear that $x^1 \notin \mathbf{R}(A)$. Since x^1 is orthogonal to $\{x^2, x^3, \ldots\}$, the closure of the subspace $\mathbf{R}(A)$ is not equal to $\mathbf{V}$. Thus $\mathbf{R}(A)$ is not dense in $\mathbf{V}$. Therefore from Def. 2.2 we have $0 \in \sigma_r(A)$. It can be shown that if $\lambda \neq 0$, then λ belongs to the resolvent set $\rho(A)$. Thus the spectrum of the destruction operator A can be summarized as follows:

$$\begin{aligned} \sigma_p(A) &= \varnothing, \\ \sigma_r(A) &= \{0\}, \\ \sigma_c(A) &= \varnothing. \end{aligned}$$

■

Example 2.3 (Point Spectrum). Let $\mathbf{V} = \mathbf{P}^n(\mathbf{R})$, and let A denote the *derivative* operator on $\mathbf{V}$. That is, for each $x \in \mathbf{V}$

$$(Ax)(t) = \frac{d}{dt} x(t), \qquad t \in \mathbf{R}.$$

Then A is linear and for each polynomial $x \in \mathbf{V}$ of degree n, Ax will be a polynomial of degree $n - 1$. Hence A can be regarded as an operator on $\mathbf{V}$. Because $\mathbf{V}$ is finite-dimensional, it follows from Exer. 2.1 that the spectrum of A is the point spectrum, and

$$\lambda \in \sigma_p(A) \Leftrightarrow \mathbf{N}(\lambda I - A) \neq \{\mathbf{0}\}.$$

Thus $\lambda \in \sigma_p(A) \Leftrightarrow Ax = \lambda x$ for some $x \neq 0$. The only way this equation can be satisfied is if $x(t)$ is a constant in which case $\lambda = 0$. Thus for the derivative operator defined on the space of nth degree polynomials,

$$\begin{aligned} \sigma_p(A) &= \{0\}, \\ \sigma_r(A) &= \varnothing, \\ \sigma_c(A) &= \varnothing. \end{aligned}$$

■

6.3. BOUNDED OPERATORS

To further characterize the spectrum of a linear operator, we introduce some topological structure to the space over which the operator is defined. The spectrum, in general, can be an unbounded subset of the complex plane **C**. However, when A is a bounded linear operator on **V** and **V** is a Banach space, then $\sigma(A)$ is also bounded. To confirm this, we begin with the following useful result.

Proposition 3.1 (Inverse Expansion). Let $A \in \mathbf{B}(\mathbf{V}, \mathbf{V})$ where **V** is a Banach space, and let $\lambda \in \mathbf{C}$. If $|\lambda| > \|A\|$, then $(\lambda I - A)^{-1} \in \mathbf{B}(\mathbf{V}, \mathbf{V})$. Moreover,

$$(\lambda I - A)^{-1} = \sum_{k=1}^{\infty} \lambda^{-k} A^{k-1}.$$

Proof. Let $\lambda \in \mathbf{C}$ with $|\lambda| > \|A\|$. Then, using the convergence property of a geometric series of scalars, we have

$$\begin{aligned}
\left\| \sum_{k=1}^{\infty} \lambda^{-k} A^{k-1} \right\| &\leqslant \sum_{k=1}^{\infty} \|\lambda^{-k} A^{k-1}\| \\
&= |\lambda^{-1}| \sum_{k=1}^{\infty} \|\lambda^{1-k} A^{k-1}\| \\
&= |\lambda^{-1}| \sum_{k=0}^{\infty} \|\lambda^{-k} A^{k}\| \\
&= |\lambda^{-1}| \sum_{k=0}^{\infty} \|A^{k}\|/|\lambda^{k}| \\
&\leqslant |\lambda^{-1}| \sum_{k=0}^{\infty} (\|A\|/|\lambda|)^{k} \\
&\leqslant |\lambda^{-1}|/(1 - \|A\|/|\lambda|) \\
&\leqslant 1/(|\lambda| - \|A\|).
\end{aligned}$$

Thus the series converges absolutely. Next, using the absolute convergence we have

$$\begin{aligned}
(\lambda I - A) \sum_{k=1}^{\infty} \lambda^{-k} A^{k-1} &= (\lambda I - A)\lambda^{-1} \sum_{k=1}^{\infty} \lambda^{1-k} A^{k-1} \\
&= (I - \lambda^{-1} A) \sum_{k=0}^{\infty} \lambda^{-k} A^{k} \\
&= (I - \lambda^{-1} A)\left(I + \sum_{k=1}^{\infty} \lambda^{-k} A^{k} \right)
\end{aligned}$$

$$= I - \lambda^{-1}A + \sum_{k=1}^{\infty} \lambda^{-k}A^{k} - \lambda^{-1}A\sum_{k=1}^{\infty} \lambda^{-k}A^{k}$$

$$= I - \left(\lambda^{-1}A + \sum_{k=2}^{\infty} \lambda^{-k}A^{k}\right) + \sum_{k=1}^{\infty} \lambda^{-k}A^{k}$$

$$= I.$$

Hence the series in Prop. 3.1 is a right inverse of $\lambda I - A$. Using a similar analysis we can show that it is also a left inverse of $\lambda I - A$. Thus

$$(\lambda I - A)^{-1} = \sum_{k=1}^{\infty} \lambda^{-k}A^{k-1}. \qquad \blacksquare$$

Exercise 3.1 (Inverse). Let $A \in \mathbf{B}(\mathbf{V}, \mathbf{V})$ where $\mathbf{V}$ is a Banach space. Show that if $\|A\| < 1$, then $A^{-1} \in \mathbf{B}(\mathbf{V}, \mathbf{V})$, and

$$A^{-1} = \sum_{k=0}^{\infty} (I - A)^{k}. \qquad \blacksquare$$

As a direct consequence of Prop. 3.1, we can make the following statement regarding the spectrum of a bounded linear operator on a Banach space.

Proposition 3.2 (Spectral Bound). Let $A \in \mathbf{B}(\mathbf{V}, \mathbf{V})$ where $\mathbf{V}$ is a Banach space. Then

$$\lambda \in \sigma(A) \Rightarrow |\lambda| \leqslant \|A\|.$$

Proof. Suppose $\lambda \in \sigma(A)$ and $|\lambda| > \|A\|$. Then from Prop. 3.1, $(\lambda I - A)^{-1} \in \mathbf{B}(\mathbf{V}, \mathbf{V})$. It then follows from Def. 2.2 that $\lambda \in \rho(A)$, which is a contradiction. $\blacksquare$

Thus if $\mathbf{V}$ is a Banach space and $A \in \mathbf{B}(\mathbf{V}, \mathbf{V})$, then $\sigma(A)$ is contained in the closed ball of radius $\|A\|$ centered at the origin. In view of Prop. 3.2, we are now in position to define something called the spectral radius.

Definition 3.1 (Spectral Radius). Let $A \in \mathbf{B}(\mathbf{V}, \mathbf{V})$ where $\mathbf{V}$ is a Banach space. Then the *spectral radius* of A is denoted $r(A)$ and is defined

$$r(A) \triangleq \sup_{\lambda \in \sigma(A)}\{|\lambda|\}. \qquad \blacksquare$$

Another way of stating Prop. 3.2 is to say that for bounded linear operators on Banach spaces, the spectral radius always exists and is bounded from above by the operator norm. Refer to Figure 6.3 for a pictorial description of the spectrum of a bounded linear operator on a Banach space.

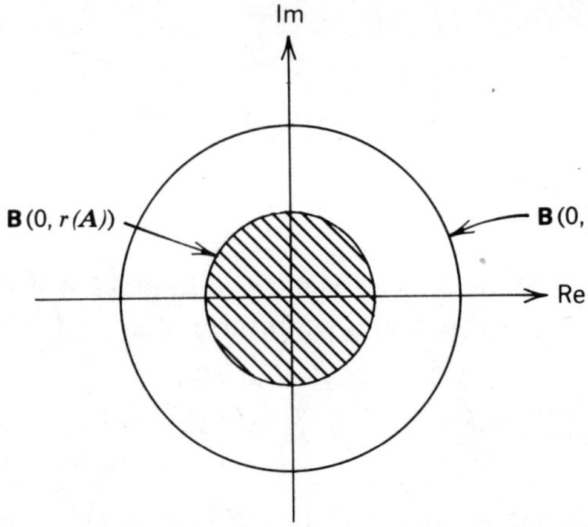

Figure 6.3. The spectral radius of a bounded linear operator A.

Depending on the operator, the exact spectrum of A can sometimes be difficult to determine. However, we can easily approximate $\sigma(A)$ by identifying a subset of it called the approximate spectrum.

Definition 3.2 (Approximate Spectrum). Let $A \in \mathbf{L}(\mathbf{V}, \mathbf{V})$ where $\mathbf{V}$ is a normed vector space, and let $\lambda \in \mathbf{C}$. Then λ belongs to the *approximate spectrum* of A, denoted $\pi(A) \Leftrightarrow$ for each $\alpha > 0$ there exists an $x \in \mathbf{V}$ such that

$$\|(\lambda I - A)x\| < \alpha \|x\| \qquad \blacksquare$$

The approximate spectrum is an approximation to the exact spectrum in the sense that every point in $\pi(A)$ is also a point in $\sigma(A)$, as can be seen from the following exercise.

Exercise 3.2 (Approximate Spectrum). Let $A \in \mathbf{L}(\mathbf{V}, \mathbf{V})$ where $\mathbf{V}$ is a normed vector space. Show that

$$\pi(A) \subseteq \sigma(A). \qquad \blacksquare$$

Thus the approximate spectrum of A is a subset of the actual spectrum $\sigma(A)$. As we shall see, for an important class of linear operators, the approximate spectrum is in fact equal to the spectrum. As an example of an infinite-dimensional bounded linear operator, consider the following.

Example 3.1 (Time-Varying Convolution). Let $\mathbf{V} = \mathbf{L}_2(\mathbf{T})$ where $T = [\alpha, \beta]$, and let $h: T \to \mathbf{C}$ be bounded in the following sense:

$$\int_T \int_T |h(t, \tau)|^2 \, d\tau \, dt < \infty.$$

Consider the following linear operator on **V**. For each $x \in \mathbf{V}$

$$(Ax)(t) \triangleq \int_\alpha^t h(t, \tau)x(\tau)\, d\tau, \qquad t \in T.$$

Using the Cauchy–Schwarz inequality, we have

$$\begin{aligned}
\|Ax\|^2 &= \int_T |(Ax)(t)|^2\, dt \\
&= \int_T \left| \int_\alpha^t h(t, \tau)x(\tau)\, d\tau \right|^2 dt \\
&\leqslant \int_T \int_\alpha^t |h(t, \tau)x(\tau)|^2\, d\tau\, dt \\
&= \int_T \int_\alpha^t |h(t, \tau)|^2 |x(\tau)|^2\, d\tau\, dt \\
&\leqslant \int_T \left\{ \int_\alpha^t |h(t, \tau)|^2\, d\tau \cdot \int_\alpha^t |x(\tau)|^2\, d\tau \right\} dt \\
&\leqslant \int_T \left\{ \int_\alpha^\beta |h(t, \tau)|^2\, d\tau \int_\alpha^\beta |x(\tau)|^2\, d\tau \right\} dt \\
&= \int_T \int_\alpha^\beta |h(t, \tau)|^2\, d\tau\, dt \cdot \int_\alpha^\beta |x(\tau)|^2\, d\tau \\
&= \int_T \int_\alpha^\beta |h(t, \tau)|^2\, d\tau\, dt \|x\|^2.
\end{aligned}$$

Hence A is bounded and

$$\|A\|^2 \leqslant \int_T \int_T |h(t, \tau)|^2\, d\tau\, dt \qquad \blacksquare$$

The bounded linear operator depicted in Exam. 3.1 is a generalization of the convolution operator discussed in Exam. 1.2. It appears frequently in engineering applications as it can be used to model the input–output behavior of linear systems whose parameters vary with time. We refer to the bounded linear operator in Exam. 3.1 as the *time-varying convolution* operator.

6.4. THE ADJOINT OPERATOR

For the remainder of this chapter we restrict our attention to bounded linear operators defined on Hilbert spaces. Each such operator $A \in \mathbf{B(V, V)}$ induces a dual operator $A^* \in \mathbf{B(V, V)}$ called its adjoint. We define the adjoint in a somewhat more general context as follows.

Definition 4.1 (Adjoint). Let $A \in \mathbf{L}(\mathbf{V}, \mathbf{V})$ where $\mathbf{V}$ is an inner product space. If there exists a unique $A^* \in \mathbf{L}(\mathbf{V}, \mathbf{V})$ such that for each $(x, y) \in \mathbf{V}^2$

$$\langle Ax, y \rangle = \langle x, A^*y \rangle,$$

then A^* is the *adjoint* of A. ∎

Thus when it exists, the adjoint operator is defined *implicitly* by means of the inner product. If the linear operator A is bounded and the inner product space $\mathbf{V}$ is a Hilbert space, then the existence of the adjoint operator A^* is assured, and we have the following result.

Proposition 4.1 (Adjoints). Let $A \in \mathbf{B}(\mathbf{V}, \mathbf{V})$ where $\mathbf{V}$ is a Hilbert space. Then the adjoint A^* exists. Moreover, $A^* \in \mathbf{B}(\mathbf{V}, \mathbf{V})$ and

$$\|A^*\| = \|A\|.$$

Proof. Let $y \in \mathbf{V}$, and consider the following functional on $\mathbf{V}$. For each $x \in \mathbf{V}$

$$f(x) \triangleq \langle Ax, y \rangle.$$

Clearly, f is linear. Moreover, using the Cauchy–Schwartz inequality, we have

$$\begin{aligned} |f(x)| &= |\langle Ax, y \rangle| \\ &\leqslant \|Ax\| \, \|y\| \\ &\leqslant \|A\| \, \|x\| \, \|y\| \\ &= (\|A\| \, \|y\|)\|x\|. \end{aligned}$$

Thus f is also bounded with $\|f\| \leqslant \|A\| \, \|y\|$. Hence from the Riesz representation theorem (Prop. 4.6.2), there exists a unique $z \in \mathbf{V}$ such that

$$f(x) = \langle x, z \rangle.$$

Let $A^*: \mathbf{V} \to \mathbf{V}$ be defined as $A^*y = z$ for each $y \in \mathbf{V}$. Then for each $(x, y) \in \mathbf{V}^2$ we have

$$\begin{aligned} \langle Ax, y \rangle &= \langle x, z \rangle \\ &= \langle x, A^*y \rangle. \end{aligned}$$

Since the inner product is linear in its first argument, it follows that $A^* \in \mathbf{L}(\mathbf{V}, \mathbf{V})$. Furthermore, by the Cauchy–Schwartz inequality we have $|f(x)| \leqslant \|x\| \, \|z\|$ for all x and $|f(x)| = \|x\| \, \|z\|$ when $x = z$. Thus $\|f\| = \|z\|$.

Hence

$$\begin{aligned}\|A^*y\| &= \|z\| \\ &= \|f\| \\ &\leqslant \|A\|\,\|y\|.\end{aligned}$$

Thus $A^* \in \mathbf{B}(\mathbf{V}, \mathbf{V})$ and $\|A^*\| \leqslant \|A\|$. Finally,

$$\begin{aligned}\|Ax\|^2 &= |\langle Ax, Ax\rangle| \\ &= |\langle x, A^*Ax\rangle| \\ &\leqslant \|x\|\,\|A^*Ax\| \\ &\leqslant \|x\|\,\|A^*\|\,\|Ax\|.\end{aligned}$$

Dividing both sides by $\|Ax\|$, we find that $\|A\| \leqslant \|A^*\|$. Hence $\|A^*\| = \|A\|$. ■

Adjoints of bounded linear operators on Hilbert spaces satisfy a number of properties. Among these the following are fundamental.

Proposition 4.2 (Adjoint Properties). Let $(A, B) \in \mathbf{B}(\mathbf{V}, \mathbf{V})^2$ where $\mathbf{V}$ is a Hilbert space, and let $\alpha \in \mathbf{F}$. Then

1. $(\alpha A + B)^* = \bar{\alpha}A^* + B^*$,
2. $(AB)^* = B^*A^*$,
3. $A^{**} = A$.

Proof. Let $(x, y) \in \mathbf{B}(\mathbf{V}, \mathbf{V})$, and let $\alpha \in \mathbf{F}$. Then from the defining properties of the inner product we have

$$\begin{aligned}\langle(\alpha A + B)x, y\rangle &= \alpha\langle Ax, y\rangle + \langle Bx, y\rangle \\ &= \alpha\langle x, A^*y\rangle + \langle x, B^*y\rangle \\ &= \langle x, \bar{\alpha}A^*y\rangle + \langle x, B^*y\rangle \\ &= \langle x, (\bar{\alpha}A^* + B^*)y\rangle.\end{aligned}$$

Hence $(\alpha A + B)^* = \bar{\alpha}A^* + B^*$. Next, for each $(x, y) \in \mathbf{B}(\mathbf{V}, \mathbf{V})$ we have $\langle ABx, y\rangle = \langle Bx, A^*y\rangle = \langle x, B^*A^*y\rangle$. Thus $(AB)^* = B^*A^*$. Finally, for each $(x, y) \in \mathbf{B}(\mathbf{V}, \mathbf{V})$ we have

$$\begin{aligned}\langle x, Ay\rangle &= \overline{\langle Ay, x\rangle}, \\ &= \overline{\langle y, A^*x\rangle} \\ &= \langle A^*x, y\rangle \\ &= \langle x, A^{**}y\rangle.\end{aligned}$$

Thus $A^{**} = A$. ■

We can summarize Prop. 4.2 in words by saying that (1) the adjoint operation is *conjugate linear*, (2) the adjoint of the composition is the composition of the adjoints in reverse order, and (3) the adjoint of the adjoint is the original operator. Note that the composition property makes the adjoint operation behave in a manner similar to the inverse operation. The relationship between the adjoint and its inverse is a simple one as can be seen from the following result.

Proposition 4.3 (Adjoint Inverse). Let $A \in \mathbf{B(V, V)}$ where $\mathbf{V}$ is a Hilbert space. Then if A is invertible, so is A^*. Moreover,

$$(A^{-1})^* = (A^*)^{-1}.$$

Proof. Clearly, for each $(x, y) \in \mathbf{V}^2$ we have $\langle \boldsymbol{I}x, y\rangle = \langle x, \boldsymbol{I}y\rangle$. Hence $\boldsymbol{I}^* = \boldsymbol{I}$. Thus using property 2 of Prop. 4.2 we have

$$\begin{aligned} \boldsymbol{I} &= \boldsymbol{I}^* \\ &= (A^{-1}A)^* \\ &= A^*(A^{-1})^*. \end{aligned}$$

Similarly,

$$\begin{aligned} \boldsymbol{I} &= \boldsymbol{I}^* \\ &= (AA^{-1})^* \\ &= (A^{-1})^*A^*. \end{aligned}$$

It follows that $(A^{-1})^* = (A^*)^{-1}$. ∎

Thus the adjoint operation and the inverse operation commute. As an example of a bounded linear operator on a Hilbert space and its adjoint, consider the following.

Example 4.1 (Time-Varying Convolution). Let $\mathbf{V} = \mathbf{L_2(T)}$ where $T = [\alpha, \beta]$, and let A be the time-varying convolution operator defined in Exam. 3.1. That is, for each $x \in \mathbf{V}$

$$(Ax)(t) = \int_\alpha^t h(t, \tau)x(\tau)\, d\tau, \qquad t \in T.$$

In Exam. 3.1 it was shown that $A \in \mathbf{B(V, V)}$. To find the adjoint of A, note that

$$\langle Ax, y\rangle = \int_\alpha^\beta \int_\alpha^t h(t, \tau)x(\tau)\, d\tau\; y(t)\, dt$$

$$
\begin{aligned}
&= \int_\alpha^\beta \int_\alpha^t y(t)h(t, \tau)x(\tau)\, d\tau\, dt \\
&= \int_\alpha^\beta \int_\tau^\beta y(t)h(t, \tau)x(\tau)\, dt\, d\tau \\
&= \int_\alpha^\beta \int_t^\beta y(\tau)h(\tau, t)x(t)\, d\tau\, dt \\
&= \int_\alpha^\beta \int_t^\beta h(\tau, t)y(\tau)\, d\tau\, x(t)\, dt \\
&= \langle x, A^*y \rangle.
\end{aligned}
$$

Here we have made use of the observation that integrating over the triangular region $(t, \tau) \in [\alpha, \beta] \times [\alpha, t]$ is equivalent to integrating over the region $(\tau, t) \in [\alpha, \beta] \times [\tau, \beta]$. Thus we have

$$(A^*y)(t) = \int_t^\beta h(\tau, t)y(\tau)\, d\tau. \qquad \blacksquare$$

The notions of adjoints, null spaces, range spaces, and orthogonal complements can all be combined. In doing so we get the following useful identities.

Proposition 4.4 (Adjoint Identities). Let $A \in \mathbf{B}(\mathbf{V}, \mathbf{V})$ where $\mathbf{V}$ is a Hilbert space. Then

1. $\mathbf{N}(A^*) = \mathbf{R}(A)^\perp$
2. $\mathbf{N}(A^*)^\perp = \tilde{\mathbf{R}}(A)$,
3. $\mathbf{N}(A) = \mathbf{R}(A^*)^\perp$,
4. $\mathbf{N}(A)^\perp - \tilde{\mathbf{R}}(A^*)$. $\blacksquare$

The interested reader is referred to Luenberger (1969) for a proof of the identities in Prop. 4.4. Note that because the orthogonal complement of a set is always a closed subspace, the closures of $\mathbf{R}(A)$ and $\mathbf{R}(A^*)$ are needed in properties 2 and 4, respectively.

The spectrum of the adjoint of a linear operator is related in a direct way to the spectrum of the original operator. Specifically, we have the following relationship that is left as an exercise.

Exercise 4.1 (Adjoint Spectrum). Let $A \in \mathbf{B}(\mathbf{V}, \mathbf{V})$ where $\mathbf{V}$ is a Hilbert space. Show that

$$\lambda \in \sigma_r(A) \Rightarrow \bar{\lambda} \in \sigma_p(A^*). \qquad \blacksquare$$

Recall that the spectrum of a bounded linear operator A defined on a Banach space $\mathbf{V}$ lies within the closed ball of radius $\|A\|$ centered at the origin. If the space $\mathbf{V}$ is a Hilbert space, then another way of bounding the spectrum of an operator in $\mathbf{B(V, V)}$ is by introducing something called the numerical range of A.

Definition 4.2 (Numerical Range). Let $A \in \mathbf{B(V, V)}$ where $\mathbf{V}$ is a Hilbert space. Then the *numerical range* of A is denoted $W(A)$ and is defined as the following subset of $\mathbf{C}$:

$$W(A) \triangleq \{\langle Ax, x\rangle \in \mathbf{C} : \|x\| = 1\}. \qquad \blacksquare$$

Thus the numerical range is the image of the unit circle under the quadratic form transformation $\langle Ax, x\rangle$. The following result tells us that the spectrum of A is restricted to the closure of the numerical range, thus narrowing down its location still further.

Proposition 4.5 (Numerical Range). Let $A \in \mathbf{B(V, V)}$ where $\mathbf{V}$ is a Hilbert space, and let $W(A)$ be the numerical range of A. Then

$$\sigma(A) \subseteq \widetilde{W(A)}.$$

Proof. Suppose $\lambda \notin \widetilde{W(A)}$. Then from Def. 4.2, there exists an $\alpha > 0$ such that for each $x \in \mathbf{V}$ with $\|x\| = 1$:

$$|\lambda - \langle Ax, x\rangle| > \alpha \Rightarrow |\lambda\langle x, x\rangle - \langle Ax, x\rangle| > \alpha$$

$$\Rightarrow |\langle(\boldsymbol{\lambda} \boldsymbol{I} - A)x, x\rangle| > \alpha.$$

Let $y \in \mathbf{V}$. Then

$$|\langle(\boldsymbol{\lambda} \boldsymbol{I} - A)y, y\rangle| = \|y\|^2 \langle(\boldsymbol{\lambda} \boldsymbol{I} - A)y/\|y\|, y/\|y\|\rangle$$

$$> \|y\|^2 \alpha.$$

Thus from the Cauchy–Schwarz inequality we have

$$\alpha\|y\|^2 < |\langle(\boldsymbol{\lambda} \boldsymbol{I} - A)y, y\rangle|$$

$$\leqslant \|(\boldsymbol{\lambda} \boldsymbol{I} - A)y\| \, \|y\|.$$

Hence $\|(\boldsymbol{\lambda} \boldsymbol{I} - A)y\| \geqslant \alpha\|y\|$ for each $y \in \mathbf{V}$. Thus from Prop. 4.6.2 it follows that $\boldsymbol{\lambda} \boldsymbol{I} - A$ has a bounded inverse on $\mathbf{R}(\boldsymbol{\lambda} \boldsymbol{I} - A)$. Hence either $\lambda \in \sigma_r(A)$ or $\lambda \in \rho(A)$. Suppose $\lambda \in \sigma_r(A)$. Then from Exer. 4.1, $\bar{\lambda} \in \sigma_p(4^*)$. Thus there exists an

$x \in \mathbf{V}$ with $\|x\| = 1$ such that $(\bar{\lambda} I - A^*)x = 0$. Hence

$$\begin{aligned} \langle Ax, x \rangle &= \langle x, A^*x \rangle \\ &= \langle x, \bar{\lambda} x \rangle \\ &= \lambda \langle x, x \rangle \\ &= \lambda. \end{aligned}$$

Thus $\lambda = W(A)$. But this is a contradiction. Hence $\lambda \notin \widetilde{W(A)} \Rightarrow \lambda \in \rho(A)$. Because $\rho(A) = \mathbf{V} - \sigma(A)$, it follows that $\lambda \in \sigma(A) \Rightarrow \lambda \in \widetilde{W(A)}$. ■

6.5. NORMAL OPERATORS

Recall that every bounded linear operator on a Hilbert space has an adjoint operator associated with it. Given the adjoint, we are now in a position to classify operators into a number of useful categories. One of the broader categories involves those operators that commute with their adjoints. These are called normal operators.

Definition 5.1 (Normal Operator). Let $A \in \mathbf{B}(\mathbf{V}, \mathbf{V})$ where $\mathbf{V}$ is a Hilbert space. Then A is a *normal* operator $\Leftrightarrow$

$$A^*A = AA^*.$$ ■

Normal operators can be further decomposed into several important subcategories. However, before we do this, let us examine some of the characteristics of normal operators in general.

Exercise 5.1 (Normal Properties). Let $A \in \mathbf{B}(\mathbf{V}, \mathbf{V})$ where $\mathbf{V}$ is a Hilbert space. Show that if A is normal, then for each $x \in \mathbf{V}$

1. $\|\mathbf{A}x\| = \|A^*x\|$,
2. $\mathbf{N}(A^*) = \mathbf{N}(A)$,
3. $\mathbf{N}(A) = \mathbf{R}(A)^\perp$. ■

One of the useful features of normal operators lies in the relative ease with which the spectrum can be determined. The spectrum of a normal operator is identical to the approximate spectrum, as can be seen from the following result.

Proposition 5.1 (Normal Spectrum). Let $A \in \mathbf{B}(\mathbf{V}, \mathbf{V})$ where $\mathbf{V}$ is a Hilbert space. If A is normal, then $\lambda \in \sigma(A) \Leftrightarrow$ for each $\alpha > 0$ there exists an $x \in \mathbf{V}$ such that

$$\|(\lambda I - A)x\| < \alpha \|x\|.$$

Proof. Suppose that for each $\alpha > 0$ there exists an $x \in \mathbf{V}$ satisfying the inequality in Prop. 5.1. Then from Def. 3.2 and Exer. 3.2, $\lambda \in \sigma(A)$. Next, let $\lambda \in \mathbf{C}$, and suppose there exists an $\alpha > 0$ such that for every $x \in \mathbf{V}$

$$\|(\lambda I - A)\| \geqslant \alpha \|x\|.$$

We will show that this implies $\lambda \notin \sigma(A)$. From Prop. 4.6.2 we know that $\lambda I - A$ has a bounded inverse on $\mathbf{R}(\lambda I - A)$. It will be demonstrated that $\mathbf{R}(\lambda I - A)$ is dense in $\mathbf{V}$. Suppose $\mathbf{R}(\lambda I - A)^{\perp} = \{\mathbf{0}\}$. Then using Exer. 3.6.2 we have

$$\begin{aligned} \tilde{\mathbf{R}}(\lambda I - A) &= \mathbf{R}(\lambda I - A)^{\perp\perp} \\ &= \{\mathbf{0}\}^{\perp} \\ &= \mathbf{V}. \end{aligned}$$

Hence it is sufficient to show that $\mathbf{R}(\lambda I - A)^{\perp} = \{\mathbf{0}\}$. From Prop. 4.4 we have $\mathbf{R}(\lambda I - A)^{\perp} = \mathbf{N}(\bar{\lambda} I - A^*)$. Suppose there exists an $x \in \mathbf{V}$ with $x \neq 0$ such that

$$(\bar{\lambda} I - A^*)x = 0.$$

Then, since A is normal, $\bar{\lambda} I - A^*$ is normal. Thus from Exer. 5.1 we have

$$\begin{aligned} 0 &= \|(\bar{\lambda} I - A^*)x\| \\ &= \|(\lambda I - A)x\|. \end{aligned}$$

But this is a contradiction since $\|(\lambda I - A)x\| \geqslant \alpha \|x\|$ for every $x \in \mathbf{V}$. Hence $\mathbf{N}(\bar{\lambda} I - A^*) = \{\mathbf{0}\}$. Thus $\mathbf{R}(\lambda I - A)$ is dense in $\mathbf{V}$. It follows that $\lambda \notin \sigma(A)$. Hence $\pi(A) = \sigma(A)$. ■

Exercise 5.2 (Normal Operators). Let $A \in \mathbf{B}(\mathbf{V}, \mathbf{V})$ where $\mathbf{V}$ is a Hilbert space. Show that if A is normal,

$$\sigma_r(A) = \varnothing.$$ ■

The class of normal operators can be decomposed in a number of ways. The following two subcategories are particularly important special cases of normal operators.

Definition 5.2 (Self-Adjoint Operator). Let $A \in \mathbf{B}(\mathbf{V}, \mathbf{V})$ where $\mathbf{V}$ is a Hilbert space. Then A is *self-adjoint* $\Leftrightarrow$

$$A^* = A.$$ ■

Definition 5.3 (Unitary Operator). Let $A \in \mathbf{B}(\mathbf{V}, \mathbf{V})$ where $\mathbf{V}$ is a Hilbert space. Then A is *unitary* $\Leftrightarrow$

$$A^* = A^{-1}. \qquad \blacksquare$$

Clearly, self-adjoint and unitary operators are normal. Refer to Figure 6.4 for a pictorial description of the relationship between the various classes of normal operators.

Exercise 5.3 (Self-Adjoint Operator). Let $A \in \mathbf{B}(\mathbf{V}, \mathbf{V})$ where $\mathbf{V}$ is a Hilbert space. Show that if A is self-adjoint, then for each $x \in \mathbf{V}$

$$\langle Ax, x \rangle \in \mathbf{R}. \qquad \blacksquare$$

If an operator A is self-adjoint, then the spectrum of A is restricted to a particular region of the complex plane. More specifically, we have the following result.

Proposition 5.2 (Self-Adjoint Spectrum). Let $A \in \mathbf{B}(\mathbf{V}, \mathbf{V})$ where $\mathbf{V}$ is a Hilbert space. If A is self-adjoint, then

$$\sigma(A) \subseteq \mathbf{R}.$$

Proof. Let $\lambda \in \mathbf{C}$ with $\mathrm{Im}(\lambda) \neq 0$, and let $x \in \mathbf{V} - \{\mathbf{0}\}$. Then because A is self-adjoint,

$$\begin{aligned}
|\lambda - \bar{\lambda}| \|x\|^2 &= |\langle (\lambda - \bar{\lambda})x, x \rangle| \\
&= |\langle (\lambda \boldsymbol{I} - A)x, x \rangle - \langle (\bar{\lambda} \boldsymbol{I} - A^*)x, x \rangle| \\
&= |\langle (\lambda \boldsymbol{I} - A)x, x \rangle - \langle x, (\lambda \boldsymbol{I} - A)x \rangle| \\
&\leqslant |\langle (\lambda \boldsymbol{I} - A)x, x \rangle| + |\langle x, (\lambda \boldsymbol{I} - A)x \rangle| \\
&\leqslant 2\|(\lambda \boldsymbol{I} - A)x\| \|x\|.
\end{aligned}$$

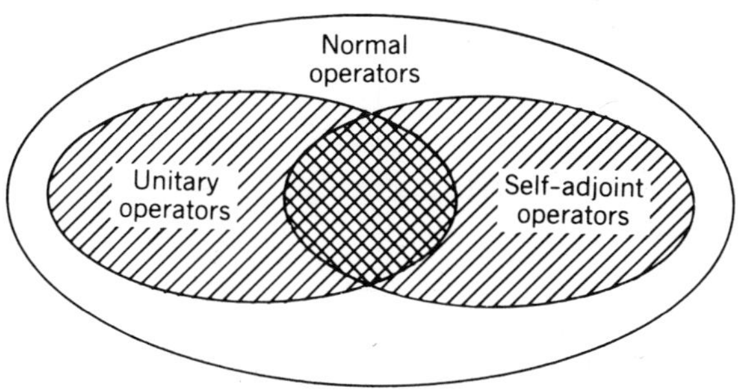

Figure 6.4. Normal operators: $A^*A = AA^*$.

Hence for each $x \in \mathbf{V}$

$$\|(\lambda I - A)x\| \geqslant |\lambda - \bar{\lambda}| \|x\|/2.$$

Thus from Prop. 5.1 it follows that $\lambda \notin \sigma(A)$. Consequently $\lambda \in \sigma(A) \Rightarrow \mathrm{Im}(\lambda) = 0$. ■

Thus the spectrum of a self-adjoint operator is real. In view of a previous result concerning the spectral radius, the spectrum of a bounded self-adjoint operator is, in fact, restricted to a finite segment of the real line as can be seen from the following exercise.

Exercise 5.4 (Self-Adjoint Spectrum). Let $A \in \mathbf{B}(\mathbf{V}, \mathbf{V})$ where $\mathbf{V}$ is a Hilbert space. Show that if A is self-adjoint, then

$$\sigma(A) \subseteq [-\|A\|, \|A\|].$$ ■

Just as the spectrum of a self-adjoint operator is restricted to a particular region of the complex plane, so is the spectrum of a unitary operator, as seen from the following exercise.

Exercise 5.5 (Unitary Spectrum). Let $A \in \mathbf{B}(\mathbf{V}, \mathbf{V})$ where $\mathbf{V}$ is a Hilbert space. Show that if A is unitary, then

$$\sigma(A) \subseteq \{\lambda \in \mathbf{C}: |\lambda| = 1\}.$$ ■

Thus the spectrum of a unitary operator is restricted to the unit circle. Refer to Figure 6.5 for a pictorial summary of the spectra of self-adjoint and unitary operators.

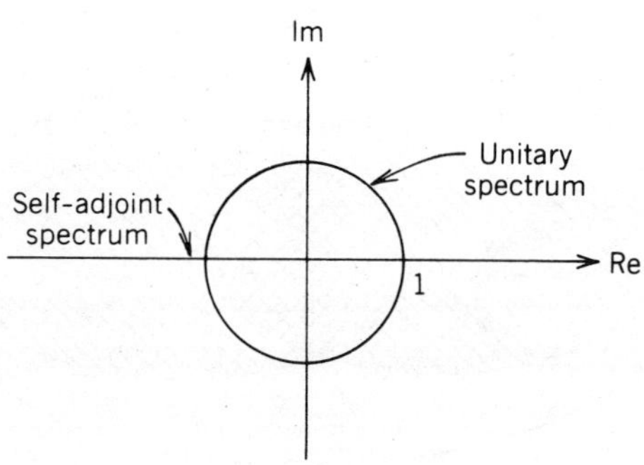

Figure 6.5. The spectra of self-adjoint and unitary operators.

6.6. PROJECTION OPERATORS

Earlier in Chapter 3 we examined the notion of orthogonal projections from the point of view of vector spaces. Recall that if **U** is a closed subspace of a Hilbert space **V**, then

$$\mathbf{V} = \mathbf{U} \oplus \mathbf{U}^{\perp}.$$

Thus each vector $x \in \mathbf{V}$ can be decomposed uniquely into a vector $x^0 \in \mathbf{U}$ plus a vector $x^1 \in \mathbf{U}^{\perp}$. These component vectors were referred to as the orthogonal projections of x onto **U** and $\mathbf{U}^{\perp}$, respectively. We now examine this direct sum decomposition of **V** from the point of view of linear operators.

Definition 6.1 (Orthogonal Projection Operator). Let **U** be a closed subspace of a Hilbert space **V**. For each $x \in \mathbf{V}$ let $x^0 \in \mathbf{U}$ be the orthogonal projection of x onto **U**. Then the *orthogonal projection operator* induced by **U** is defined as

$$Ex \triangleq x^0. \qquad \blacksquare$$

Exercise 6.1 (Projection Operators). Let **V** be a Hilbert space, and let E be the orthogonal projection operator induced by a closed subspace **U**. Show that $E \in \mathbf{B}(\mathbf{V}, \mathbf{V})$ and

1. $\|E\| = 1$,
2. $\mathbf{R}(E) - \mathbf{U}$,
3. $\mathbf{N}(E) = \mathbf{U}^{\perp}$,
4. $\mathbf{R}(E) = \{x : Ex = x\}$ ■

In view of Exer. 6.1, a projection operator is a bounded linear operator. As such it is meaningful to speak of the adjoint of a projection operator. Every orthogonal projection operator is, in fact, self-adjoint. In addition, orthogonal projection operators are idempotent operators, which are defined as follows.

Definition 6.2 (Idempotent Operator). Let $A \in \mathbf{L}(\mathbf{V}, \mathbf{V})$ where **V** is a vector space. Then A is an *idempotent* operator $\Leftrightarrow A^2 = A$. ■

Here the notation A^2 denotes the composition AA. The self-adjoint and idempotent properties of orthogonal projection operators are not only necessary, they are also sufficient. As such, an alternative way to characterize orthogonal projection operators is by means of the following result.

Proposition 6.1 (Orthogonal Projection Operator). Let $E \in \mathbf{B}(\mathbf{V}, \mathbf{V})$ where **V** is a Hilbert space. Then E is an orthogonal projection operator $\Leftrightarrow$

1. E is idempotent,
2. E is self-adjoint.

Proof. Suppose E is an orthogonal projection operator. Then from Exer. 6.1, E is the orthogonal projection operator induced by the closed subspace $\mathbf{R}(E)$ and

$$\mathbf{V} = \mathbf{R}(E) \oplus \mathbf{N}(E).$$

Let $x \in \mathbf{V}$. Then there exist unique $(x^0, x^1) \in \mathbf{R}(E) \times \mathbf{N}(E)$ such that $x = x^0 + x^1$. Hence from Def. 6.1,

$$\begin{aligned} E^2 x &= E(Ex) \\ &= Ex^0 \\ &= Ex^0 + Ex^1 \\ &= E(x^0 + x^1) \\ &= Ex. \end{aligned}$$

Thus E is idempotent. Next, let $(x, y) \in \mathbf{V}^2$. Then there exist unique $(x^0, y^0) \in \mathbf{R}(E)^2$ and $(x^1, y^1) \in \mathbf{N}(E)^2$ such that $x = x^0 + x^1$ and $y = y^0 + y^1$. Hence, using Exer. 6.1, we have

$$\begin{aligned} \langle Ex, y\rangle &= \langle x^0, y\rangle \\ &= \langle x^0, y^0 + y^1\rangle \\ &= \langle x^0, y^0\rangle \\ &= \langle x^0 + x^1, y^0\rangle \\ &= \langle x, y^0\rangle \\ &= \langle x, Ey\rangle. \end{aligned}$$

Thus E is self-adjoint. Next, suppose E is idempotent and self-adjoint. First, we will show that $\mathbf{R}(E)$ is closed. From property 4 of Exer. 6.1 we have

$$\mathbf{R}(E) = \{y \in \mathbf{V} : Ey = y\}.$$

Let $\{y^k\}$ be a sequence in $\mathbf{R}(E)$ and suppose $y^k \to y$ as $k \to \infty$ where $y \in \mathbf{V}$. Since $E \in \mathbf{B}(\mathbf{V}, \mathbf{V})$, E is continuous. Hence

$$\begin{aligned} Ey &= E\left(\lim_{k\to\infty} \{y^k\}\right) \\ &= \lim_{k\to\infty} \{Ey^k\} \\ &= \lim_{k\to\infty} \{y^k\} \\ &= y. \end{aligned}$$

Thus $y \in \mathbf{R}(\boldsymbol{E})$. Hence $\mathbf{R}(\boldsymbol{E})$ is closed. It follows that $\mathbf{V} = \mathbf{R}(\boldsymbol{E}) \oplus \mathbf{R}(\boldsymbol{E})^{\perp}$. Now for each $x \in \mathbf{V}$,

$$\begin{aligned} x &= \boldsymbol{I}x \\ &= (\boldsymbol{I} - \boldsymbol{E} + \boldsymbol{E})x \\ &= \boldsymbol{E}x + (\boldsymbol{I} - \boldsymbol{E})x. \end{aligned}$$

Let $y \in \mathbf{R}(\boldsymbol{E})$. Then $\boldsymbol{E}y = y$ and, since $\boldsymbol{E}$ is self-adjoint, for each $x \in \mathbf{V}$ we have

$$\begin{aligned} \langle(\boldsymbol{I} - \boldsymbol{E})x, y\rangle &= \langle x, (\boldsymbol{I} - \boldsymbol{E})y\rangle \\ &= \langle x, y - \boldsymbol{E}y\rangle \\ &= \langle x, y - y\rangle \\ &= \langle x, 0\rangle \\ &= 0. \end{aligned}$$

Thus for each $x \in \mathbf{V}$, $(\boldsymbol{E}x, (\boldsymbol{I} - \boldsymbol{E})x) \in \mathbf{R}(\boldsymbol{E}) \times \mathbf{R}(\boldsymbol{E})^{\perp}$. Hence $\boldsymbol{E}$ is the orthogonal projection operator induced by the closed subspace $\mathbf{R}(\boldsymbol{E})$. ■

Thus an orthogonal projection operator is a bounded, linear, self-adjoint, idempotent operator. In certain cases, the composition of two orthogonal projection operators is again an orthogonal projection operator.

Proposition 6.2 (Composition). Let $\mathbf{V}$ be a Hilbert space, and let $\boldsymbol{E}_1$ and $\boldsymbol{E}_2$ be orthogonal projection operators induced by the closed subspaces U_1 and U_2, respectively. Then $\boldsymbol{E}_1\boldsymbol{E}_2$ is an orthogonal projection operator $\Leftrightarrow$ $\boldsymbol{E}_1$ and $\boldsymbol{E}_2$ commute, in which case

$$\mathbf{R}(\boldsymbol{E}_1\boldsymbol{E}_2) = U_1 \cap U_2.$$

Proof. Suppose $\boldsymbol{E}_1\boldsymbol{E}_2$ is an orthogonal projection operator. Then from Prop. 6.1, $\boldsymbol{E}_1\boldsymbol{E}_2$ is self-adjoint. Thus

$$\begin{aligned} \boldsymbol{E}_1\boldsymbol{E}_2 &= (\boldsymbol{E}_1\boldsymbol{E}_2)^* \\ &= \boldsymbol{E}_2^*\boldsymbol{E}_1^* \\ &= \boldsymbol{E}_2\boldsymbol{E}_1. \end{aligned}$$

Hence $\boldsymbol{E}_1$ and $\boldsymbol{E}_2$ commute. Next, suppose $\boldsymbol{E}_1$ and $\boldsymbol{E}_2$ commute. Then

$$\begin{aligned} (\boldsymbol{E}_1\boldsymbol{E}_2)^* &= \boldsymbol{E}_2^*\boldsymbol{E}_1^* \\ &= \boldsymbol{E}_2\boldsymbol{E}_1 \\ &= \boldsymbol{E}_1\boldsymbol{E}_2. \end{aligned}$$

Hence E_1E_2 is self-adjoint. Furthermore,

$$\begin{aligned}(E_1E_2)^2 &= E_1E_2E_1E_2\\ &= E_1E_2E_2E_1\\ &= E_1E_2E_1\\ &= E_1E_1E_2\\ &= E_1E_2.\end{aligned}$$

Thus E_1E_2 is idempotent. Hence from Prop. 6.1, E_1E_2 is an orthogonal projection operator. It remains to establish the range space of E_1E_2. Let $x \in U_1 \cap U_2$. Then $E_k x = x$ for $1 \leqslant k \leqslant 2$. Hence

$$\begin{aligned}E_1E_2x &= E_1x\\ &= x.\end{aligned}$$

Thus $U_1 \cap U_2$ is a subset of $\mathbf{R}(E_1E_2)$. Next, let $x \in \mathbf{R}(E_1E_2)$. Then as E_1E_2 is an orthogonal projection operator,

$$\begin{aligned}x &= E_1E_2x\\ &= E_1(E_2x),\end{aligned}$$

and

$$\begin{aligned}x &= E_1E_2x\\ &= E_2E_1x\\ &= E_2(E_1x).\end{aligned}$$

Hence $x \in \mathbf{R}(E_1)$ and $x \in \mathbf{R}(E_2)$. But from Exer. 6.1, we have $\mathbf{R}(E_k) = U_k$ for $1 \leqslant k \leqslant 2$. Thus $\mathbf{R}(E_1E_2) = U_1 \cap U_2$. ■

6.7. SIGN-DEFINITE OPERATORS

Next, we introduce a scheme for establishing an *ordering* of operators within $\mathbf{B}(\mathbf{V}, \mathbf{V})$. To do this, we isolate a subset of $\mathbf{B}(\mathbf{V}, \mathbf{V})$ called sign-definite operators.

Definition 7.1 (Positive-Semidefinite Operator). Let $A \in \mathbf{B}(\mathbf{V}, \mathbf{V})$ where $\mathbf{V}$ is a Hilbert space and A is self-adjoint. Then A is a *positive-semidefinite* operator, and we write $A \geqslant 0 \Leftrightarrow$

$$\langle Ax, x\rangle \geqslant 0, \qquad x \in \mathbf{V}.$$ ■

Recall from Exer. 5.3 that $\langle Ax, x\rangle$ is real for all $x \in \mathbf{V}$ when A is self-adjoint. If we strengthen Def. 7.1 by replacing the inequality by a strict inequality, then we say that A is *positive-definite*. It is also possible to define *negative-definite* and *negative-semidefinite* operators by replacing A with $-A$.

The characterization of a self-adjoint operator in terms of its sign is really a characterization of the location of the spectrum of that operator. The following exercise illustrates this point for the case of positive-semidefinite operators.

Exercise 7.1 (Spectrum Bound). Let $A \in \mathbf{B}(\mathbf{V}, \mathbf{V})$ where $\mathbf{V}$ is a Hilbert space and A is self-adjoint. Show that if $A \geqslant \boldsymbol{0}$, then

$$\sigma(A) \subseteq \{\lambda \in \mathbf{R} : \lambda \geqslant 0\}. \qquad \blacksquare$$

Within the class of self-adjoint operators in $\mathbf{B}(\mathbf{V}, \mathbf{V})$ we are now in a position to define an ordering of operators in the following sense.

Definition 7.2 (Operator Order). Let $(A, B) \in \mathbf{B}(\mathbf{V}, \mathbf{V})^2$ where $\mathbf{V}$ is a Hilbert space and A and B are self-adjoint. Then $A \geqslant B \Leftrightarrow A - B \geqslant \boldsymbol{0}$. ■

Thus, a self-adjoint operator A is greater than or equal to a self-adjoint operator B if and only if the difference operator $A - B$ is positive-semidefinite. The notions of sign-definite operators and orthogonal projection operators can now be combined to produce the following result.

Proposition 7.1 (Projection Operator Order). Let $(E_1, E_2) \in \mathbf{B}(\mathbf{V}, \mathbf{V})^2$ be orthogonal projection operators induced by the closed subspaces $\mathbf{U}_1$ and $\mathbf{U}_2$ of a Hilbert space $\mathbf{V}$. Then $E_1 \leqslant E_2 \Leftrightarrow \mathbf{U}_1 \subseteq \mathbf{U}_2$.

Proof. Suppose $E_1 \leqslant E_2$. Then for each $x \in \mathbf{V}$,

$$\begin{aligned}
\|E_1 x\|^2 &= \langle E_1 x, E_1 x\rangle \\
&= \langle E_1 x, E_1^* x\rangle \\
&= \langle E_1^2 x, x\rangle \\
&= \langle E_1 x, x\rangle \\
&\leqslant \langle E_2 x, x\rangle \\
&= \langle E_2^2 x, x\rangle \\
&= \langle E_2 x, E_2^* x\rangle \\
&= \langle E_2 x, E_2 x\rangle \\
&= \|E_2 x\|^2.
\end{aligned}$$

Let $x \in \mathbf{U}_1$. Then

$$\begin{aligned} \|x\| &= \|E_1 x\| \\ &\leqslant \|E_2 x\| \\ &\leqslant \|E_2\| \, \|x\| \\ &= \|x\|. \end{aligned}$$

Thus $\|E_2 x\| = \|x\|$. Hence $x \in \mathbf{R}(E_2)$. But $\mathbf{R}(E_2) = \mathbf{U}_2$. It follows that $\mathbf{U}_1 \subseteq \mathbf{U}_2$. Next, suppose $\mathbf{U}_1 \subseteq \mathbf{U}_2$. Let $x \in \mathbf{V}$. Then there exist unique $(x^0, x^1) \in \mathbf{U}_1 \times \mathbf{U}_1^{\perp}$, such that $x = x^0 + x^1$. Thus for each $x \in \mathbf{V}$,

$$\begin{aligned} E_2 E_1 x &= E_2 x^0 \\ &= x^0 \\ &= E_1 x. \end{aligned}$$

Hence $E_2 E_1 = E_1$. Using adjoints, we have

$$\begin{aligned} E_2 E_1 &= E_1 \\ &= E_1^* \\ &= (E_2 E_1)^* \\ &= E_1^* E_2^* \\ &= E_1 E_2. \end{aligned}$$

Let $x \in \mathbf{V}$. Then

$$\begin{aligned} \langle (E_2 - E_1)x, x \rangle &= \langle E_2 x, x \rangle - \langle E_1 x, x \rangle \\ &= \langle E_2 x, x \rangle - \langle E_2 E_1 x, x \rangle \\ &= \langle E_2 (I - E_1) x, x \rangle. \end{aligned}$$

Now E_2 and $(I - E_1)$ commute. Thus from Prop. 6.2, $E_2(I - E_1)$ is an orthogonal projection operator. Let $E \triangleq E_2(I - E_1)$. Then for each $x \in \mathbf{V}$

$$\begin{aligned} \langle (E_2 - E_1)x, x \rangle &= \langle (E_2 - E_2 E_1)x, x \rangle \\ &= \langle Ex, x \rangle \\ &= \langle E^2 x, x \rangle \\ &= \langle Ex, E^* x \rangle \\ &= \langle Ex, Ex \rangle \\ &= \|Ex\|^2 \\ &\geqslant 0. \end{aligned}$$

It follows that $E_2 \geqslant E_1$. ■

Exercise 7.2. Let $(E_1, E_2) \in \mathbf{B}(\mathbf{V}, \mathbf{V})^2$ be orthogonal projection operators induced by the closed subspaces $\mathbf{U}_1$ and $\mathbf{U}_2$ of a Hilbert space $\mathbf{V}$. Show that $E_2 - E_1$ is an orthogonal projection operator $\Leftrightarrow E_2 \geqslant E_1$, in which case

$$\mathbf{R}(E_2 - E_1) = \mathbf{U}_1 \cap \mathbf{U}_2^{\perp} \qquad \blacksquare$$

One question which arises in applications involving positive-semidefinite operators is the following. If A and B are positive-semidefinite, is the composition operator AB positive-semidefinite? This question cannot always be answered in the affirmative. However, in the event that A and B commute, then the composition AB is indeed positive-semidefinite and we have the following result.

Proposition 7.2 (Composition). Let $(A, B) \in B(V, V)^2$ where $\mathbf{V}$ is a Hilbert space. Suppose A and B are self-adjoint and commute. Then

$$A \geqslant 0,\ B \geqslant 0 \Rightarrow AB \geqslant 0. \qquad \blacksquare$$

The interested reader can refer to Bachman and Narici (1966) for a proof of Prop. 7.2.

Exercise 7.3. Let $(A, B) \in B(V, V)^2$ be orthogonal projection operators induced by the closed subspaces $\mathbf{U}_1$ and $\mathbf{U}_2$ of a Hilbert space $\mathbf{V}$. Show that $E_1 \leqslant E_2 \Leftrightarrow$ for each $x \in \mathbf{V}$

$$\|\mathbf{E}_1 x\| \leqslant \|\mathbf{E}_2 x\|. \qquad \blacksquare$$

6.8. THE SPECTRAL REPRESENTATION

Many operators which occur in mathematical models of physical systems can, with an appropriate choice of inner product, be viewed as self-adjoint. In order to solve the equations within which these operators appear, it is useful to have a standard way of representing such operators. With this in mind we now introduce an integral representation for self-adjoint operators called the spectral representation. The material in this section is based on a more detailed treatment of the subject that can be found in Bachman and Narici (1966). We begin with the unbounded infinite-dimensional case.

Proposition 8.1 (Spectral Theorem: Unbounded Case). Let $A \in \mathbf{L}(\mathbf{V}, \mathbf{V})$ where $\mathbf{V}$ is a Hilbert space. If A is self-adjoint, then there exists a family of orthogonal projection operators $\{E(\lambda)\}$ such that

1. $\lambda_1 \leqslant \lambda_2 \Rightarrow E(\lambda_1) \leqslant E(\lambda_2)$,
2. $E(\lambda) \to 0$ as $\lambda \to -\infty$,
3. $E(\lambda) \to I$ as $\lambda \to \infty$,
4. $E(\lambda + |\alpha|) \to E(\lambda)$ as $\alpha \to 0$.

In addition, the operator A can be represented as follows:

$$A = \int_R \lambda \, dE(\lambda). \qquad \blacksquare$$

A proof of Prop. 8.1 can be found in Bachman and Narici (1966). Note from property 1 that $\{E(\lambda)\}$ is a *monotone* family of orthogonal projection operators parameterized by λ. In light of the results from Section 6, it follows that $E(\lambda)$ and $E(\mu)$ commute and $E(\lambda)E(\mu)$ is again an orthogonal projection operator. We call the family of orthogonal projection operators $\{E(\lambda)\}$ the *resolution of the identity* induced by the operator A. The name arises from the fact that from properties 2 and 3 we have

$$\int_{\mathrm{R}} dE(\lambda) = I.$$

Property 4 says that the operator $E(\lambda)$ is continuous from the right. The principal result of Prop. 8.1 is the integral representation of A, which is called *spectral representation* of the operator A.

To further investigate the properties of the spectral representation, we restrict our attention to bounded self-adjoint operators. Recall from Exer. 5.3 that if $A \in (\mathbf{V}, \mathbf{V})$ is self-adjoint, then $\langle Ax, x\rangle$ is real for each $x \in \mathbf{V}$. This observation allows us to consider something we will call the spectral limits of A.

Definition 8.1 (Spectral Limits). Let $A \in \mathbf{B}(\mathbf{V}, \mathbf{V})$ where $\mathbf{V}$ is a Hilbert space and A is self-adjoint. The lower and upper *spectral limits* of A are denoted m and M, respectively, and are defined

$$m \triangleq \inf_{\|x\|=1}\{\langle Ax, x\rangle\},$$
$$M \triangleq \sup_{\|x\|=1}\{\langle \mathbf{A}x, x\rangle\}. \qquad \blacksquare$$

Because A is bounded and self-adjoint, the lower and upper spectral limits of A are well defined. The names arise from the fact that they provide us with a small interval in $\mathbf{R}$ within which the spectrum of A must lie.

Proposition 8.2 (Spectral Bounds). Let $A \in \mathbf{B}(\mathbf{V}, \mathbf{V})$ where $\mathbf{V}$ is a Hilbert space. If A is self-adjoint and m and M are the spectral limits, then

$$\sigma(A) \subseteq [m, M].$$

Proof. Since $A \in \mathbf{B}(\mathbf{V}, \mathbf{V})$, from Prop. 4.5 we know that $\sigma(A)$ is a subset of $\tilde{W}(A)$, the closure of the numerical range of A where

$$W(A) = \{\langle Ax, x\rangle \in \mathbf{C} : \|x\| = 1\}.$$

From Exer. 5.3 we know that the numerical range $W(A)$ is real, as A is self-adjoint. Finally, because $A \in \mathbf{B}(\mathbf{V}, \mathbf{V})$, using the Cauchy–Schwarz inequality for each $x \in \mathbf{V}$ we have

$$\begin{aligned}|\langle Ax, x\rangle| &\leqslant \|Ax\|\,\|x\| \\ &\leqslant \|A\|\,\|x\|^2.\end{aligned}$$

It follows that $\tilde{W}(A)$ is a subset of the interval $[-\|A\|, \|A\|]$. Hence the lower and upper spectral limits are well defined and from Def. 8.1,

$$\sigma(A) \subseteq [m, M]. \qquad \blacksquare$$

Given the notion of the spectral limits, we are now in a position to state the spectral representation theorem for a bounded infinite-dimensional self-adjoint operator on a Hilbert space.

Proposition 8.3 (Spectral Theorem: Bounded Case). Let $A \in \mathbf{B}(\mathbf{V}, \mathbf{V})$ where $\mathbf{V}$ is a Hilbert space. If A is self-adjoint, then there exists a family of orthogonal projection operators $\{E(\lambda)\}$ such that

1. $\lambda_1 \leqslant \lambda_2 \Rightarrow E(\lambda_1) \leqslant E(\lambda_2)$,
2. $\lambda < m \Rightarrow E(\lambda) = 0$,
3. $\lambda \geqslant M \Rightarrow E(\lambda) = I$,
4. $E(\lambda + |\alpha|) \to E(\lambda)$ as $\alpha \to 0$,
5. $E(\lambda)A = AE(\lambda)$,

In addition, if $\alpha < m$ and $\beta \geqslant M$, then for each polynomial $p \in \mathbf{P}(\mathbf{C})$ the operator $\mathbf{p}(A)$ can be represented as follows:

$$\mathbf{p}(A) = \int_\alpha^\beta p(\lambda)\, dE(\lambda). \qquad \blacksquare$$

A proof of Prop. 8.3 can be found in Bachman and Narici (1966). By comparing the spectral representation of a bounded self-adjoint operator with the spectral representation of its unbounded counterpart, we see that properties 2 and 3 have now allowed us to restrict the limits of integration to $\alpha < m$ and $\beta \geqslant M$. The strict inequality involving m arises from the fact that $E(\lambda)$ is not necessarily continuous from the left. Note also from property 5 that the operators $E(\lambda)$ and A commute. It is this feature which allows us to generalize the spectral representation to polynomials of A.

As with the unbounded case, no statement has been made claiming that the projection operator $E(\lambda)$ is continuous from the left and therefore continuous. Indeed, $E(\lambda)$ is discontinuous on the point spectrum of A. More specifically, we can characterize the spectrum of A in terms of the family of orthogonal projection operators $\{E(\lambda)\}$ as follows.

Proposition 8.4 (Resolution of Identity). Let $A \in \mathbf{B}(\mathbf{V}, \mathbf{V})$ where $\mathbf{V}$ is a Hilbert space. If A is self-adjoint and $\{E(\lambda)\}$ is the resolution of the identity induced by A, then the following properties hold:

1. $\lambda \in \sigma_p(A) \Leftrightarrow E(\lambda) \neq E(\lambda^-)$ in which case $\mathbf{N}(\lambda I - A) = \mathbf{R}\{E(\lambda) - E(\lambda^-)\}$.
2. $\lambda \in \rho(A) \Leftrightarrow E(\lambda) = E(\lambda^-)$, and there exists an $\alpha > 0$ such that $E(\mu)$ is constant for $\mu \in (\lambda - \alpha, \lambda + \alpha)$.
3. $\lambda \in \sigma_c(A) \Leftrightarrow E(\lambda) = E(\lambda^-)$ and $\lambda \notin \rho(A)$. ■

The interested reader can refer to Bachman and Narici (1966) for a proof of Prop. 8.4. The interpretation of Prop. 8.4 is that the family of orthogonal projection operators $E(\lambda)$ is continuous except at the eigenvalues where jump discontinuities occur. Furthermore, $E(\lambda)$ is constant on the resolvent set of A. It follows that the integration involved in the spectral representation can be confined to the spectrum. Note that Prop. 8.4 agrees with a previous result that indicated that the residual spectrum of a self-adjoint operator is empty.

As an example of a bounded self-adjoint linear operator and its spectral representation, consider the following.

Example 8.1 [Multiplication Operator (Bachman and Narici, 1966)]. Let $\mathbf{V} = \mathbf{L}_2([-1, 1])$, and let A be the following *multiplication* operator on $\mathbf{V}$. For each $x \in \mathbf{V}$

$$(Ax)(t) \triangleq tx(t), \qquad -1 \leqslant t \leqslant 1.$$

Clearly, A is linear. To show that A is bounded, let $x \in \mathbf{V}$. Then

$$\begin{aligned}
\|Ax\|^2 &= \int_{-1}^{1} |(Ax)(t)|^2 \, dt \\
&= \int_{-1}^{1} |tx(t)|^2 \, dt \\
&= \int_{-1}^{1} t^2 |x(t)|^2 \, dt \\
&\leqslant \int_{-1}^{1} |x(t)|^2 \, dt \\
&= \|x\|^2.
\end{aligned}$$

Thus $A \in \mathbf{B}(\mathbf{V}, \mathbf{V})$ with $\|A\| \leqslant 1$. Next, to show that A is self-adjoint, let $(x, y) \in \mathbf{V}^2$. Then

$$\begin{aligned}\langle Ax, y\rangle &= \int_{-1}^{1} tx(t)y(t)\, dt \\ &= \int_{-1}^{1} x(t)ty(t)\, dt \\ &= \langle x, Ay\rangle.\end{aligned}$$

Thus A is self-adjoint. Next, we must find the resolution of the identity induced by A. For each $\lambda \in \mathbf{R}$ let

$$(\boldsymbol{E}(\lambda)x)(t) = x(t)1(\lambda - t), \qquad -1 \leqslant t \leqslant 1.$$

Here $1(\alpha)$ denotes the *unit step* function, which is zero for $\alpha < 0$ and unity for $\alpha \geqslant 0$. Then $\boldsymbol{E}(\lambda)$ is the orthogonal projection operator induced by the following closed subspace of $\mathbf{V}$:

$$\mathbf{R}(\boldsymbol{E}(\lambda)) = \{x \in \mathbf{V} : x(t) = 0,\ t > \lambda\}.$$

It is clear that $\boldsymbol{E}(\lambda) = \boldsymbol{0}$ for $\lambda \in (-\infty, -1)$ and also $\boldsymbol{E}(\lambda) = \boldsymbol{I}$ for $\lambda \in [1, \infty)$. Finally, using integration by parts, we have the following spectral representation of A:

$$\begin{aligned}(Ax)(t) &= \left(\int_R \lambda\, \boldsymbol{dE}(\lambda)x\right)(t) \\ &= \left(\int_{-2}^{1} \lambda\, \boldsymbol{dE}(\lambda)x\right)(t) \\ &= \lambda \boldsymbol{E}(\lambda)|_{-2}^{1}\, x(t) - \int_{-2}^{1} (\boldsymbol{E}(\lambda)x)(t)\, d\lambda \\ &= x(t) - \int_{-2}^{1} (\boldsymbol{E}(\lambda)x)(t)\, d\lambda \\ &= x(t) - \int_{-2}^{1} x(t)1(\lambda - t)\, d\lambda \\ &= x(t) - \int_{t}^{1} x(t)\, d\lambda \\ &= x(t) - x(t)(1 - t) \\ &= tx(t).\end{aligned}$$

■

PROBLEMS

1. Let $A \in \mathbf{B}(\mathbf{V}, \mathbf{V})$ where $\mathbf{V}$ is a normed vector space. Show that $\sigma_p(A) \cup \sigma_c(A) \subseteq \pi(A)$.

2. Let $A \in \mathbf{B}(\mathbf{V}, \mathbf{V})$ where $\mathbf{V}$ is a Banach space. Show that

$$\lambda \in \rho(A) \Rightarrow \mathbf{R}(\lambda I - A) = \mathbf{V}.$$

3. Let $\mathbf{V} = \mathbf{C}_\infty([0, 1])$, and let A be the integral operator on $\mathbf{V}$

$$(Ax)(t) \triangleq \int_0^t x(\tau)\, d\tau, \qquad 0 \leqslant t \leqslant 1.$$

Show that $\sigma_r(A) = \{0\}$ and $\rho(A) = \mathbf{V} - \{0\}$.

4. Let $A \in \mathbf{B}(\mathbf{V}, \mathbf{V})$ where $\mathbf{V}$ is a Hilbert space. Show that if A is invertible and $A^{-1} \in \mathbf{B}(\mathbf{V}, \mathbf{V})$, then

$$\lambda \in \sigma(A) \Leftrightarrow \lambda^{-1} \in \sigma(A^{-1}).$$

5. Let $A \in \mathbf{B}(\mathbf{V}, \mathbf{V})$ where $\mathbf{V}$ is a Hilbert space. Show that

$$\lambda \in \sigma(A^*) \Rightarrow \bar{\lambda} \in \sigma(A).$$

6. Let $A \in \mathbf{B}(\mathbf{V}, \mathbf{V})$ where $\mathbf{V}$ is a Hilbert space. Show that if A is normal, then

$$\mathbf{V} = \mathbf{N}(A) \oplus \tilde{\mathbf{R}}(A).$$

7. Let $A \in \mathbf{B}(\mathbf{V}, \mathbf{V})$ where $\mathbf{V}$ is a Hilbert space. Show that A is an orthogonal projection operator $\Leftrightarrow$

$$A = A^*A.$$

8. Let E_1 and E_2 be orthogonal projection operators. Show that $E_1 + E_2$ is an orthogonal projection operator $\Leftrightarrow$

$$E_1E_2 = 0.$$

9. Let $A \in \mathbf{B}(\mathbf{V}, \mathbf{V})$ where $\mathbf{V}$ is a Hilbert space. Show that

$$A^*A \geqslant 0.$$

10. Let $A \in \mathbf{B}(\mathbf{V}, \mathbf{V})$ where $\mathbf{V}$ is a Hilbert space. Find $\sigma(A)$ if A is an orthogonal projection operator

11. Let $A \in \mathbf{B}(\mathbf{V}, \mathbf{V})$ where $\mathbf{V}$ is a Hilbert space. Show that if A is self-adjoint, then there exist $(B, C) \in \mathbf{B}(\mathbf{V}, \mathbf{V})^2$ with $B \geqslant 0$, $C \geqslant 0$ such that

$$A = B - C.$$

12. Let $A \in \mathbf{B(V, V)}$ where $\mathbf{V}$ is a Hilbert space. Show that if A is normal, then A is self-adjoint $\Leftrightarrow \sigma(A)$ is a subset of $\mathbf{R}$.

13. Let $A \in \mathbf{B(V, V)}$ where $\mathbf{V}$ is a Hilbert space. Show that if A is normal, then A is unitary $\Leftrightarrow \sigma(A)$ is a subset of the unit circle in $\mathbf{C}$.

14. Let $A \in \mathbf{B(V, V)}$ where $\mathbf{V}$ is a Hilbert space. Show that if A is self-adjoint, then A is positive-semidefinite $\Leftrightarrow \sigma(A)$ is nonnegative.

15. Let $\mathbf{V} = \mathbf{L_p(R)}$, and for each $x \in \mathbf{V}$ let Ax be defined by the following convolution operation (Desoer and Vidyasagar, 1975)

$$(Ax)(t) = \int_R h(t - \tau)x(\tau)\, d\tau.$$

Show that if $h \in \mathbf{L_1(R)}$, then $A \in \mathbf{B(V, V)}$ for $1 \leqslant p \leqslant \infty$.

16. Let $A \in \mathbf{B(V, V)}$ where $\mathbf{V}$ is a Hilbert space and A is self-adjoint. Show that if $A > 0$, then

$$\lambda \in (A) \Rightarrow \lambda > 0.$$

REFERENCES

Bachman G., and L. Narici, *Functional Analysis*, Academic, New York, 1966.

Desoer C. A., and M. Vidyasagar, *Feedback Systems: Input-Output Properties*, Academic, New York, 1975.

Luenberger, D. G., Luenberger, *Optimization by Vector Space Methods*, Wiley, New York, 1969.

7 FINITE-DIMENSIONAL LINEAR OPERATORS

Considering that a finite-dimensional linear operator is an important special case of a finite-dimensional linear transformation, all of the results concerning finite-dimensional linear transformations apply to finite-dimensional linear operators as well. In particular, every finite-dimensional linear operator is bounded, continuous, and can be represented by a matrix. However, finite-dimensional linear operators have certain additional properties which set them apart from finite-dimensional linear transformations. It is with these properties that we will be concerned.

7.1. FINITE-DIMENSIONAL LINEAR OPERATORS

In the discussion of linear operators in general, considerable emphasis was placed upon the notion of the adjoint operator. For a finite-dimensional linear operator A defined on an inner product space, the adjoint operator A^* always exists. Moreover, in certain coordinate systems the matrix representation of the adjoint operator A^* is related in a simple way to the matrix representation of the operator A. To see this, it is useful to first consider the following generalization of the transpose operation.

Definition 1.1 (Hermitian Transpose). Let $A \in \mathbf{F}^{m \times n}$. The *Hermitian transpose* of A is the matrix $A^H \in \mathbf{F}^{n \times m}$ defined as follows:

$$A^H \triangleq \bar{A}^T. \qquad \blacksquare$$

Thus the Hermitian transpose of A is simply the transpose of the complex conjugate of A. In particular, if the field $\mathbf{F}$ is real, then the Hermitian transpose is synonymous with the transpose. The fundamental properties of the Hermitian transpose operation are analogous to those of the regular transpose operation, as can be seen from the following exercise.

Exercise 1.1 (Hermitian Transpose Properties). Let $(A, B, C) \in \mathbf{F}^{m \times k} \times \mathbf{F}^{k \times n} \times \mathbf{F}^{m \times k}$, and let $\alpha \in \mathbf{F}$. Show that

1. $(\alpha A + C)^{H} = \bar{\alpha} A^{H} + C^{H}$,
2. $(AB)^{H} = B^{H} A^{H}$,
3. $(A^{H})^{H} = A$. ■

Note that the fundamental properties of the Hermitian transpose A^{H} are identical to those of the adjoint operator A^*. Indeed, the relationship between these two operations is an intimate one. To see this, it is useful to first consider the following preliminary result that is left as an exercise.

Exercise 1.2 (Inner Product). Let $\mathbf{B} = \{b^1, b^2, \ldots, b^n\}$ be an orthonormal basis for an inner product space $\mathbf{V}$. Show that for each $(x, y) \in \mathbf{V}^2$

$$\langle x, y \rangle = ([y]^{\mathbf{B}})^{H}[x]^{\mathbf{B}}.$$ ■

In an orthonormal coordinate system, the operation of forming the Hermitian transpose is identical to the operation of computing the adjoint as can be seen from the following result.

Proposition 1.1 (Adjoint Representation). Let $\mathbf{B} = \{b^1, b^2, \ldots, b^n\}$ be an orthonormal basis for an inner product space $\mathbf{V}$. Then for each $A \in \mathbf{L(V, V)}$

$$[A^*]^{\mathbf{B,B}} = ([A]^{\mathbf{B,B}})^{H}.$$

Proof. From Exer. 1.2, for each $(x, y) \in \mathbf{V}^2$ we have

$$\begin{aligned}\langle Ax, y \rangle &= ([y]^{\mathbf{B}})^{H}[Ax]^{\mathbf{B}} \\ &= ([y]^{\mathbf{B}})^{H}[A]^{\mathbf{B,B}}[x]^{\mathbf{B}}.\end{aligned}$$

Similarly, using Exer. 1.1 and Exer. 1.2, for each $(x, y) \in \mathbf{V}^2$

$$\begin{aligned}\langle x, A^*y \rangle &= ([A^*y]^{\mathbf{B}})^{H}[x]^{\mathbf{B}} \\ &= ([A^*]^{\mathbf{B,B}}[y]^{\mathbf{B}})^{H}[x]^{\mathbf{B}} \\ &= ([y]^{\mathbf{B}})^{H}([A^*]^{\mathbf{B,B}})^{H}[x]^{\mathbf{B}}.\end{aligned}$$

It follows that $([A^*]^{\mathbf{B,B}})^{H} = [A]^{\mathbf{B,B}}$. Taking the Hermitian transpose of both sides and using Exer. 1.1 then yields

$$[A^*]^{\mathbf{B,B}} = ([A]^{\mathbf{B,B}})^{H}.$$ ■

Thus when an orthonormal basis for **V** is used, the matrix representation of the adjoint of $\boldsymbol{A}$ is the Hermitian transpose of the matrix representation of $\boldsymbol{A}$. A matrix which is equal to its own Hermitian transpose is called a *Hermitian matrix*. Thus Prop. 1.1 says that a finite-dimensional linear operator $\boldsymbol{A}$ is self-adjoint if and only if its matrix representation, in an orthonormal coordinate system, is a Hermitian matrix.

7.2. MATRIX REPRESENTATIONS

Because a finite-dimensional linear operator is a special case of a finite-dimensional linear transformation, every finite-dimensional linear operator can be represented by a matrix. In the case of a linear operator $\boldsymbol{A}$ defined on an n-dimensional space **V**, the representation is a square $n \times n$ matrix denoted A. We compute the matrix representation of $\boldsymbol{A}$ as follows.

Proposition 2.1 (Matrix Representation). Let $\boldsymbol{A} \in \mathbf{L}(\mathbf{V}, \mathbf{V})$, and let $\mathbf{B} = \{b^1, b^2, \ldots, b^n\}$ be a basis for **V**. Suppose A is the $n \times n$ matrix whose kth column is defined $a^k = [\boldsymbol{A}b^k]^{\mathbf{B}}$ for $1 \leqslant k \leqslant n$. Then $\mathbf{L}(\mathbf{V}, \mathbf{V})$ is isomorphic to $\mathbf{F}^{\mathbf{n} \times \mathbf{n}}$ and for each $x \in \mathbf{V}$,

$$[\boldsymbol{A}x]^{\mathbf{B}} = A[x]^{\mathbf{B}}. \qquad \blacksquare$$

The proof of Prop. 2.1 follows directly from Prop. 5.2.1 with $\mathbf{W} = \mathbf{V}$ and $\mathbf{C} = \mathbf{B}$. We call $A \in \mathbf{F}^{\mathbf{n} \times \mathbf{n}}$ the *matrix representation* of $\boldsymbol{A}$ induced by the basis **B**, and we write $A = [\boldsymbol{A}]^{\mathbf{B},\mathbf{B}}$. The matrix representation of a finite-dimensional linear operator is general in the sense that Prop. 2.1 is valid for any n-dimensional vector space **V**. However, if the space **V** is an inner product space and we choose an orthonormal basis for **V**, then the matrix representation of $\boldsymbol{A}$ can be simplified somewhat. More specifically, if $A = [\boldsymbol{A}]^{\mathbf{B},\mathbf{B}}$, then the components of A can be expressed explicitly in terms of the basis vectors, as can be seen from the following exercise.

Exercise 2.1 (Orthonormal Representation). Let **V** be an inner product space, and let $\mathbf{B} = \{b^1, b^2, \ldots, b^n\}$ be an orthonormal basis for **V**. Show that for each $\boldsymbol{A} \in \mathbf{L}(\mathbf{V}, \mathbf{V})$ if $A \triangleq [\boldsymbol{A}]^{\mathbf{B},\mathbf{B}}$, then

$$A_{kj} = \langle \boldsymbol{A}b^j, b^k \rangle, \qquad 1 \leqslant k, j \leqslant n. \qquad \blacksquare$$

The problem of transforming from one matrix representation to another can also be viewed as a special case of its more general counterpart for linear transformations.

Proposition 2.2 (Coordinate Transformation). Let $\boldsymbol{A} \in \mathbf{L}(\mathbf{V}, \mathbf{V})$, and let $\mathbf{B} = \{b^1, b^2, \ldots, b^n\}$ and $\mathbf{C} = \{c^1, c^2, \ldots, c^n\}$ be bases for **V**. Suppose P is the

$n \times n$ matrix whose kth column is defined $p^k = [c^k]^{\mathbf{B}}$ for $1 \leqslant k \leqslant n$. Then

$$[A]^{\mathbf{C},\mathbf{C}} = P^{-1}[A]^{\mathbf{B},\mathbf{B}}P.$$

Proof. Applying Prop. 5.3.1 with $\mathbf{E} = \mathbf{C}$ and $\mathbf{D} = \mathbf{B}$,

$$[A]^{\mathbf{C},\mathbf{C}} = Q[A]^{\mathbf{B},\mathbf{B}}P.$$

Here P is as in Prop. 2.2, and Q is the $n \times n$ matrix whose kth column is $q^k = [b^k]^{\mathbf{C}}$ for $1 \leqslant k \leqslant n$. Recalling Prop. 1.10.1, for each $x \in \mathbf{V}$

$$[x]^{\mathbf{B}} = P[x]^{\mathbf{C}},$$
$$[x]^{\mathbf{C}} = Q[x]^{\mathbf{B}}.$$

Thus the coordinate transformation matrix P is invertible and $Q = P^{-1}$. ■

Note that the coordinate transformation problem for finite-dimensional linear operators is somewhat simpler than the more general coordinate transformation problem for finite-dimensional linear transformations as only two bases and one coordinate transformation matrix are involved. Refer to Figure 7.1 for a pictorial description of the relationships between a finite-dimensional linear operator A and its various matrix representations.

Any two square matrices that represent the same linear operator, but in different coordinate systems, are said to be similar. More specifically, we have the following definition.

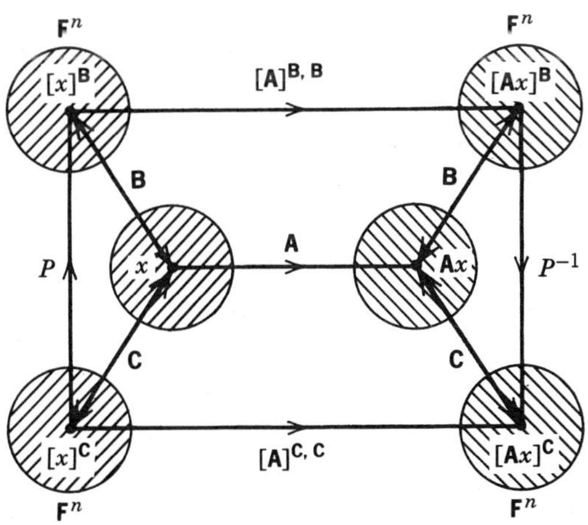

Figure 7.1. Matrix representations of a linear operator A.

Definition 2.1 (Similar Matrices). Let $(A, B) \in \mathbf{F}^{n \times n} \times \mathbf{F}^{n \times n}$. Then A and B are *similar matrices* $\Leftrightarrow$ there exists an invertible $P \in \mathbf{F}^{n \times n}$ such that

$$B = P^{-1}AP. \qquad \blacksquare$$

The transformation from A to B in Def. 2.1 is called a *similarity transformation.* In view of Prop. 2.2, a similarity transformation is in fact a coordinate transformation. If the operator $\boldsymbol{A}$ happens to already be a matrix represented with respect to the standard basis $\boldsymbol{I}$, then the problem of transforming to a new coordinate system $\mathbf{B}$ is particularly simple, as can be seen from the following exercise.

Exercise 2.2 (Standard Basis). Let $A \in \mathbf{F}^{n \times n}$, and let $\mathbf{B} = \{b^1, b^2, \ldots, b^n\}$ be a basis for $\mathbf{F}^n$. Suppose B is the $n \times n$ matrix whose kth column is b^k for $1 \leqslant k \leqslant n$. Show that

$$[A]^{B,B} = B^{-1}AB. \qquad \blacksquare$$

7.3. THE SPECTRUM

Recall from Chapter 6 that each linear operator $\boldsymbol{A} \in \mathbf{L}(\mathbf{V}, \mathbf{V})$ induces a decomposition of the complex plane into the spectrum $\sigma(\boldsymbol{A})$ and the resolvent set $\rho(\boldsymbol{A})$. The spectrum is further decomposed into the point spectrum $\sigma_p(\boldsymbol{A})$, the residual spectrum $\sigma_r(\boldsymbol{A})$, and the continuous spectrum $\sigma_c(\boldsymbol{A})$. If $\boldsymbol{A}$ is finite-dimensional, then from Exer. 6.2.1 both $\sigma_r(\boldsymbol{A})$ and $\sigma_c(\boldsymbol{A})$ are empty. Thus for finite-dimensional linear operators, the spectrum reduces to the point spectrum. The following is therefore an equivalent definition of the spectrum for finite-dimensional linear operators.

Definition 3.1 (Spectrum). Let $\boldsymbol{A} \in \mathbf{L}(\mathbf{V}, \mathbf{V})$ where $\mathbf{V}$ is finite-dimensional. Then $\lambda \in \mathbf{C}$ is an *eigenvalue* of $\boldsymbol{A} \Leftrightarrow \mathbf{N}(\boldsymbol{\lambda I} - \boldsymbol{A}) \neq \{\mathbf{0}\}$. The *spectrum* of $\boldsymbol{A}$, denoted $\sigma(\boldsymbol{A})$, is the set of all eigenvalues of $\boldsymbol{A}$. If $\lambda \in \sigma(\boldsymbol{A})$, then the *geometric multiplicity* of λ is denoted $n(\lambda)$ and is defined

$$n(\lambda) = \dim\{\mathbf{N}(\boldsymbol{\lambda I} - \boldsymbol{A})\}. \qquad \blacksquare$$

Thus the spectrum of $\boldsymbol{A}$ is the set of all complex scalars λ such that $\boldsymbol{B}(\boldsymbol{\lambda}) = \boldsymbol{\lambda I} - \boldsymbol{A}$ is not one-to-one. Note that if $\lambda \in \sigma(\boldsymbol{A})$, then the geometric multiplicity of λ must be greater than or equal to one. When $\lambda \in \sigma(\boldsymbol{A})$, the space $\mathbf{N}(\boldsymbol{\lambda I} - \boldsymbol{A})$ is given a special name. It is called an eigenspace of $\boldsymbol{A}$.

Definition 3.2 (Eigenspace). Let $\boldsymbol{A} \in \mathbf{L}(\mathbf{V}, \mathbf{V})$ where $\mathbf{V}$ is finite-dimensional, and let $\lambda \in \sigma(\boldsymbol{A})$. Then the *eigenspace* of $\boldsymbol{A}$ associated with λ is denoted $\mathbf{G}(\boldsymbol{\lambda})$ and is defined

$$\mathbf{G}(\lambda) \triangleq \mathbf{N}(\lambda I - A).$$

A vector $x \in \mathbf{V}$ is an *eigenvector* of A associated with the eigenvalue $\lambda \Leftrightarrow x \in \mathbf{G}(\lambda) - \{\mathbf{0}\}$. ■

Thus if λ is an eigenvalue of A, then $x \in \mathbf{V}$ is an eigenvector associated with λ if and only if x is nonzero and $x \in \mathbf{N}(\lambda I - A)$. Note that the geometric multiplicity $n(\lambda)$ is simply the dimension of the eigenspace associated with λ. In view of Def. 3.2, the eigenvectors of a linear operator A are special vectors. They are special in the sense that they are *directionally invariant* under A. That is, if $\lambda \in \sigma(A)$ and $x \in \mathbf{G}(\lambda)$, then $Ax = \lambda x$. Hence, apart from scaling by the eigenvalue λ, every eigenvector is its *own image* under A. Refer to Figure 7.2 for a simple geometric interpretation of the directional invariance of an eigenvector of A for the case when $\mathbf{V} = \mathbf{R}^2$.

As an example of a finite-dimensional linear operator and its eigenvalues and eigenvectors, consider the following.

Example 3.1 (Eigenspace). Let $\mathbf{V} = \mathbf{P}^n(\mathbf{R})$, and let A denote the *derivative* operator on $\mathbf{V}$. That is, for each $x \in \mathbf{V}$

$$(Ax)(t) = \frac{d}{dt}\, x(t), \qquad t \in \mathbf{R}.$$

In this case we have $\sigma(A) = \{0\}$ and $\mathbf{G}(0) = \mathbf{P}^0(\mathbf{R})$. That is, every polynomial of degree zero gets mapped into the zero polynomial by the derivative operator. ■

The spectrum of a linear operator is a fundamental attribute of the operator itself. In particular, the spectrum does not depend upon the coordinate system used to represent the operator, as can be seen from the following exercise.

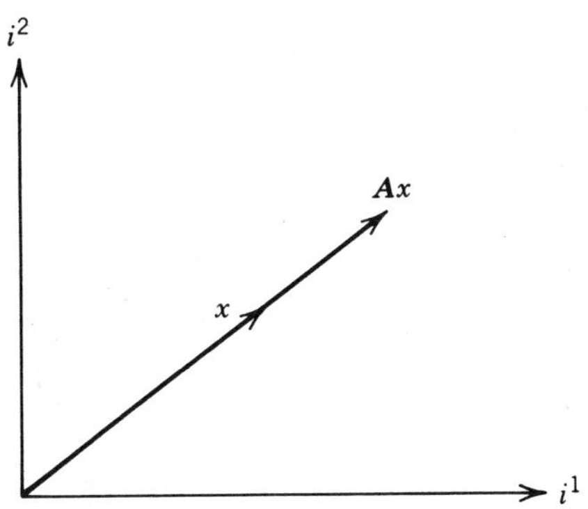

Figure 7.2. Directional invariance of eigenvectors of A.

Exercise 3.1 (Matrix Spectrum). Let $\mathcal{A} \in \mathbf{L}(\mathbf{V}, \mathbf{V})$ where $\mathbf{V}$ is n-dimensional, and let $A \in \mathbf{F}^{\mathbf{n}\times\mathbf{n}}$ be a representation of $\mathcal{A}$. Show that $\sigma(\mathcal{A}) = \sigma(A)$. ■

As similar matrices represent the same linear operator, but in different coordinate systems, it follows from Exer. 3.1 that similar matrices have the same spectrum. Finding the spectrum of a finite-dimensional linear operator $\mathcal{A}$ therefore reduces to the problem of finding the spectrum of any one of its matrix representations. In order to investigate this problem further, we introduce the following familiar functional on the space of $n \times n$ matrices.

Definition 3.3 (Matrix Determinant). Let $A \in \mathbf{F}^{\mathbf{n}\times\mathbf{n}}$. Then the *determinant* of A is a function mapping $\mathbf{F}^{\mathbf{n}\times\mathbf{n}}$ into $\mathbf{F}$ denoted $\det(A)$ defined recursively as follows:

1. $\det(A) = A, \qquad n = 1,$
2. $\det(A) = \sum_{k=1}^{n} (-1)^{k+1} A_{1k} \det(A^{1k}), \qquad n \geqslant 2.$

Here A^{1k} denotes the $(n-1) \times (n-1)$ matrix obtained by eliminating row 1 and column k of A. ■

The expression for the determinant of A in Def. 3.3 is called an expansion of the determinant along row 1. One of the fundamental properties of the determinant is that it is also possible to expand it along any row or, for that matter, any column. Determinants can also be defined in terms of permutation operations. Computing the determinant of an $n \times n$ matrix A requires a total of $n!$ scalar multiplications. Consequently, the determinant function is not particularly attractive from a computational viewpoint. However, the determinant is quite useful from a theoretical one, as we will soon see. Among the numerous qualitative properties that characterize the determinant function, the following two are quite useful for our purposes.

Proposition 3.1 (Determinant Properties). Let $(A, B) \in \mathbf{F}^{\mathbf{n}\times\mathbf{n}} \times \mathbf{F}^{\mathbf{n}\times\mathbf{n}}$. Then the following are properties of the determinant:

1. $\det(A^{\mathrm{T}}) = \det(A)$,
2. $\det(AB) = \det(A)\det(B)$. ■

A proof of Prop. 3.1, along with a discussion of other useful properties of the matrix determinant, can be found in Noble and Deniel (1977). Recall from Exer. 3.1 that similar matrices have the same spectrum. They also have the same determinant, as can be seen from the following exercise.

Exercise 3.2 (Similar Determinant). Let $(A, B) \in \mathbf{F}^{\mathbf{n}\times\mathbf{n}} \times \mathbf{F}^{\mathbf{n}\times\mathbf{n}}$. Show that if A and B are similar, then $\det(A) = \det(B)$. ■

Since similar matrices represent the same linear operator, but in different coordinate systems, it is meaningful in light of Exer. 3.2 to define the determinant of a finite-dimensional linear operator A in terms of the determinant of one of its matrix representations.

Definition 3.4 (Operator Determinant). Let $A \in \mathbf{L}(\mathbf{V}, \mathbf{V})$ where $\mathbf{V}$ is finite-dimensional, and let $A \in \mathbf{F}^{n \times n}$ be a representation of $\boldsymbol{A}$. Then

$$\det(\boldsymbol{A}) \triangleq \det(A).$$

We say $\boldsymbol{A}$ is *singular* $\Leftrightarrow \det(\boldsymbol{A}) = 0$. Otherwise $\boldsymbol{A}$ is *nonsingular*. ■

For finite-dimensional linear operators the notions of invertability and nonsingularity are synonymous. This is a direct consequence of the following result, which is called Laplace's classical expansion of the inverse.

Proposition 3.2 (Laplace). Let $\boldsymbol{A} \in \mathbf{L}(\mathbf{V}, \mathbf{V})$ where $\mathbf{V}$ is finite-dimensional. Then $\boldsymbol{A}$ is invertible $\Leftrightarrow \det(\boldsymbol{A})$ is invertible, in which case we can write

$$\boldsymbol{A}^{-1} \triangleq \{\det(\boldsymbol{A})\}^{-1}\boldsymbol{B}.$$

For a proof of Prop. 3.2 refer to Noble and Deniel (1977). The operator $\boldsymbol{B}$, defined implicitly in Prop. 3.2, is called the *classical adjoint* of $\boldsymbol{A}$. Laplace's expansion of the inverse is important in so much as it provides us with a means of further characterizing the eigenvalues of a finite-dimensional linear operator $\boldsymbol{A}$. To see this we introduce a special polynomial associated with $\boldsymbol{A}$ called the characteristic polynomial.

Definition 3.5 (Characteristic Polynomial). Let $\boldsymbol{A} \in \mathbf{L}(\mathbf{V}, \mathbf{V})$ where $\mathbf{V}$ is finite-dimensional. Then the *characteristic polynomial* of $\boldsymbol{A}$ is defined as

$$\Delta(\lambda) \triangleq \det(\lambda \boldsymbol{I} - \boldsymbol{A}), \qquad \lambda \in \mathbf{C}.$$ ■

If $\boldsymbol{A} \in \mathbf{L}(\mathbf{V}, \mathbf{V})$ where $\mathbf{V}$ is n-dimensional, then Def. 3.5 implies that Δ is in fact a polynomial in $\mathbf{P}^n(\mathbf{C})$. Moreover, the coefficient of the highest power of λ is unity. As such, Δ is called a *monic* polynomial. Finding the zeros of this monic polynomial is tantamount to finding the spectrum of $\boldsymbol{A}$, as the following result shows.

Proposition 3.3 (Spectrum). Let $\boldsymbol{A} \in \mathbf{L}(\mathbf{V}, \mathbf{V})$ where $\mathbf{V}$ is finite-dimensional. Then

$$\sigma(\boldsymbol{A}) = \{\lambda \in \mathbf{C} : \Delta(\lambda) = 0\}.$$

Proof. From Defs. 3.1, 3.5, and Prop. 3.2,

$$\begin{aligned} \lambda \in \sigma(\boldsymbol{A}) &\Leftrightarrow \mathbf{N}(\lambda \boldsymbol{I} - \boldsymbol{A}) \neq \{\mathbf{0}\} \\ &\Leftrightarrow \det(\lambda \boldsymbol{I} - \boldsymbol{A}) = 0 \\ &\Leftrightarrow \Delta(\lambda) = 0. \end{aligned}$$ ■

According to the fundamental theorem of algebra, every polynomial in $\mathbf{P}^n(\mathbf{C})$ has exactly n zeros in $\mathbf{C}$. If we factor the characteristic polynomial into its zeros, we can further characterize the eigenvalues of A as follows.

Definition 3.6 (Eigenvalue Multiplicity). Let $A \in \mathbf{L}(\mathbf{V}, \mathbf{V})$ where $\mathbf{V}$ is finite-dimensional, and let Δ be the characteristic polynomial of A. Then λ_k is an eigenvalue of A of *multiplicity* $m(k)$ for $1 \leqslant k \leqslant q \Leftrightarrow$

$$\Delta(\lambda) = (\lambda - \lambda_1)^{m(1)}(\lambda - \lambda_2)^{m(2)} \cdots (\lambda - \lambda_q)^{m(q)}. \qquad \blacksquare$$

Thus λ is an eigenvalue of A of multiplicity m if and only if λ occurs exactly m times as a zero of the characteristic polynomial of A. The multiplicity of λ is sometimes called the *algebraic multiplicity*. It is distinct from the geometric multiplicity, which is the dimension of the eigenspace $\mathbf{N}(\boldsymbol{\lambda} \boldsymbol{I} - \boldsymbol{A})$. An eigenvalue λ with unit multiplicity is called a *simple eigenvalue*. Otherwise, λ is a referred to as a *multiple eigenvalue*.

Exercise 3.3 (Multiplicity). Let $A \in \mathbf{L}(\mathbf{V}, \mathbf{V})$ where $\mathbf{V}$ is n-dimensional. Let λ_k be an eigenvalue of A of multiplicity $m(k)$ for $1 \leqslant k \leqslant q$. Show that

1. $1 \leqslant q \leqslant n$,
2. $1 \leqslant m(k) \leqslant n$, $1 \leqslant k \leqslant q$,
3. $\sum_{k=1}^{q} m(k) = n$. ■

Given Prop. 3.3, we are now in a position to develop an algorithm for finding the eigenvalues and eigenvectors of a finite-dimensional linear operator A.

Algorithm 3.1 (Eigenvectors)

1. Pick a basis $\mathbf{B} = \{b^1, b^2, \ldots, b^n\}$ for $\mathbf{V}$ and apply Prop. 2.1 to find a matrix representation $A = [\mathbf{A}]^{\mathbf{B},\mathbf{B}}$.
2. Compute the characteristic polynomial $\Delta(\lambda) = \det(\lambda I - A)$ and factor it into its zeros,

$$\Delta(\lambda) = (\lambda - \lambda_1)^{m(1)}(\lambda - \lambda_2)^{m(2)} \cdots (\lambda - \lambda_q)^{m(q)}.$$

3. Set $k = 1$.
4. Apply Alg. 5.5.2 to find a basis $\mathbf{X}^k = \{x^{k1}, x^{k2}, \ldots, x^{kn(k)}\}$ for $\mathbf{N}(\boldsymbol{\lambda}_{\mathbf{k}} \boldsymbol{I} - \boldsymbol{A})$.
5. Compute $\mathbf{C}^{\mathbf{k}} = \{c^{k1}, c^{k2}, \ldots, c^{kn(k)}\}$ from $\mathbf{X}^{\mathbf{k}}$ as follows:

$$c^{kj} = \sum_{i=1}^{n} x_i^{kj} b^i, \qquad 1 \leqslant j \leqslant n(k).$$

6. Set $k = k + 1$. If $k \leqslant q$, go to 4. ■

Proposition 3.4 (Eigenvectors). Let $A \in \mathbf{L(V, V)}$ where $\mathbf{V}$ is n-dimensional. Then the vectors $\mathbf{C^k} = \{c^{k1}, c^{k2}, \ldots, c^{kn(k)}\}$ computed in Alg. 3.1 form a basis for the kth eigenspace $\mathbf{G(\lambda_k)}$ for $1 \leqslant k \leqslant q$.

Proof. From steps 1 and 2 of Alg. 3.1, and Prop. 3.3, it follows that $\sigma(A) = \{\lambda_1, \lambda_2, \ldots \lambda_q\}$. Then from step 4 of Alg. 3.1, $\mathbf{X^k} = \{x^{k1}, x^{k2}, \ldots, x^{kn(k)}\}$ forms a basis for $\mathbf{N(\lambda_k I - A)}$ where $A = [A]^{\mathbf{B,B}}$ with $\mathbf{B} = \{b^1, b^2, \ldots, b^n\}$. Thus for each $1 \leqslant k \leqslant q$ and each $1 \leqslant j \leqslant n(k)$ we have, from step 5,

$$x^{kj} = [c^{kj}]^{\mathbf{B}}.$$

Hence $\mathbf{C^k} = \{c^{k1},\ c^{k2},\ \ldots,\ c^{kn(k)}\}$ is a basis for the eigenspace $\mathbf{G(\lambda_k)} = \mathbf{N}(\lambda_k I - A)$ for $1 \leqslant k \leqslant q$. ∎

As an example of an application of Alg. 3.1 to compute the eigenvalues and eigenvectors of a finite-dimensional linear operator, consider the following example.

Example 3.2 (Eigenvectors). Let $\mathbf{V} = \mathbf{P^2(R)}$, and let A be the following linear operator on $\mathbf{V}$. For each $x \in \mathbf{V}$

$$(Ax)(t) = \frac{d^2}{dt^2}\, x(t) - x(t), \qquad t \in \mathbf{R}.$$

Let $\mathbf{B} = \{b^1, b^2, b^3\}$ be the basis of elementary polynomials on $\mathbf{V}$. That is, $b^k(t) = t^{k-1}$ for $1 \leqslant k \leqslant 3$. If $A = [A]^{\mathbf{B,B}}$, then from Prop. 2.1 we have $a^k = [Ab^k]^{\mathbf{B}}$ for $1 \leqslant k \leqslant 3$. Now $Ab^1 = -b^1$, $Ab^2 = -b^2$, and $Ab^3 = 2b^1 - b^3$. Thus

$$A = \begin{bmatrix} -1 & 0 & 2 \\ 0 & -1 & 0 \\ 0 & 0 & -1 \end{bmatrix}.$$

Next to find the spectrum of A, we compute and factor the characteristic polynomial of A. Thus

$$\begin{aligned} \Delta(\lambda) &= \det(\lambda I - A) \\ &= (\lambda + 1)^3 \\ &\triangleq (\lambda - \lambda_1)^{m(1)}. \end{aligned}$$

Hence $\sigma(A) = \{-1\}$ with $\lambda_1 = -1$ and $m(1) = 3$. To find a basis $\mathbf{X_1}$ for

$\mathbf{N(\lambda_1 I - A)}$, we apply Alg. 5.5.2 using row reduction

$$B(\lambda_1) = [(\lambda_1 I - A)^T, I]$$

$$= \begin{bmatrix} 0 & 0 & 0 & 1 & 0 & 0 \\ 0 & 0 & 0 & 0 & 1 & 0 \\ -2 & 0 & 0 & 0 & 0 & 1 \end{bmatrix}$$

$$\rightarrow \begin{bmatrix} -2 & 0 & 0 & 0 & 0 & 1 \\ 0 & 0 & 0 & 0 & 1 & 0 \\ 0 & 0 & 0 & 1 & 0 & 0 \end{bmatrix}$$

$$\rightarrow \left[\begin{array}{ccc|ccc} 1 & 0 & 0 & 0 & 0 & -\frac{1}{2} \\ \hline 0 & 0 & 0 & 0 & 1 & 0 \\ 0 & 0 & 0 & 1 & 0 & 0 \end{array}\right].$$

Thus $n(1) = 2$ and $\mathbf{X}^1 = \{x^{11}, x^{12}\}$ is a basis for $\mathbf{N(\lambda_1 I - A)}$ where $x^{11} = [0, 1, 0]^T$ and $x^{12} = [1, 0, 0]^T$. From Alg. 3.1, we then have

$$c^{11}(t) = b^2(t) = t.$$

$$c^{12}(t) = b^1(t) = 1.$$

It follows that $\mathbf{C}^1 = \{t, 1\}$ is a basis for the eigenspace $\mathbf{G(\lambda_1)}$. Hence $\sigma(A) = \{-1\}$ and $\mathbf{G(-1)} = \mathbf{P^1(R)}$. That is, each polynomial in $\mathbf{P^1(R)}$ is its own image under the operator A, apart from scaling by -1. ■

The composition of a linear operator with itself is again a linear operator. This fact allows us to define a polynomial of an operator.

Definition 3.7 (Polynomial of *A*). Let $A \in \mathbf{L(V, V)}$, and let $p \in \mathbf{P^m(F)}$. If $a \in \mathbf{F}^{m+1}$ is the vector of coefficients of p, then the operator $\mathbf{p}(A) \in \mathbf{L(V, V)}$ is defined

$$p(A) \triangleq \sum_{k=0}^{m} a_{k+1} A^k.$$ ■

Here $A^0 \triangleq I$ by convention. Since the characteristic polynomial Δ is an example of a polynomial, it should be possible to use Def. 3.7 to compute $\Delta(A)$. Interestingly enough, every finite-dimensional linear operator A is a *zero* of its own characteristic polynomial. This important result is called the Cayley–Hamilton theorem.

Proposition 3.5 (Cayley–Hamilton). Let $A \in \mathbf{L(V, V)}$ where $\mathbf{V}$ is finite-dimensional. If Δ denotes the characteristic polynomial of A then

$$\Delta(A) = 0.$$

Proof. Let $\mathbf{B}(\lambda)$ denote the classical adjoint of the operator $\lambda\mathbf{I} - \mathbf{A}$. Then from Prop. 3.2,

$$\mathbf{B}(\lambda) = \Delta(\lambda)(\lambda\mathbf{I} - \mathbf{A})^{-1}.$$

Suppose $\mathbf{V}$ is n-dimensional. Then since Δ is a polynomial of degree n, there exist operators $\{\mathbf{B}_1, \mathbf{B}_2, \ldots, \mathbf{B}_n\}$ such that

$$\mathbf{B}(\lambda) = \sum_{k=1}^{n} \lambda^{k-1}\mathbf{B}_k.$$

Let $a \in \mathbf{F}^{n+1}$ be the vector of coefficients of the characteristic polynomial Δ. Then

$$\begin{aligned}
\left(\sum_{k=0}^{n} a_{k+1}\lambda^k\right)\mathbf{I} &= \Delta(\lambda)\mathbf{I} \\
&= \Delta(\lambda)(\lambda\mathbf{I} - \mathbf{A})^{-1}(\lambda\mathbf{I} - \mathbf{A}) \\
&= \mathbf{B}(\lambda)(\lambda\mathbf{I} - \mathbf{A}) \\
&= \left(\sum_{k=1}^{n} \lambda^{k-1}\mathbf{B}_k\right)(\lambda\mathbf{I} - \mathbf{A}) \\
&= \sum_{k=1}^{n} \lambda^{k}\mathbf{B}_k - \sum_{k=1}^{n} \lambda^{k-1}\mathbf{B}_k\mathbf{A} \\
&= \sum_{k=2}^{n+1} \lambda^{k-1}\mathbf{B}_{k-1} - \sum_{k=1}^{n} \lambda^{k-1}\mathbf{B}_k\mathbf{A} \\
&= \lambda^n\mathbf{B}_n + \left\{\sum_{k=2}^{n} \lambda^{k-1}(\mathbf{B}_{k-1} - \mathbf{B}_k\mathbf{A})\right\} - \mathbf{B}_1\mathbf{A}.
\end{aligned}$$

Since this must hold for each $\lambda \in \rho(A)$, it follows that

$$\begin{aligned}
a_1\mathbf{I} &= -\mathbf{B}_1\mathbf{A}, \\
a_k\mathbf{I} &= \mathbf{B}_{k-1} - \mathbf{B}_k\mathbf{A}, \qquad 2 \leqslant k \leqslant n, \\
a_{n+1}\mathbf{I} &= \mathbf{B}_n.
\end{aligned}$$

Finally, the characteristic polynomial, evaluated at $\mathbf{A}$, is

$$\begin{aligned}
\Delta(\mathbf{A}) &= \sum_{k=0}^{n} a_{k+1}\mathbf{A}^k \\
&= a_1\mathbf{I} + \left\{\sum_{k=1}^{n-1} a_{k+1}\mathbf{A}^k\right\} + a_{n+1}\mathbf{A}^n
\end{aligned}$$

$$= -B_1A + \left\{\sum_{k=1}^{n-1} (B_k - B_{k+1}A)A^k\right\} + B_nA^n$$

$$= -B_1A + (B_1 - B_2A)A + \cdots + (B_{n-1} - B_nA)A^{n-1} + B_nA^n$$

$$= 0.$$ ■

A polynomial which maps an operator A into the zero operator is said to *annihilate* A. Thus the Cayley–Hamilton theorem says that every finite-dimensional linear operator gets annihilated by its own characteristic polynomial. The characteristic polynomial is not necessarily the polynomial of least degree with this property. The monic polynomial m_Δ of least degree which annihilates A is called the *minimal polynomial* of A.

Exercise 3.4 (Minimal Polynomial). Let $A \in \mathbf{L}(\mathbf{V}, \mathbf{V})$ where $\mathbf{V}$ is finite-dimensional, and let Δ and m_Δ denote the characteristic and minimal polynomials of A, respectively. Find examples of linear operators such that

1. $\Delta = m_\Delta$,
2. $\Delta \neq m_\Delta$. ■

Exercise 3.5. Let $A \in \mathbf{L}(\mathbf{V}, \mathbf{V})$ where $\mathbf{V}$ is n-dimensional, and let $p \in \mathbf{P}(\mathbf{F})$. Show that there exists a polynomial $g \in \mathbf{P}^{n-1}(\mathbf{F})$ such that $p(A) = g(A)$. ■

7.4. INVARIANT SUBSPACES

To investigate the internal structure of a linear operator $A \in \mathbf{L}(\mathbf{V}, \mathbf{V})$, it is useful to consider the restriction of A to various subspaces of the space $\mathbf{V}$.

Definition 4.1 (Restriction). Let $A \in \mathbf{L}(\mathbf{V}, \mathbf{V})$, and let $\mathbf{U}$ be a subspace of $\mathbf{V}$. Then the *restriction* of A to $\mathbf{U}$ is denoted A_u and defined

$$A_u x \triangleq Ax, \qquad x \in \mathbf{U}.$$ ■

Thus the restriction A_u is actually a linear transformation from $\mathbf{U}$ into $\mathbf{V}$. On the subspace $\mathbf{U}$, the operator A and its restriction A_u are synonymous. However, outside the subspace $\mathbf{U}$, the transformation $A_u \in \mathbf{L}(\mathbf{U}, \mathbf{V})$ is not defined. An alternative way to view the restriction of an operator to a subspace is in terms of the dual concept, an extension. That is, we can say that the operator A is an *extension* of the transformation A_u to the entire space $\mathbf{V}$. Refer to Figure 7.3 for a pictorial description of the restriction of A to $\mathbf{U}$.

There are certain cases when the restriction of an operator to a subspace is itself an operator on that subspace. Here the relationship between the operator and the subspace is said to be one of invariance in the following sense.

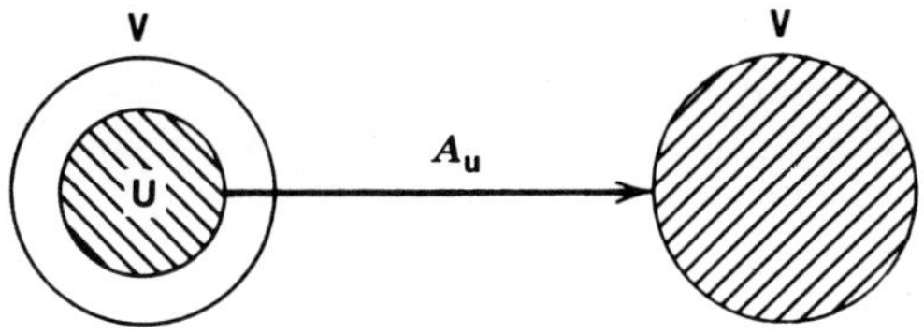

Figure 7.3. The restriction of A to a subspace **U**.

Definition 4.2 (Invariant Subspace). Let $A \in \mathbf{L}(\mathbf{V}, \mathbf{V})$, and let **U** be a subspace of **V**. Then **U** is *invariant* under $A \Leftrightarrow$

$$x \in \mathbf{U} \Rightarrow Ax \in \mathbf{U}. \qquad \blacksquare$$

Thus if **U** is invariant under A, then the image of any vector in **U** under A is again a vector in **U**. Hence the restriction of A to **U** is an operator on **U**. Refer to Figure 7.4 for a pictorial description of an invariant subspace of an operator.

Exercise 4.1 (Invariant Subspaces). Let $A \in \mathbf{L}(\mathbf{V}, \mathbf{V})$. Show that the subspaces $\mathbf{N}(A)$ and $\mathbf{R}(A)$ are invariant under A. ■

The notions of direct sums and invariant subspaces can now be combined. The combination allows us to develop a simplified representation for a linear operator A. It is here that the essential features of the internal structure of A begin to become clearer.

Proposition 4.1 (Direct Sums). Let $A \in \mathbf{L}(\mathbf{V}, \mathbf{V})$ where **V** is finite-dimensional. Suppose $\mathbf{V} = \mathbf{U} \oplus \mathbf{W}$ where the subspaces **U** and **W** are invariant under A. Let $A_u = [A_\mathbf{u}]^{\mathbf{B},\mathbf{B}}$ and $A_w = [A_\mathbf{w}]^{\mathbf{C},\mathbf{C}}$ where **B** and **C** are bases for **U** and **W**, respectively. Then $\mathbf{D} = \{\mathbf{B}, \mathbf{C}\}$ is a basis for **V** and if $A = [A]^{\mathbf{D},\mathbf{D}}$ then

$$A = \begin{bmatrix} A_u & 0 \\ 0 & A_w \end{bmatrix}.$$

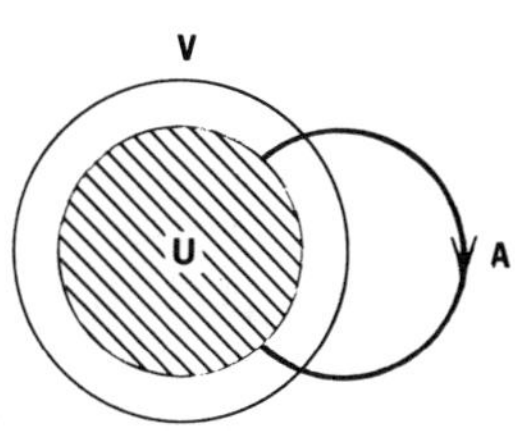

Figure 7.4. A subspace **U** that is invariant under the operator A.

Proof. Suppose $\mathbf{V}$ is n-dimensional and $\mathbf{U}$ is m-dimensional. Let $\mathbf{B} = \{d^1, d^2, \ldots, d^m\}$ and $\mathbf{C} = \{d^{m+1}, d^{m+2}, \ldots, d^n\}$ be bases for $\mathbf{U}$ and $\mathbf{W}$, respectively. Then since $\mathbf{V} = \mathbf{U} \oplus \mathbf{W}$, it follows that $\mathbf{D} = \{\mathbf{B}, \mathbf{C}\} = \{d^1, d^2, \ldots, d^n\}$ is a basis for $\mathbf{V}$. If $A = [A]^{\mathbf{D},\mathbf{D}}$, then from Prop. 2.1

$$a^k = [Ad^k]^{\mathbf{D}}, \qquad 1 \leqslant k \leqslant n.$$

Since $\mathbf{U}$ is invariant under A, it follows that $Ad^k \in \mathbf{U}$ for $1 \leqslant k \leqslant m$. Thus for $1 \leqslant k, j \leqslant m$

$$\begin{aligned} a_j^k &= [Ad^k]_j^{\mathbf{D}} \\ &= [Ad^k]_j^{\mathbf{B}} \\ &= [A_{\mathbf{u}}d^k]_j^{\mathbf{B}}. \end{aligned}$$

Furthermore, since $Ad^k \notin \mathbf{W}$ for $1 \leqslant k \leqslant m$, it follows that $a_j^k = 0$ for $1 \leqslant k \leqslant m$ and $m + 1 \leqslant j \leqslant n$. Next, since $\mathbf{W}$ is invariant under A, it follows that $Ad^k \in \mathbf{W}$ for $m + 1 \leqslant k \leqslant n$. Thus for $m + 1 \leqslant k, j \leqslant n$,

$$\begin{aligned} a_j^k &= [Ad^k]_j^{\mathbf{D}} \\ &= [Ad^k]_j^{\mathbf{C}} \\ &= [A_{\mathbf{w}}d^k]_j^{\mathbf{C}}. \end{aligned}$$

Furthermore, since $Ad^k \notin \mathbf{U}$ for $m + 1 \leqslant k \leqslant n$, it follows that $a_j^k = 0$ for $m + 1 \leqslant k \leqslant n$ and $1 \leqslant j \leqslant m$. Thus if $A_u = [A_{\mathbf{u}}]^{\mathbf{B},\mathbf{B}}$, $A_w = [A_{\mathbf{w}}]^{\mathbf{C},\mathbf{C}}$, and $A = [A]^{\mathbf{D},\mathbf{D}}$, then

$$A = \begin{bmatrix} A_u & 0 \\ 0 & A_w \end{bmatrix}. \qquad \blacksquare$$

In view of Prop. 4.1, it is worthwhile to search for an invariant direct sum decomposition of the space $\mathbf{V}$ that can be associated with each linear operator A. With this in mind, we generalize the notion of an eigenvector in the following manner.

Definition 4.3 (Generalized Eigenvector). Let $A \in \mathbf{L}(\mathbf{V}, \mathbf{V})$ where $\mathbf{V}$ is finite-dimensional, and let $\lambda \in \sigma(A)$ be an eigenvalue of A of multiplicity m. Then the *generalized eigenspace* of A associated with λ is denoted $\mathbf{H}(\lambda)$ and is defined

$$\mathbf{H}(\lambda) \triangleq \mathbf{N}\{(\lambda I - A)^{m}\}.$$

A vector x is a *generalized eigenvector* of A associated with $\lambda \Leftrightarrow x \in \mathbf{H}(\lambda) - \{\mathbf{0}\}$. $\blacksquare$

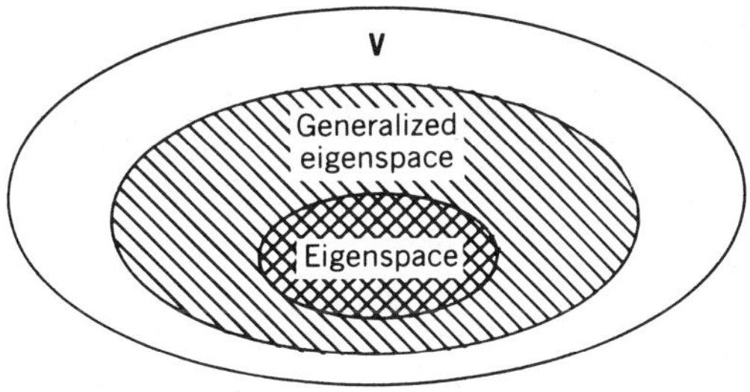

Figure 7.5. Eigenspaces and generalized eigenspaces of A.

Clearly, for each $\lambda \in \sigma(A)$ the eigenspace $\mathbf{G}(\boldsymbol{\lambda})$ is a subspace of the generalized eigenspace $\mathbf{H}(\boldsymbol{\lambda})$. That is, every eigenvector is also a generalized eigenvector. For simple eigenvalues, the generalized eigenspace is the eigenspace, but for multiple eigenvalues the generalized eigenspace sometimes includes vectors that are not eigenvectors. Refer to Figure 7.5 for a pictorial description of the relationship between the eigenspaces and generalized eigenspaces of a linear operator.

Exercise 4.2 (Generalized Eigenspace). Let $A \in \mathbf{L}(\mathbf{V}, \mathbf{V})$ where $\mathbf{V}$ is finite-dimensional. Show that for each $\lambda \in \sigma(A)$ the generalized eigenspace, $\mathbf{H}(\boldsymbol{\lambda})$, is invariant under A. ■

For each finite-dimensional linear operator $A \in \mathbf{L}(\mathbf{V}, \mathbf{V})$, the vector space $\mathbf{V}$ can be decomposed into the direct sum of the generalized eigenspaces of A. This useful result is called the decomposition theorem.

Proposition 4.2 (Decomposition). Let $A \in \mathbf{L}(\mathbf{V}, \mathbf{V})$ where $\mathbf{V}$ is finite-dimensional. If $\sigma(A) = \{\lambda_1, \lambda_2, \ldots, \lambda_q\}$, and $\mathbf{H}(\boldsymbol{\lambda}_k)$ is the generalized eigenspace of A associated with λ_k, then

$$\mathbf{V} = \mathbf{H}(\boldsymbol{\lambda}_1) \oplus \mathbf{H}(\boldsymbol{\lambda}_2) \oplus \cdots \oplus \mathbf{H}(\boldsymbol{\lambda}_q).$$

Proof. Let Δ be the characteristic polynomial of A. That is,

$$\Delta(\lambda) = (\lambda - \lambda_1)^{m(1)}(\lambda - \lambda_2)^{m(2)} \cdots (\lambda - \lambda_q)^{m(q)}.$$

We can expand $1/\Delta(\lambda)$ into partial fractions as follows:

$$\frac{1}{\Delta(\lambda)} = \sum_{k=1}^{q} \sum_{j=1}^{m(k)} \frac{R_{kj}(\lambda_k)}{(\lambda - \lambda_k)^j}.$$

Here the coefficients for the terms of the partial fraction expansion for $1 \leqslant k \leqslant q$ and $0 \leqslant j \leqslant (m(k) - 1)$ are

$$R_{k,m(k)-j} = \frac{1}{j!} \frac{d^j}{d\lambda^j} \frac{(\lambda - \lambda_k)^{m(k)}}{\Delta(\lambda)}$$

For each $1 \leqslant k \leqslant q$, we define three functions f_k, g_k, and h_k as follows:

$$f_k(\lambda) \triangleq \sum_{j=1}^{m(k)} (\lambda - \lambda_k)^{m(k)-j} R_{kj}(\lambda_k),$$

$$g_k(\lambda) \triangleq \frac{\Delta(\lambda)}{(\lambda - \lambda_k)^{m(k)}},$$

$$h_k(\lambda) \triangleq f_k(\lambda) g_k(\lambda).$$

From the partial fraction expansion of $1/\Delta(\lambda)$,

$$\begin{aligned}
\sum_{k=1}^{q} h_k(\lambda) &= \sum_{k=1}^{q} f_k(\lambda) g_k(\lambda) \\
&= \Delta(\lambda) \sum_{k=1}^{q} \frac{f_k(\lambda)}{(\lambda - \lambda_k)^{m(k)}} \\
&= \Delta(\lambda) \sum_{k=1}^{q} \sum_{j=1}^{m(k)} \frac{R_{kj}(\lambda_k)}{(\lambda - \lambda_k)^{j}} \\
&= \frac{\Delta(\lambda)}{\Delta(\lambda)} \\
&= 1.
\end{aligned}$$

Thus $\sum_{k=1}^{q} \mathbf{h}_k(A) = I$. Let $x^k \triangleq \mathbf{h}_k(A)x$ for $1 \leqslant k \leqslant q$. Then for each $x \in \mathbf{V}$,

$$\begin{aligned}
x &= Ix \\
&= \left\{ \sum_{k=1}^{q} \mathbf{h}_k(A) \right\} x \\
&= \sum_{k=1}^{q} x^k.
\end{aligned}$$

Recalling the definitions of f_k, g_k, and h_k and using the Cayley–Hamilton theorem,

$$\begin{aligned}
(\lambda_k I - A)^{m(k)} x^k &= (\lambda_k I - A)^{m(k)} \mathbf{h}_k(A) x \\
&= (\{(\lambda_k - \lambda)^{m(k)} h_k(\lambda)\}|_{\lambda = A}) x \\
&= (\{(\lambda_k - \lambda)^{m(k)} f_k(\lambda) g_k(\lambda)\}|_{\lambda = A}) x \\
&= (\{(-1)^{m(k)} f_k(\lambda)(\lambda - \lambda_k)^{m(k)} g_k(\lambda)\}|_{\lambda = A}) x \\
&= (\{(-1)^{m(k)} f_k(\lambda) \Delta(\lambda)\}|_{\lambda = A}) x \\
&= (-1)^{m(k)} \mathbf{f}_k(A) \Delta(A) x \\
&= 0.
\end{aligned}$$

It follows that $x^k \in \mathbf{H}(\lambda_k)$ for $1 \leqslant k \leqslant q$. It remains to demonstrate that this decomposition of x is unique. Suppose there exist $y^k \in \mathbf{H}(\lambda_k)$ for $1 \leqslant k \leqslant q$ such that

$$x = \sum_{k=1}^{q} y^k.$$

Let $z^k \triangleq x^k - y^k$ for $1 \leqslant k \leqslant q$. Then

$$\begin{aligned}\sum_{k=1}^{q} z^k &= \sum_{k=1}^{q} (x^k - y^k) \\ &= \sum_{k=1}^{q} x^k - \sum_{k=1}^{q} y^k \\ &= x - x \\ &= 0.\end{aligned}$$

It follows that for each $1 \leqslant j \leqslant q$

$$z^j = -\sum_{k \neq j} z^k.$$

Because $z^k \in \mathbf{H}(\lambda_k)$ for $k \neq j$, we have

$$\begin{aligned}\mathbf{g}_j(A)z^j &= -\sum_{k \neq j} \mathbf{g}_j(A)z^k \\ &= 0.\end{aligned}$$

The polynomials $g_j(\lambda)$ and $(\lambda - \lambda_j)^{m(j)}$ have no common factors. Hence by Euclid's algorithm, there exist polynomials p and q such that

$$p(\lambda)g_j(\lambda) + q(\lambda)(\lambda - \lambda_j)^{m(j)} = 1.$$

It then follows that

$$\begin{aligned}z^j &= Iz^j \\ &= \{\mathbf{p}(A)\mathbf{g}_j(A) + \mathbf{q}(A)(A - \lambda_j I)^{\mathbf{m(j)}}\}z^j \\ &= \mathbf{p}(A)\mathbf{g}_j(A)z^j + (-1)^{m(j)}\mathbf{q}(A)(\lambda_j I - A)^{\mathbf{m(j)}}z^j \\ &= 0.\end{aligned}$$

Thus $z^j = 0$ for $1 \leqslant j \leqslant q$. Hence $y^k = x^k$ for $1 \leqslant k \leqslant q$. Therefore the decomposition of $\mathbf{V}$ is unique, and

$$\mathbf{V} = \mathbf{H}(\lambda_1) \oplus \mathbf{H}(\lambda_2) \oplus \cdots \oplus \mathbf{H}(\lambda_q). \qquad \blacksquare$$

Exercise 4.3 (Linear Independence). Show that generalized eigenvectors associated with distinct eigenvalues are linearly independent. ■

7.5. BLOCK DIAGONAL REPRESENTATION

The invariant direct sum decomposition of the space **V** can now be put to use to obtain a simplified standard representation for a linear operator on **V**. The structure of this representation is as follows.

Definition 5.1 (Block Diagonal). Let $A \in \mathbf{F}^{n \times n}$. Then A is a *block diagonal matrix* $\Leftrightarrow$ there exist square submatrices $\{A_1, A_2, \ldots, A_p\}$ called *diagonal blocks* such that $A = \text{diag}\{A_1, A_2, \ldots, A_p\}$ where

$$\text{diag}\{A_1, A_2, \ldots, A_p\} \triangleq \begin{bmatrix} A_1 & 0 & \cdots & 0 \\ 0 & A_2 & \cdots & 0 \\ \vdots & \vdots & & \vdots \\ 0 & 0 & \cdots & A_p \end{bmatrix}.$$ ■

The most important consequence of the decomposition theorem is that every finite-dimensional linear operator $\boldsymbol{A}$ can be represented by a matrix with the block diagonal structure.

Proposition 5.1 (Block Diagonal Representation). Let $\boldsymbol{A} \in \mathbf{L}(\mathbf{V}, \mathbf{V})$ where **V** is finite-dimensional, and let $\sigma(\boldsymbol{A}) = \{\lambda_1, \lambda_2, \ldots, \lambda_q\}$. Let $\boldsymbol{A}_k$ be the restriction of $\boldsymbol{A}$ to the generalized eigenspace $\mathbf{H}(\boldsymbol{\lambda}_k)$, and let A_k be the representation of $\boldsymbol{A}_k$ with respect to a basis $\mathbf{B}_k$ for $1 \leqslant k \leqslant q$. Then $\mathbf{B} = \{\mathbf{B}_1, \mathbf{B}_2, \ldots, \mathbf{B}_q\}$ is a basis for **V** and if $A = [\boldsymbol{A}]^{\mathbf{B},\mathbf{B}}$,

$$A = \text{diag}\{A_1, A_2, \ldots, A_q\}.$$

Proof. Let **U** and **W** be the following subspaces of **V**:

$$\mathbf{U} = \mathbf{H}(\boldsymbol{\lambda}_1),$$
$$\mathbf{W} = \mathbf{H}(\boldsymbol{\lambda}_2) \oplus \cdots \oplus \mathbf{H}(\boldsymbol{\lambda}_q).$$

Then from the decomposition theorem, Prop. 4.2, we have $\mathbf{V} = \mathbf{U} \oplus \mathbf{W}$. From Exer. 4.2, the subspaces **U** and **W** are invariant under $\boldsymbol{A}$. It then follows from Prop. 4.1 that

$$A = \text{diag}\{A_1, A_w\}.$$

Here A_1 is the representation of the restriction $\boldsymbol{A}_1$ induced by the basis $\mathbf{B}_1$. Similarly, A_w is the representation of the restriction of $\boldsymbol{A}$ to **W** induced by the basis

$$\mathbf{B_w} = \{\mathbf{B_2}, \mathbf{B_3}, \ldots, \mathbf{B_q}\}.$$

Clearly, the same procedure which was applied to $A \in \mathbf{L(V, V)}$ can now be applied to $A_w \in \mathbf{L(W, W)}$. Repeating this process q − 1 times yields

$$A = \text{diag}\{A_1, A_2, \ldots, A_q\}. \qquad \blacksquare$$

Thus every finite-dimensional linear operator A admits a block diagonal representation. This block diagonal representation has as many diagonal blocks as A has distinct eigenvalues. The size of the kth diagonal block is the dimension of the kth generalized eigenspace.

Exercise 5.1 (Characteristic Polynomial). Let $A \in \mathbf{F}^{n \times n}$ be block diagonal with diagonal blocks $\{A_1, A_2, \ldots, A_p\}$. Show that if Δ and Δ_k denote the characteristic polynomials of A and A_k, respectively, then

$$\Delta(\lambda) = \Delta_1(\lambda)\Delta_2(\lambda) \cdots \Delta_p(\lambda). \qquad \blacksquare$$

Exercise 5.2 (Multiplicity). Let $A \in \mathbf{L(V, V)}$ where $\mathbf{V}$ is finite-dimensional, and let λ_k be an eigenvalue of A of multiplicity $m(k)$. Show that

$$\dim\{\mathbf{H}(\lambda_k)\} = m(k). \qquad \blacksquare$$

In view of Exer. 5.2, the size of the kth diagonal block of the block diagonal representation of A is equal to the multiplicity of the kth eigenvalue of A. Proposition 5.1 is a constructive result in the sense that it not only establishes the existence of a block diagonal representation for A, but it also provides a means for constructing this representation.

Algorithm 5.1 (Block Diagonal Representation)

1. Pick a basis $\mathbf{B} = \{b^1, b^2, \ldots, b^n\}$ for $\mathbf{V}$ and apply Prop. 2.1 to find a matrix representation $A = [A]^{\mathbf{B},\mathbf{B}}$.
2. Compute the characteristic polynomial $\Delta(\lambda) = \det(\lambda I - A)$ and factor it into its zeros:

$$\Delta(\lambda) = (\lambda - \lambda_1)^{m(1)}(\lambda - \lambda_2)^{m(2)} \cdots (\lambda - \lambda_q)^{m(q)}.$$

3. Set $k = 1$.
4. Apply Alg. 5.5.2 to find a basis $\mathbf{x}^k = \{x^{k1}, x^{k2}, \ldots, x^{km(k)}\}$ for $\mathbf{N}\{(\lambda_k \mathbf{I} - \mathbf{A})^{m(k)}\}$.
5. Compute $\mathbf{C}^k = \{c^{k1}, c^{k2}, \ldots, c^{km(k)}\}$ from $\mathbf{X}^k$ as follows:

$$c^{kj} = \sum_{m=1}^{n} x_m^{kj} b^m, \qquad 1 \leqslant j \leqslant m(k).$$

6. Set $k = k + 1$. If $k \leqslant q$, go to 4.
7. Set $\mathbf{C} = \{\mathbf{C}^1, \mathbf{C}^2, \ldots, \mathbf{C}^q\}$, and compute $P \in \mathbf{F}^{n \times n}$ where $p^k = [b^k]^{\mathbf{C}}$ for $1 \leqslant k \leqslant n$.
8. Compute $B = [A]^{\mathbf{C},\mathbf{C}}$ using $B = P^{-1}AP$. ■

Proposition 5.2 (Block Diagonalization). Let $A \in \mathbf{L}(\mathbf{V}, \mathbf{V})$ where $\mathbf{V}$ is finite-dimensional. Then the matrix B obtained by applying Alg. 5.1 to A is a block diagonal representation of A with the number of diagonal blocks equal to the number of distinct eigenvalues, and the size of the kth block equal to the multiplicity of the kth eigenvalue.

Proof. From steps 1–4 of Alg. 5.1, $\mathbf{X}^k = \{x^{k1}, x^{k2}, \ldots, x^{km(k)}\}$ is a basis for $\mathbf{N}\{(\lambda_k\mathbf{I} - \mathbf{A})^{m(k)}\}$ where $A = [A]^{\mathbf{B},\mathbf{B}}$ with $\mathbf{B} = \{b^1, b^2, \ldots, b^n\}$. Thus for each $1 \leqslant k \leqslant q$ and each $1 \leqslant j \leqslant m(k)$ we have from step 5

$$x^{kj} = [c^{kj}]^{\mathbf{B}}.$$

Hence $\mathbf{C}^k = \{c^{k1}, c^{k2}, \ldots, c^{km(k)}\}$ is a basis for the generalized eigenspace $\mathbf{H}(\lambda_k) = \mathbf{N}\{(\lambda_k\mathbf{I} - \mathbf{A})^{m(k)}\}$ for $1 \leqslant k \leqslant q$. It then follows from Prop. 4.2 that $\mathbf{C} = \{\mathbf{C}^1, \mathbf{C}^2, \ldots, \mathbf{C}^q\}$ is a basis for $\mathbf{V}$. If $B = [A]^{\mathbf{C},\mathbf{C}}$, then from Prop. 5.1

$$B = \operatorname{diag}(A_1, A_2, \ldots, A_q).$$

Here A_k is the representation of the restriction of A to $\mathbf{H}(\lambda_k)$ induced by the basis $\mathbf{C}^k$ for $1 \leqslant k \leqslant q$. Thus from Exer. 5.2, A_k is a square matrix of order $m(k)$. ■

As an example of an application of Alg. 5.1 to find the block diagonal representation of a finite-dimensional linear operator, consider the following.

Example 5.1 (Block Diagonal Representation). Let $\mathbf{V} = \mathbf{R}^3$, and let A be the following linear operator on $\mathbf{V}$. For each $x \in \mathbf{V}$

$$(Ax)_1 = 5x_1 + 4x_2 + 3x_3,$$

$$(Ax)_2 = -x_1 - 3x_3,$$

$$(Ax)_3 = x_1 - 2x_2 + x_3.$$

If $A \in \mathbf{R}^{3 \times 3}$ is the representation of A induced by the standard basis $\mathbf{B} = \mathbf{I}$, then

$$A = \begin{bmatrix} 5 & 4 & 3 \\ -1 & 0 & -3 \\ 1 & -2 & 1 \end{bmatrix},$$

To determine the spectrum of A, we compute and factor the characteristic polynomial as follows:

$$\begin{aligned}
\Delta(\lambda) &= \det(\lambda I - A) \\
&= (\lambda - 5)\{\lambda(\lambda - 1) - 6\} + 4\{(\lambda - 1) + 3\} - 3(2 + \lambda) \\
&= (\lambda - 5)(\lambda^2 - \lambda - 6) + 4(\lambda + 2) - 3(\lambda + 2) \\
&= (\lambda - 5)(\lambda - 3)(\lambda + 2) + (\lambda + 2) \\
&= (\lambda + 2)\{(\lambda - 5)(\lambda - 3) + 1\} \\
&= (\lambda + 2)(\lambda^2 - 8\lambda + 16) \\
&= (\lambda + 2)(\lambda - 4)^2 \\
&\triangleq (\lambda - \lambda_1)^{m(1)}(\lambda - \lambda_2)^{m(2)}.
\end{aligned}$$

Thus $\sigma(A) = \{-2, 4\}$ with $\lambda_1 = -2$, $m(1) = 1$, and $\lambda_2 = 4$, $m(2) = 2$. To find a basis for the generalized eigenspace associated with $\lambda_1 = -2$, we form

$$(\lambda_1 I - A)^{m(1)} = \begin{bmatrix} -7 & -4 & -3 \\ 1 & -2 & 3 \\ -1 & 2 & -3 \end{bmatrix}.$$

We then apply Alg. 5.2.2 and row reduction to find a basis for the null space of this matrix. Thus

$$\begin{bmatrix} -7 & 1 & -1 & 1 & 0 & 0 \\ -4 & -2 & 2 & 0 & 1 & 0 \\ -3 & 3 & -3 & 0 & 0 & 1 \end{bmatrix} \to \begin{bmatrix} 1 & -1 & 1 & 0 & 0 & -\frac{1}{3} \\ -4 & -2 & 2 & 0 & 1 & 0 \\ -7 & 1 & -1 & 1 & 0 & 0 \end{bmatrix}$$

$$\to \begin{bmatrix} 1 & -1 & 1 & 0 & 0 & -\frac{1}{3} \\ 0 & -6 & 6 & 0 & 1 & -\frac{4}{3} \\ 0 & -6 & 6 & 1 & 0 & -\frac{7}{3} \end{bmatrix}$$

$$\to \begin{bmatrix} 1 & -1 & 1 & 0 & 0 & -\frac{1}{3} \\ 0 & 1 & -1 & 0 & -\frac{1}{6} & \frac{2}{9} \\ 0 & -6 & 6 & 1 & 0 & -\frac{7}{3} \end{bmatrix}$$

$$\to \left[\begin{array}{ccc|ccc} 1 & 0 & 0 & 0 & -\frac{1}{6} & -\frac{1}{9} \\ 0 & 1 & -1 & 0 & -\frac{1}{6} & \frac{2}{9} \\ \hline 0 & 0 & 0 & 1 & -1 & -1 \end{array}\right].$$

Thus $\mathbf{X}^1 = \{x^{11}\} = [1, -1, -1]^T$ is a basis for the null space of $(\lambda_1 I - A)^{m(1)}$.

Next, to find a basis for the generalized eigenspace associated with $\lambda_2 = 4$, we form

$$(\lambda_2 I - A)^{m(2)} = \begin{bmatrix} -1 & -4 & -3 \\ 1 & 4 & 3 \\ -1 & 2 & 3 \end{bmatrix}^2 = \begin{bmatrix} 0 & -18 & -18 \\ 0 & 18 & 18 \\ 0 & 18 & 18 \end{bmatrix}.$$

We then apply Alg. 5.2.2 and row reduction to find a basis for the null space of this matrix. Here

$$\begin{bmatrix} 0 & 0 & 0 & 1 & 0 & 0 \\ -18 & 18 & 18 & 0 & 1 & 0 \\ -18 & 18 & 18 & 0 & 0 & 1 \end{bmatrix} \rightarrow \begin{bmatrix} 1 & -1 & -1 & 0 & 0 & -\frac{1}{18} \\ -18 & 18 & 18 & 0 & 1 & 0 \\ 0 & 0 & 0 & 1 & 0 & 0 \end{bmatrix}$$

$$\rightarrow \left[\begin{array}{ccc:ccc} 1 & -1 & -1 & 0 & 0 & -\frac{1}{18} \\ \hdashline 0 & 0 & 0 & 0 & 1 & -1 \\ 0 & 0 & 0 & 1 & 0 & 0 \end{array}\right].$$

Thus $\mathbf{X}^2 = \{x^{21}, x^{22}\}$ is a basis for the null space of $(\lambda_2 I - A)^{m(2)}$ where $x^{21} = [0, 1, -1]^{\mathrm{T}}$ and $x^{22} = [1, 0, 0]^{\mathrm{T}}$. Since $\mathbf{B} = \mathbf{I}$, we have $c^{kj} = x^{kj}$ for $1 \leqslant k \leqslant 2$ and $1 \leqslant j \leqslant m(k)$. Hence

$$\mathbf{C} = \{x^{11}, x^{21}, x^{22}\}.$$

Now the coordinate transformation matrix P is defined in terms of its columns as $p^k = [c^k]^{\mathbf{B}}$ for $1 \leqslant k \leqslant 3$. Since $\mathbf{B} = \mathbf{I}$, it follows that $P = [x^{11}, x^{21}, x^{22}]$ or

$$P = \begin{bmatrix} 1 & 0 & 1 \\ -1 & 1 & 0 \\ -1 & -1 & 0 \end{bmatrix}.$$

Applying Alg. 5.5.1 to find P^{-1} we form $[P, I]$ and use row reduction. Here

$$\left[\begin{array}{ccc:ccc} 1 & 0 & 1 & 1 & 0 & 0 \\ -1 & 1 & 0 & 0 & 1 & 0 \\ -1 & -1 & 0 & 0 & 0 & 1 \end{array}\right] \rightarrow \begin{bmatrix} 1 & 0 & 1 & 1 & 0 & 0 \\ 0 & 1 & 1 & 1 & 1 & 0 \\ 0 & -1 & 1 & 1 & 0 & 1 \end{bmatrix}$$

$$\rightarrow \begin{bmatrix} 1 & 0 & 1 & 1 & 0 & 0 \\ 0 & 1 & 1 & 1 & 1 & 0 \\ 0 & 0 & 2 & 2 & 1 & 1 \end{bmatrix}$$

$$\rightarrow \begin{bmatrix} 1 & 0 & 1 & 1 & 0 & 0 \\ 0 & 1 & 1 & 1 & 1 & 0 \\ 0 & 0 & 1 & 1 & \frac{1}{2} & \frac{1}{2} \end{bmatrix}$$

$$\rightarrow \left[\begin{array}{ccc|ccc} 1 & 0 & 0 & 0 & -\frac{1}{2} & -\frac{1}{2} \\ 0 & 1 & 0 & 0 & \frac{1}{2} & -\frac{1}{2} \\ 0 & 0 & 1 & 1 & \frac{1}{2} & \frac{1}{2} \end{array}\right].$$

Finally, the block diagonal representation of A is

$$B = P^{-1}AP$$

$$= \begin{bmatrix} 0 & -1 & -1 \\ 0 & 1 & -1 \\ 2 & 1 & 1 \end{bmatrix} \begin{bmatrix} 5 & 4 & 3 \\ -1 & 0 & -3 \\ 1 & -2 & 1 \end{bmatrix} \begin{bmatrix} 1 & 0 & 1 \\ -1 & 1 & 0 \\ -1 & -1 & 0 \end{bmatrix} \left(\frac{1}{2}\right)$$

$$= \begin{bmatrix} 0 & -1 & -1 \\ 0 & 1 & -1 \\ 2 & 1 & 1 \end{bmatrix} \begin{bmatrix} -2 & 1 & 5 \\ 2 & 3 & -1 \\ 2 & -3 & 1 \end{bmatrix} \left(\frac{1}{2}\right)$$

$$= \left[\begin{array}{c|cc} -2 & 0 & 0 \\ \hline 0 & 3 & -1 \\ 0 & 1 & 5 \end{array}\right].$$

■

The block diagonal representation theorem ensures that every finite-dimensional linear operator A can be represented by a block diagonal matrix. The kth diagonal block is a representation of the restriction of A to the kth generalized eigenspace. No constraints have been placed on the way the basis vectors for the generalized eigenspaces are chosen. If the bases for the generalized eigenspaces are selected in the proper way, the representation of A can be simplified still further. In particular, each diagonal block $A(\lambda)$ associated with an eigenvalue λ can itself be represented by a block diagonal matrix of the form $A(\lambda) = \text{diag}\{J_1(\lambda), J_2(\lambda), \ldots, J_r(\lambda)\}$. Here the subblocks $\{J_k(\lambda)\}$ are called *Jordan blocks*. For example, a Jordan block of order 3 is of the form

$$J(\lambda) = \begin{bmatrix} \lambda & 1 & 0 \\ 0 & \lambda & 1 \\ 0 & 0 & \lambda \end{bmatrix}.$$

More generally, a Jordan block of order m associated with λ is an $m \times m$ upper triangular matrix with λ on the diagonal, ones on the first superdiagonal, and zeros elsewhere. A Jordan block of order 1 is simply the scalar λ, viewed as 1×1 matrix. The number of Jordan blocks associated with a given eigenvalue is equal to the number of linearly independent eigenvectors associated with that eigenvalue, that is, the geometric multiplicity of the eigenvalue. The matrix representation based on Jordan blocks is called the Jordan representation or the *Jordan canonical form* (Noble and Deniel, 1977). The simplest special case of the Jordan canonical form occurs when the number of blocks is n. In this case we say the operator A is diagonalizable.

Definition 5.2 (Diagonal Matrix). Let $A \in \mathbf{F}^{\mathbf{n} \times \mathbf{n}}$. Then A is a *diagonal matrix* $\Leftrightarrow A_{kj} = 0$ for $j \neq k$. A finite-dimensional linear operator $\boldsymbol{A}$ is *diagonalizable* $\Leftrightarrow$ it can be represented by a diagonal matrix. ■

Thus a diagonal representation is a block diagonal representation with all of the blocks being 1×1 blocks. A linear operator $\boldsymbol{A}$ is diagonalizable if and only if the generalized eigenspaces are identical to the eigenspaces, as can be seen from the following exercise.

Exercise 5.3 (Diagonalizability). Let $\boldsymbol{A} \in \mathbf{L}(\mathbf{V}, \mathbf{V})$ where $\mathbf{V}$ is finite-dimensional, and let $\sigma(A) = \{\lambda_1, \lambda_2, \ldots, \lambda_q\}$. Show that $\boldsymbol{A}$ is diagonalizable $\Leftrightarrow$

$$\mathbf{H}(\boldsymbol{\lambda}_k) = \mathbf{G}(\boldsymbol{\lambda}_k), \qquad 1 \leqslant k \leqslant q.$$ ■

Recall that for each eigenvalue λ, the eigenspace $\mathbf{G}(\boldsymbol{\lambda})$ is a subspace of the generalized eigenspace $\mathbf{H}(\boldsymbol{\lambda})$. Thus Exer. 5.3 says that if the converse is also true, then and only then is $\boldsymbol{A}$ diagonalizable.

Exercise 5.4 (Diagonalization). Let $\boldsymbol{A} \in \mathbf{L}(\mathbf{V}, \mathbf{V})$ where $\mathbf{V}$ is finite-dimensional, and let A be a block diagonal representation of $\boldsymbol{A}$. Show that $\boldsymbol{A}$ is diagonalizable $\Leftrightarrow A$ is diagonal. ■

In light of Exer. 5.4, Alg. 5.1 provides us with a means of obtaining a diagonal representation of $\boldsymbol{A}$ whenever $\boldsymbol{A}$ is diagonalizable. Moreover, Exer. 5.3 asserts that the basis relative to which this representation is expressed will consist solely of eigenvectors of $\boldsymbol{A}$. There are two important types of linear operators that are always diagonalizable.

Proposition 5.3 (Diagonalizability). Let $\boldsymbol{A} \in \mathbf{L}(\mathbf{V}, \mathbf{V})$ where $\mathbf{V}$ is finite-dimensional. Then $\boldsymbol{A}$ is diagonalizable if either of the following conditions is satisfied:

1. $\boldsymbol{A}$ has only simple eigenvalues,
2. $\boldsymbol{A}$ is normal. ■

The proof of part 1 of Prop. 5.3 follows directly from the observation that the block diagonal representation will have n blocks when the eigenvalues of $\boldsymbol{A}$ are simple. Hence the size of the largest block is 1×1. The interested reader is referred to Noble and Deniel (1977) for a proof of part 2 of Prop. 5.3. Recall that normal operators include self-adjoint operators and unitary operators as special cases. Although normal operators and operators with simple eigenvalues are diagonalizable, there are diagonalizable operators which satisfy neither of these criteria. Refer to Figure 7.6 for a pictorial summary of the family of diagonalizable linear operators.

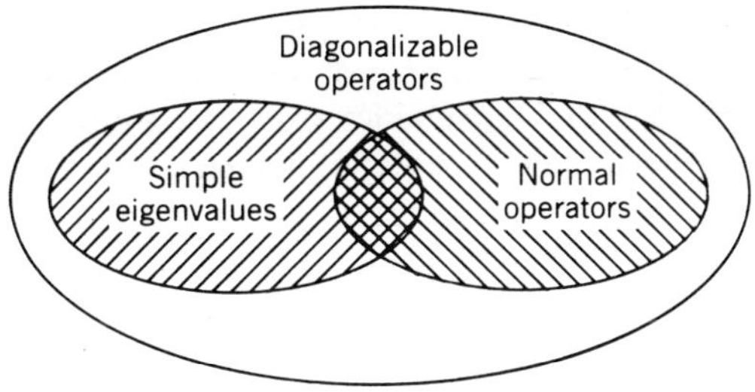

Figure 7.6. Diagonalizable linear operators.

Exercise 5.5. Let $A \in \mathbf{L(V, V)}$ where $\mathbf{V}$ is a finite-dimensional inner product space. Show that if A is self-adjoint, then the eigenvectors of A associated with distinct eigenvalues are orthogonal. ■

7.6. FUNCTIONS OF OPERATORS

Recall that whenever $A \in \mathbf{L(V, V)}$, the composition $A^{\mathrm{k}} \in \mathbf{L(V, V)}$ for each $k \geqslant 0$. It is this observation that allowed us to define the operator $\mathbf{p}(A) \in \mathbf{L(V, V)}$ where p is a polynomial in $\mathbf{P(F)}$. Indeed if Δ is the characteristic polynomial of A, then the Cayley–Hamilton theorem says that $\Delta(A)$ is the zero operator in $\mathbf{L(V, V)}$. We now use this important result to extend the definition of functions of operators to functions other than polynomials. For this purpose we assume that the field $\mathbf{F}$ over which the space $\mathbf{V}$ is defined is the complex plane $\mathbf{C}$. The one constraint we must place on the function $f \in \mathbf{F(C, C)}$ is that it be analytic on $\sigma(A)$.

Definition 6.1 (Analytic). Let $f: \mathbf{C} \to \mathbf{C}$, and let $\lambda \in \mathbf{C}$. Then f is *analytic* at $\lambda \Leftrightarrow f$ is differentiable in an open ball $B(\lambda, r)$ for some $r > 0$. For each $k \geqslant 0$, the kth *derivative* of f at λ is denoted $f^{(k)}(\lambda)$. ■

One of the fundamental features of analytic functions is that they are infinitely differentiable. Indeed, if f is analytic at λ then $f^{(k)}(\lambda)$ not only exists for $k \geqslant 1$, but it can be represented in terms of an integral (Hille, 1973). Given the infinite differentiability of an analytic function f on $\sigma(A)$, we are now in a position to introduce the interpolating polynomial of f on $\sigma(A)$.

Definition 6.2 (Interpolating Polynomial). Let $A \in \mathbf{L(V, V)}$ where $\mathbf{V}$ is n-dimensional, and let Δ be the characteristic polynomial of A. Suppose $\sigma(A) = \{\lambda_1, \lambda_2, \ldots, \lambda_q\}$ where $m(k)$ is the multiplicity of λ_k for $1 \leqslant k \leqslant q$. Let $f: \mathbf{C} \to \mathbf{C}$ be analytic on $\sigma(A)$. Then the *interpolating polynomial* of f on $\sigma(A)$ is the polynomial $g \in \mathbf{P}^{\mathrm{n}-1}(\mathbf{C})$ such that for $1 \leqslant k \leqslant q$ and $1 \leqslant j \leqslant m(k)$

$$g^{((j-1)}(\lambda_k) = f^{(j-1)}(\lambda_k).$$ ■

Note that the definition of the interpolating polynomial involves n equations in the n unknown polynomial coefficients. Thus the interpolating polynomial is unique. In the event that the eigenvalues of A are simple, the interpolating polynomial is simply that the polynomial of degree $n-1$, which agrees with f on $\sigma(A)$. The importance of the interpolating polynomial arises from the following consequence of the Cayley–Hamilton theorem.

Proposition 6.1 (Polynomial of A). Let $A \in \mathbf{L}(\mathbf{V}, \mathbf{V})$ where $\mathbf{V}$ is finite-dimensional, and let $f \in \mathbf{P}(\mathbf{C})$. If g is the interpolating polynomial of f on $\sigma(A)$, then

$$\mathbf{f}(A) = \mathbf{g}(A).$$

Proof. Let $f \in \mathbf{P}(\mathbf{C})$, and let g be the interpolating polynomial of f on $\sigma(A)$. Define $h \triangleq f - g$. Then from Def. 6.2, for each $1 \leqslant k \leqslant q$ and $1 \leqslant j \leqslant m(k)$

$$\begin{aligned} h^{(j-1)}(\lambda_k) &= f^{(j-1)}(\lambda_k) - g^{(j-1)}(\lambda_k) \\ &= 0. \end{aligned}$$

Hence there exists a polynomial p such that

$$\begin{aligned} h(\lambda) &= p(\lambda)(\lambda - \lambda_1)^{m(1)}(\lambda - \lambda_2)^{m(2)} \cdots (\lambda - \lambda_q)^{m(q)} \\ &= p(\lambda)\Delta(\lambda). \end{aligned}$$

It then follows from the Cayley–Hamilton theorem that

$$\begin{aligned} \mathbf{f}(A) &= \mathbf{h}(A) + \mathbf{g}(A) \\ &= \mathbf{p}(A)\boldsymbol{\Delta}(A) + \mathbf{g}(A) \\ &= \mathbf{g}(A). \end{aligned}$$ ■

Prop. 6.1 is a rather remarkable result. It says that *every* polynomial of an n-dimensional linear operation A can be replaced by a polynomial of degree $n-1$. This fact provides us with the following means of extending the definition of $\mathbf{f}(A)$ to functions other than polynomials.

Definition 6.3 (Function of A). Let $A \in \mathbf{L}(\mathbf{V}, \mathbf{V})$ where $\mathbf{V}$ is finite-dimensional, and let $f: \mathbf{C} \to \mathbf{C}$ be analytic on $\sigma(A)$. If g is the interpolating polynomial of f on $\sigma(A)$, then

$$\mathbf{f}(A) \triangleq \mathbf{g}(A).$$ ■

Thus to compute $\mathbf{f}(A)$ we compute $\mathbf{g}(A)$ where g is the interpolating polynomial of f on $\sigma(A)$. Note that in light of Prop. 6.1, Def. 6.3 is consistent with Def. 3.7 and therefore can be regarded as an extension of it.

Exercise 6.1. Let $A \in \mathbf{L(V, V)}$ where $\mathbf{V}$ is finite-dimensional, and let f and g be analytic on $\sigma(A)$. Show that

1. $\mathbf{(f + g)}(A) = \mathbf{f}(A) + \mathbf{g}(A)$,
2. $\mathbf{(fg)}(A) = \mathbf{f}(A)\mathbf{g}(A)$. ■

Given a linear operator A and a function f, which is analytic on $\sigma(A)$, the problem of computing $\mathbf{f}(A)$ reduces to the problem of finding the coefficients of the interpolating polynomial g. The following algorithm serves this purpose.

Algorithm 6.1 (Interpolating Polynomial)

1. Compute the characteristic polynomial of A, and factor it into the form

$$\Delta(\lambda) = (\lambda - \lambda_1)^{m(1)}(\lambda - \lambda_2)^{m(2)} \cdots (\lambda - \lambda_q)^{m(q)}.$$

2. Set $k = 1$.
3. Compute $B^k \in \mathbf{C}^{\mathbf{m(k)} \times \mathbf{n}}$ as follows. $B^k_{ij} = 0$ for $j < i$, and for $j \geqslant i$

$$B^k_{ij} = \frac{d^{i-1}}{d\lambda^{i-1}} \lambda^{j-1}, \qquad \lambda = \lambda_k.$$

4. Compute $b^k \in \mathbf{C}^{\mathbf{m(k)}}$ where for $1 \leqslant j \leqslant m(k)$

$$b^k_j = f^{(j-1)}(\lambda_k).$$

5. Set $k = k + 1$. If $k \leqslant q$, go to 3.
6. Form $C \in \mathbf{C}^{\mathbf{n} \times (\mathbf{n}+1)}$ as follows:

$$C = \begin{bmatrix} B^1 & b^1 \\ B^2 & b^2 \\ \vdots & \vdots \\ B^q & b^q \end{bmatrix}.$$

7. Use Alg. 5.4.1 to find D, the row echelon form of C. Partition D as $D = [I, a]$. ■

Proposition 6.2 (Interpolating Polynomial). The vector $a \in \mathbf{C}^{\mathbf{n}}$ constructed in Alg. 6.1 is the coefficient vector of the interpolating polynomial of f on $\sigma(A)$. That is,

$$\mathbf{f}(A) = \sum_{k=1}^{n} a_k A^{\mathbf{k}-1}.$$

Proof. Let g be the interpolating polynomial of f on $\sigma(A)$. Then from Def. 6.2, for $1 \leqslant k \leqslant q$ and $1 \leqslant i \leqslant m(k)$

$$g^{(i-1)}(\lambda_k) = f^{(i-1)}(\lambda_k).$$

It then follows from steps 3 and 4 of Alg. 6.1 that for $1 \leqslant k \leqslant q$ and $1 \leqslant i \leqslant m(k)$

$$\begin{aligned}
b_i^k &= f^{(i-1)}(\lambda_k) \\
&= g^{(i-1)}(\lambda_k) \\
&= \frac{d^{i-1}}{d\lambda^{i-1}} \sum_{j=1}^{n} a_j \lambda^{j-1}, \qquad \lambda = \lambda_k \\
&= \sum_{j=1}^{n} a_j \frac{d^{i-1}}{d\lambda^{i-1}} \lambda^{j-1}, \qquad \lambda = \lambda_k \\
&= \sum_{j=i}^{n} a_j B_{ij}^k \\
&= \sum_{j=1}^{n} a_j B_{ij}^k \\
&= (B^k a)_i .
\end{aligned}$$

Hence $B^k a = b^k$ for $1 \leqslant k \leqslant q$. Let $B \in \mathbf{C}^{n \times n}$ and $b \in \mathbf{C}^n$ be defined as follows:

$$B = \begin{bmatrix} B^1 \\ B^2 \\ \vdots \\ B^q \end{bmatrix}, \qquad b = \begin{bmatrix} b^1 \\ b^2 \\ \vdots \\ b^q \end{bmatrix}$$

Then $Ba = b$. Because $a \in \mathbf{C}^n$ is the coefficient vector of the interpolating polynomial g, and the interpolating polynomial is unique, it follows that B is invertible and

$$a = B^{-1} b.$$

From step 7 of Alg. 6.1, D is the row echelon form of $C = [B, b]$. Thus there exists a nonsingular matrix E such that

$$\begin{aligned}
D &= E[B, b] \\
&= [EB, Eb].
\end{aligned}$$

Because B is invertible and D is a row echelon matrix, it follows that $EB = I$

or $E = B^{-1}$. Thus

$$\begin{aligned} Eb &= B^{-1}b \\ &= a. \end{aligned}$$ ■

As an example of an application of Alg. 6.1, consider the following linear operator.

Example 6.1 (Function of A). Let $\mathbf{V} = \mathbf{P}^2(\mathbf{C})$, and let A be the following linear operator on $\mathbf{V}$. For each $x \in \mathbf{V}$

$$(Ax)(\lambda) = \frac{d^2}{d\lambda^2} x(\lambda) - x(\lambda), \qquad \lambda \in \mathbf{C}.$$

From Exam. 3.2, the characteristic polynomial of A is

$$\begin{aligned} \Delta(\lambda) &= (\lambda + 1)^3 \\ &\triangleq (\lambda - \lambda_1)^{m(1)}. \end{aligned}$$

Using Alg. 6.1 with $\lambda_1 = -1$ and $m(1) = 3$,

$$B^1 = \begin{bmatrix} 1 & -1 & 1 \\ 0 & 1 & -2 \\ 0 & 0 & 2 \end{bmatrix}.$$

Suppose we compute $\mathbf{f}(A)$ where $f(\lambda) = 1/\lambda$. Now, $\sigma(A) = \{-1\}$. Hence f is analytic on $\sigma(A)$. Using Alg. 6.1, we have $b^1 = [-1, \quad 1, \quad 2]^T$. Thus the matrix $C = [B^1, b^1]$ is

$$C = \begin{bmatrix} 1 & -1 & 1 & -1 \\ 0 & 1 & -2 & -1 \\ 0 & 0 & 2 & -2 \end{bmatrix}.$$

Then using Alg. 5.4.1 to find the row echelon form of C we have

$$C \to \begin{bmatrix} 1 & 0 & -1 & -2 \\ 0 & 1 & -2 & -1 \\ 0 & 0 & 2 & -2 \end{bmatrix} \to \begin{bmatrix} 1 & 0 & -1 & -2 \\ 0 & 1 & -2 & -1 \\ 0 & 0 & 1 & -1 \end{bmatrix}$$

$$\to \begin{bmatrix} 1 & 0 & 0 & -3 \\ 0 & 1 & 0 & -3 \\ 0 & 0 & 1 & -1 \end{bmatrix} = D.$$

Hence the coefficient vector of the interpolating polynomial of f on $\sigma(A)$ is a $[-3, -3, -1]^T$. That is,

$$\begin{aligned}
\mathbf{f}(A) &= a_1 I + a_2 A + a_3 A^2 \\
&= -(3I + 3A + A^2) \\
&= -\left(3 + 3\left(\frac{d^2}{-d\lambda^2} - 1\right) + \left(\frac{d^2}{d\lambda^2} - 1\right)^2\right) \\
&= -\left\{3\frac{d^2}{d\lambda^2} + \left(\frac{d^4}{d\lambda^4} - 2\frac{d^2}{d\lambda^2} + 1\right)\right\} \\
&= -\left(\frac{d^4}{d\lambda^4} + \frac{d^2}{d\lambda^2} + 1\right).
\end{aligned}$$

As a check, let $x \in \mathbf{P}^2(\mathbf{C})$. Then $x(\lambda) = b_1 + b_2\lambda + b_3\lambda^2$ for some $b \in \mathbf{C}^3$. Hence

$$\begin{aligned}
(\mathbf{f}(A)Ax)(\lambda) &= \mathbf{f}(A)\left(\frac{d^2}{d\lambda^2} - 1\right)x(\lambda) \\
&= \mathbf{f}(A)\left(\frac{d^2}{d\lambda^2} - 1\right)(b_1 + b_2\lambda + b_3\lambda^2) \\
&= \mathbf{f}(A)(2b_3 - b_1 - b_2\lambda - b_3\lambda^2) \\
&= -\left(\frac{d^4}{d\lambda^4} + \frac{d^2}{d\lambda^2} + 1\right)(2b_3 - b_1 - b_2\lambda - b_3\lambda^2) \\
&= -(-2b_3 + 2b_3 - b_1 - b_2\lambda - b_3\lambda^2) \\
&= -(-b_1 - b_2\lambda - b_3\lambda^2) \\
&= b_1 + b_2\lambda + b_3\lambda^2 \\
&= x(\lambda).
\end{aligned}$$

Thus $\mathbf{f}(A) = A^{-1}$. This is, of course, what one would expect, for $f(\lambda) = \lambda^{-1}$. ■

Algorithm 6.1 provides us with a direct means of computing an analytic function of a finite-dimensional linear operator. Because every finite-dimensional linear operator can be represented by a matrix, it is useful to consider functions of matrices in particular. There are a number of useful results that can be easily obtained. We begin by examining functions of similar matrices. Recall that similar matrices have the same spectrum.

Proposition 6.3 (Similar Matrices). Let $A \in \mathbf{C}^{n \times n}$, and let f be analytic on $\sigma(A)$. If $P \in \mathbf{C}^{n \times n}$ is nonsingular, then

$$f(P^{-1}AP) = P^{-1}f(A)P.$$

Proof. Let g be the interpolating polynomial of f on $\sigma(A)$, and let $B = P^{-1}AP$. Since B is similar to A, $\sigma(B) = \sigma(A)$. Thus g is also the interpolating polynomial of f on $\sigma(B)$. Hence from Def. 6.2

$$\begin{aligned}
f(P^{-1}AP) &= f(B) \\
&= g(B) \\
&= \sum_{k=1}^{n} a_k B^{k-1} \\
&= \sum_{k=1}^{n} a_k (P^{-1}AP)^{k-1} \\
&= \sum_{k=1}^{n} a_k (P^{-1}AP)(P^{-1}AP)\cdots(P^{-1}AP) \\
&= \sum_{k=1}^{n} a_k P^{-1}A(PP^{-1})A(PP^{-1})\cdots(PP^{-1})AP \\
&= \sum_{k=1}^{n} a_k P^{-1}AIAI\cdots IAP \\
&= \sum_{k=1}^{n} a_k P^{-1}A^{k-1}P \\
&= P^{-1}\left(\sum_{k=1}^{n} a_k A^{k-1}\right)P \\
&= P^{-1}g(A)P \\
&= P^{-1}f(A)P.
\end{aligned}$$

■

Thus if A and B are related by a similarity transformation, then $f(A)$ and $f(B)$ are related by the same similarity transformation. Recall that every finite-dimensional linear operator can be represented by a block diagonal matrix. The block diagonal structure is invariant under the operation of analytic function evaluation. Hence if A is block diagonal, then so is $f(A)$, as can be seen from the following result.

Proposition 6.4 (Block Diagonal Matrices). Let $A \in \mathbf{C}^{n \times n}$, and let f be analytic on $\sigma(A)$. If A is block diagonal with p blocks, then

$$f(\text{diag}\{A_1, A_2, \ldots, A_p\}) = \text{diag}\{f(A_1), f(A_2), \ldots, f(A_p)\}.$$

Proof. Because A is block diagonal with p blocks, we have $A = \text{diag}\{A_1, A_2, \ldots, A_p\}$. It then follows that when $k = 0$, we have, $A^k = \text{diag}\{A_1^k, A_2^k, \ldots, A_p^k\}$. Suppose this formulation of A^k holds when $k = j$.

Then

$$\begin{aligned}A^{j+1} &= AA^j \\ &= \operatorname{diag}\{A_1, A_2, \ldots, A_p\} \operatorname{diag}\{A_1^j, A_2^j, \ldots, A_p^j\} \\ &= \operatorname{diag}\{A_1^{j+1}, A_2^{j+1}, \ldots, A_p^{j+1}\}.\end{aligned}$$

Hence by induction $A^k = \operatorname{diag}\{A_1^k, A_2^k, \ldots, A_p^k\}$ for all $k \geqslant 0$. Let g be the interpolating polynomial of f on $\sigma(A)$. Then

$$\begin{aligned}f(A) &= g(A) \\ &= \sum_{k=1}^{n} a_k A^{k-1} \\ &= \sum_{k=1}^{n} a_k \operatorname{diag}\{A_1^{k-1}, A_2^{k-1}, \ldots, A_q^{k-1}\} \\ &= \operatorname{diag}\left\{\sum_{k=1}^{n} a_k A_1^{k-1}, \sum_{k=1}^{n} a_k A_2^{k-1}, \ldots, \sum_{k=1}^{n} a_k A_q^{k-1}\right\} \\ &= \operatorname{diag}\{g(A_1), g(A_2), \ldots, g(A_q)\} \\ &= \operatorname{diag}\{f(A_1), f(A_2), \ldots, f(A_q)\}.\end{aligned}$$

■

The importance of Prop. 6.4 lies in the fact that it allows us to break up the large computation of $f(A)$ into several smaller computations of $f(A_k)$ for $1 \leqslant k \leqslant q$. In the extreme case when A is diagonal, $f(\mathrm{A})$ can be obtained by inspection since $f(A)_{kk} = f(A_{kk})$ for $1 \leqslant k \leqslant n$ in this case. More generally, since every matrix is similar to a block diagonal matrix, we can use Props. 6.3 and 6.4 together to compute $f(A)$. In particular, if $B = P^{-1}AP$ is the block diagonal representation of A, then $f(A) = Pf(B)P^{-1}$. The following example illustrates this approach.

Example 6.2 (Function of A). Let A be the following matrix in $\mathbf{C}^{3\times 3}$, and consider the problem of computing $f(A)$ where $f(\lambda) = \exp(\lambda t)$,

$$A = \begin{bmatrix} 5 & 4 & 3 \\ -1 & 0 & -3 \\ 1 & -2 & 1 \end{bmatrix}.$$

From Exam. 5.1, the block diagonal representation of A is $B = P^{-1}AP$ where

$$B = \begin{bmatrix} -2 & 0 & 0 \\ 0 & 3 & -1 \\ 0 & 1 & 5 \end{bmatrix}, \quad P = \begin{bmatrix} 1 & 0 & 1 \\ -1 & 1 & 0 \\ -1 & -1 & 0 \end{bmatrix}, \quad P^{-1} = \begin{bmatrix} 0 & -1 & -1 \\ 0 & 1 & -1 \\ 2 & 1 & 1 \end{bmatrix}\left(\frac{1}{2}\right).$$

Thus $B = \text{diag}\{B_1, B_2\}$. Now, $f(B_1) = f(-2) = \exp(-2t)$. To compute $f(B_2)$, we apply Alg. 6.1. The characteristic polynomial is

$$\begin{aligned}\Delta(\lambda) &= \det(\lambda I - B_2)\\ &= (\lambda - 3)(\lambda - 5) + 1\\ &= \lambda^2 - 8\lambda + 16\\ &= (\lambda - 4)^2\\ &\triangleq (\lambda - \lambda_1)^{m(1)}.\end{aligned}$$

Thus $\sigma(B_2) = \{4\}$ with $\lambda_1 = 4$ and $m(l) = 2$. From Alg. 6.1 we have

$$C = \begin{bmatrix} 1 & 4 & \exp(4t) \\ 0 & 1 & t\exp(4t) \end{bmatrix} \to \begin{bmatrix} 1 & 0 & (1-4t)\exp(4t) \\ 0 & 1 & t\exp(4t) \end{bmatrix}.$$

Thus the coefficient vector of the interpolating polynomial of f on $\sigma(B_2)$ is $a = \exp(4t)[(1 - 4t), t]^T$. Hence

$$\begin{aligned}f(B_2) &= \exp(4t)\left\{(1-4t)\begin{bmatrix} 1 & 0 \\ 0 & 1 \end{bmatrix} + t\begin{bmatrix} 3 & -1 \\ 1 & 5 \end{bmatrix}\right\}\\ &= \exp(4t)\begin{bmatrix} 1-t & -t \\ t & 1+t \end{bmatrix}.\end{aligned}$$

When $B = P^{-1}AP$, we have $A = PBP^{-1}$. From Prop. 6.4

$$f(B) = \exp(4t)\begin{bmatrix} \exp(-6t) & 0 & 0 \\ 0 & (1-t) & -t \\ 0 & t & (1+t) \end{bmatrix}.$$

If we now premultiply by P and postmultiply by P^{-1}, the result, after simplification, is

$$f(A) = \exp(4t)\begin{bmatrix} (t+1) & (t+\frac{1}{2}) - \frac{1}{2}\exp(-6t) & \frac{1}{2} - \frac{1}{2}\exp(-6t) \\ -t & \frac{1}{2}\exp(-6t) - (t-\frac{1}{2}) & \frac{1}{2}\exp(-6t) - \frac{1}{2} \\ t & \frac{1}{2}\exp(-6t) + (t-\frac{1}{2}) & \frac{1}{2}\exp(-6t) + \frac{1}{2} \end{bmatrix}$$ ∎

7.7. THE SPECTRAL REPRESENTATION

The essential attributes of a function of a linear operator can be viewed most clearly by means of the spectral representation. If A is a finite-dimensional linear operator and f is analytic on $\sigma(A)$, then the operator **f(A)** depends upon both f

and A. The dependence upon f is determined entirely by the behavior of f on $\sigma(A)$, as can be seen from the following formulation of $\mathbf{f}(A)$ called the spectral representation.

Proposition 7.1 (Spectral Representation). Let $A \in \mathbf{L}(\mathbf{V}, \mathbf{V})$ where $\mathbf{V}$ is n-dimensional, and let λ_k be an eigenvalue of A of multiplicity $m(k)$ for $1 \leqslant k \leqslant q$. Then there exist polynomials $p_{kj} \in \mathbf{P}^{n-1}(\mathbf{C})$ called the *spectral polynomials* of A such that if f is analytic on $\sigma(A)$, then

$$\mathbf{f}(A) = \sum_{k=1}^{q} \sum_{j=1}^{m(k)} f^{(j-1)}(\lambda_k)\mathbf{p}_{kj}(A).$$

Proof. Let Δ be the characteristic polynomial of A, and expand $1/\Delta$ into partial fractions. Then

$$\frac{1}{\Delta(\lambda)} = \sum_{k=1}^{q} \sum_{j=1}^{m(k)} \frac{R_{kj}(\lambda_k)}{(\lambda - \lambda_k)^j}.$$

Here for $1 \leqslant k \leqslant q$ and $0 \leqslant j \leqslant m(k) - 1$ the coefficients of the partial fraction terms are

$$R_{k,m(k)-j} = \frac{1}{j!} \frac{d^j}{d\lambda^j} \frac{(\lambda - \lambda_k)^{m(k)}}{\Delta(\lambda)}.$$

Next for $1 \leqslant k \leqslant q$ let

$$u_k(\lambda) \triangleq \Delta(\lambda) \sum_{j=1}^{m(k)} \frac{R_{kj}(\lambda_k)}{(\lambda - \lambda_k)^j}.$$

Using the partial fraction expansion of $1/\Delta(\lambda)$ we then have for each $\lambda \in \mathbf{C}$

$$\begin{aligned}\sum_{k=1}^{q} u_k(\lambda) &= \Delta(\lambda) \sum_{k=1}^{q} \sum_{j=1}^{m(k)} \frac{R_{kj}(\lambda_k)}{(\lambda - \lambda_k)^j} \\ &= \frac{\Delta(\lambda)}{\Delta(\lambda)} \\ &= 1.\end{aligned}$$

The function $u_k(\lambda)$ can be written in terms of positive powers of λ as follows:

$$u_k(\lambda) = \{\Delta(\lambda)/(\lambda - \lambda_k)^{m(k)}\} \sum_{j=1}^{m(k)} R_{kj}(\lambda_k)(\lambda - \lambda_k)^{m(k)-j}.$$

It follows that $u_k^{(i-1)}(\lambda_j) = 0$ for $1 \leqslant i \leqslant m(j)$ and $j \neq k$. Furthermore, from the definition of $R_{k,m(k)}$ we have for $1 \leqslant i \leqslant m(k)$

$$u_k^{(i-1)}(\lambda_k) = I_{1i}.$$

Let the polynomials $p_{kj}(\lambda)$ for $1 \leqslant k \leqslant q$ and $1 \leqslant j \leqslant m(k)$ be defined as follows:

$$p_{kj}(\lambda) \triangleq \frac{(\lambda - \lambda_k)^{j-1} u_k(\lambda)}{(j-1)!}.$$

Then using the chain rule of differentiation,

$$p_{kj}^{(i-1)}(\lambda) = \sum_{m=0}^{i-1} \binom{i-1}{m} \frac{(\lambda - \lambda_k)^{j-i+m} u_k^{(m)}(\lambda)}{(j-i+m)!}$$

Here $\binom{i-1}{m}$ denotes the binomial coefficient. Because $u_k^{(i-1)}(\lambda_j) = 0$ for $1 \leqslant i \leqslant m(j)$ and $j \neq k$, it follows that

$$p_{kj}^{(i-1)}(\lambda_r) = 0, \qquad 1 \leqslant i \leqslant m(r), \quad r \neq k.$$

Next, if we evaluate the derivatives of $p_{kj}(\lambda)$ at $\lambda = \lambda_k$, we obtain for $1 \leqslant j \leqslant m(k)$ and $1 \leqslant i \leqslant j$

$$\begin{aligned}
p_{kj}^{(i-1)}(\lambda_k) &= \sum_{m=0}^{i-1} \binom{i-1}{m} \frac{(\lambda - \lambda_k)^{j-i+m} u_k^{(m)}(\lambda_k)}{(j-i+m)!}, \qquad && \lambda = \lambda_k \\
&= \sum_{m=0}^{i-1} \binom{i-1}{m} \frac{(\lambda - \lambda_k)^{j-i+m} I_{1,m+1}}{(j-i+m)!}, && \lambda = \lambda_k \\
&= \frac{(\lambda - \lambda_k)^{j-i}}{(j-i)!}, \qquad \lambda = \lambda_k \\
&= I_{ij}.
\end{aligned}$$

Finally, let g be the following polynomial:

$$g(\lambda) \triangleq \sum_{k=1}^{q} \sum_{j=1}^{m(k)} f^{(j-1)}(\lambda_k) p_{kj}(\lambda).$$

Then for $1 \leqslant r \leqslant q$ and $1 \leqslant i \leqslant m(r)$

$$\begin{aligned}
g^{(i-1)}(\lambda_r) &= \sum_{k=1}^{q} \sum_{j=1}^{m(k)} f^{(j-1)}(\lambda_k) p_{kj}^{(i-1)}(\lambda_r) \\
&= \sum_{j=1}^{m(r)} f^{(j-1)}(\lambda_r) p_{rj}^{(i-1)}(\lambda_r) \\
&= \sum_{j=1}^{m(r)} f^{(j-1)}(\lambda_r) I_{ij} \\
&= f^{(i-1)}(\lambda_r).
\end{aligned}$$

Hence g is the interpolating polynomial of f on $\sigma(A)$. Thus

$$\begin{aligned}\mathbf{f}(A) &= \mathbf{g}(A)\\ &= \sum_{k=1}^{q} \sum_{j=1}^{m(k)} f^{(j-1)}(\lambda_k)\mathbf{p}_{kj}(A).\end{aligned}$$ ■

The importance of Prop. 7.1 lies in the fact that the spectral polynomials, $\{p_{kj}\}$, depend only upon A. They are independent of f. We call the formulation of $\mathbf{f}(A)$ in Prop. 7.1 the *spectral representation* of $\mathbf{f}(A)$. Knowing that such a representation exists can, in itself, provide valuable information regarding the qualitative characteristics of $\mathbf{f}(A)$. The following exercise is a case in point.

Exercise 7.1 (Stability). Let A be a finite-dimensional linear operator with eigenvalues $\{\lambda_1, \lambda_2, \ldots, \lambda_q\}$. Show that $\exp(At) \to 0$ as $t \to \infty$ if and only if

$$\operatorname{Re}(\lambda_k) < 0, \qquad 1 \leqslant k \leqslant q.$$ ■

The proof of Prop. 7.1 is constructive in the sense that it provides us with a means of obtaining the spectral polynomials of A. We formalize the technique in the form of the following algorithm.

Algorithm 7.1 (Spectral Polynomials)

1. Compute the characteristic polynomial of A, and factor it into the form

$$\Delta(\lambda) = (\lambda - \lambda_1)^{m(1)}(\lambda - \lambda_2)^{m(2)} \cdots (\lambda - \lambda_q)^{m(q)}.$$

2. Set $k = 1$.

3. Compute $R_{kj}(\lambda)$ for $1 \leqslant j \leqslant m(k)$ recursively as follows:

$$R_{k,m(k)}(\lambda) = \frac{(\lambda - \lambda_k)^{m(k)}}{\Delta(\lambda)}$$

$$R_{k,m(k)-j}(\lambda) = \frac{1}{j}\frac{d}{d\lambda} R_{k,m(k)-j+1}(\lambda).$$

4. Compute $p_{kj}(\lambda)$ for $1 \leqslant j \leqslant m(k)$ using

$$p_{k1}(\lambda) = \Delta(\lambda) \sum_{j=1}^{m(k)} \frac{R_{kj}(\lambda_k)}{(\lambda - \lambda_k)^j},$$

$$p_{kj}(\lambda) = \frac{(\lambda - \lambda_k)^{j-1} p_{k1}(\lambda)}{(j-1)!}.$$

5. Set $k = k + 1$. If $k \leqslant q$, go to 3. ■

As an example of an application of Alg. 7.1, consider the following.

Example 7.1 (Spectral Polynomials). Let $\mathbf{V} = \mathbf{P}^2(\mathbf{C})$, and let A be the following linear operator on $\mathbf{V}$. For each $x \in \mathbf{V}$

$$(Ax)(\lambda) = \frac{d^2}{d\lambda^2} x(\lambda) - x(\lambda), \qquad \lambda \in C.$$

Then from Exam. 6.1, the characteristic polynomial of A is

$$\begin{aligned}\Delta(\lambda) &= (\lambda + 1)^3 \\ &\triangleq (\lambda - \lambda_1)^{m(1)}.\end{aligned}$$

Thus from step 3 of Alg. 7.1 we have

$$R_{13}(\lambda) = \frac{(\lambda + 1)^3}{\Delta(\lambda)} = 1,$$

$$R_{12}(\lambda) = \frac{d}{d\lambda} R_{13}(\lambda) = 0,$$

$$R_{11}(\lambda) = \frac{d}{d\lambda} \frac{R_{12}(\lambda)}{2} = 0.$$

Then from step 4 of Alg. 7.1

$$\begin{aligned}p_{11}(\lambda) &= \Delta(\lambda) \sum_{j=1}^{3} \frac{R_{1j}(\lambda_1)}{(\lambda - \lambda_1)^j} \\ &= \frac{\Delta(\lambda) R_{13}(\lambda_1)}{(\lambda - \lambda_1)^3} \\ &= R_{13}(\lambda_1) \\ &= 1.\end{aligned}$$

Now that $p_{11}(\lambda)$ is known, the remaining polynomials are easily obtained from step 4:

$$\begin{aligned}p_{12}(\lambda) &= (\lambda + 1), \\ p_{13}(\lambda) &= (\lambda + 1)^2/2.\end{aligned}$$

Suppose $f(\lambda) = 1/\lambda$. Then since $\sigma(A) = \{-1\}$, f is analytic on $\sigma(A)$. From

Prop. 7.1 we have

$$\begin{aligned}
\mathbf{f}(A) &= \sum_{j=1}^{3} f^{(j-1)}(\lambda_1)\mathbf{p}_{1j}(A) \\
&= -\mathbf{p}_{11}(A) - \mathbf{p}_{12}(A) - 2\mathbf{p}_{13}(A) \\
&= -\mathbf{I} - (A + I) - (A + I)^2 \\
&= -2I - A - (A^2 + 2A + I) \\
&= -3I - 3A - A^2 \\
&= -3 - 3\left(\frac{d^2}{d\lambda^2} - 1\right) - \left(\frac{d^2}{d\lambda^2} - 1\right)^2 \\
&= -3\frac{d^2}{d\lambda^2} - \left(\frac{d^4}{d\lambda^4} - 2\frac{d^2}{d\lambda^2} + 1\right) \\
&= -\left(\frac{d^4}{d\lambda^4} + \frac{d^2}{d\lambda^2} + 1\right).
\end{aligned}$$

Note that this agrees with the direct computation of $\mathbf{f}(A)$ developed in Exam. 6.1. However, there is one important difference here. Once the spectral polynomials $\{\mathbf{p}_{kj}(A)\}$ are known, the computation of $\mathbf{f}(A)$ for *any* f that is analytic on $\sigma(A)$ is a simple procedure. ■

Example 7.2 (Spectral Polynomials). Let $A \in \mathbf{C}^{3\times 3}$ be the following linear operator on $\mathbf{V} = \mathbf{C}^3$:

$$A = \begin{bmatrix} 5 & 4 & 3 \\ -1 & 0 & -3 \\ 1 & -2 & 1 \end{bmatrix}.$$

If $f(\lambda) = \exp(\lambda t)$, then from Exam. 6.2 we already know that

$$f(A) = \exp(4t)\begin{bmatrix} (t+1) & (t+\frac{1}{2}) - \frac{1}{2}\exp(-6t) & \frac{1}{2} - \frac{1}{2}\exp(-6t) \\ -t & \frac{1}{2}\exp(-6t) - (t - \frac{1}{2}) & \frac{1}{2}\exp(-6t) - \frac{1}{2} \\ t & \frac{1}{2}\exp(-6t) + (t - \frac{1}{2}) & \frac{1}{2}\exp(-6t) + \frac{1}{2} \end{bmatrix}.$$

Since the characteristic polynomial of A is $\Delta(\lambda) = (\lambda + 2)(\lambda - 4)^2$, we can see from inspection that the spectral representation of $\exp(At)$ must be

$$\exp(At) = \exp(-2t)\begin{bmatrix} 0 & -\frac{1}{2} & -\frac{1}{2} \\ 0 & \frac{1}{2} & \frac{1}{2} \\ 0 & \frac{1}{2} & \frac{1}{2} \end{bmatrix} + \exp(4t)\begin{bmatrix} 1 & \frac{1}{2} & \frac{1}{2} \\ 0 & \frac{1}{2} & -\frac{1}{2} \\ 0 & -\frac{1}{2} & \frac{1}{2} \end{bmatrix}$$

$$+t\exp(4t)\begin{bmatrix} 1 & 1 & 0 \\ -1 & -1 & 0 \\ 1 & 1 & 0 \end{bmatrix}.$$

Thus for *any* function f which is analytic on $\{-2, 4\}$, the spectral representation of $f(A)$ is

$$f(A) = f(-2)\begin{bmatrix} 0 & -\frac{1}{2} & -\frac{1}{2} \\ 0 & \frac{1}{2} & \frac{1}{2} \\ 0 & \frac{1}{2} & \frac{1}{2} \end{bmatrix} + f(4)\begin{bmatrix} 1 & \frac{1}{2} & \frac{1}{2} \\ 0 & \frac{1}{2} & -\frac{1}{2} \\ 0 & -\frac{1}{2} & \frac{1}{2} \end{bmatrix}$$

$$+ f^{(1)}(4)\begin{bmatrix} 1 & 1 & 0 \\ -1 & -1 & 0 \\ 1 & 1 & 0 \end{bmatrix}$$ ■

Recall from Chapter 6 that the spectral representation of a self-adjoint operator $\boldsymbol{A}$ was based on a family of orthogonal projection operators $\{\boldsymbol{E}(\lambda)\}$ called the *resolution of the identity* induced by $\boldsymbol{A}$. For finite-dimensional linear operators the family of orthogonal projection operators $\{\boldsymbol{E}(\lambda_k)\}$ is the set of first-order spectral polynomials $\{\mathbf{p}_{k1}(\boldsymbol{A})\}$, as can be seen from the following exercise.

Exercise 7.2 (Resolution of Identity). Let A be a finite-dimensional linear operator with eigenvalues $\{\lambda_1, \lambda_2, \ldots, \lambda_q\}$ and spectral polynomials $\{p_{kj}\}$. Show that

$$\sum_{k=1}^{q} \mathbf{p}_{k1}(A) = I.$$ ■

The spectral representation of A simplifies considerably in the event that A is diagonalizable. In this case the first-order spectral polynomials $\{p_{k1}\}$ completely characterize the behavior of A, as shown in the following result.

Proposition 7.2 (Diagonalizable Operators). Let A be a finite-dimensional diagonalizable linear operator with eigenvalues $\{\lambda_1, \lambda_2, \ldots, \lambda_q\}$ and spectral polynomials $\{p_{kj}\}$. If f is analytic on $\sigma(A)$, then

$$\mathbf{f}(A) = \sum_{k=1}^{q} f(\lambda_k)\mathbf{p}_{k1}(A).$$

Proof. Let Δ be the characteristic polynomial of A, and let $m(k)$ be the multiplicity of λ_k for $1 \leqslant k \leqslant q$. From step 4 of Alg. 7.1

$$p_{k1}(\lambda) = \Delta(\lambda) \sum_{j=1}^{m(k)} \frac{R_{kj}(\lambda_k)}{(\lambda - \lambda_k)^j}.$$

Thus for $1 \leqslant k \leqslant q$ we have

$$\begin{aligned}(\lambda_k - \lambda)^{m(k)} p_{k1}(\lambda) &= (-1)^{m(k)}(\lambda - \lambda_k)^{m(k)} p_{k1}(\lambda) \\ &= (-1)^{m(k)}\Delta(\lambda)(\lambda - \lambda_k)^{m(k)} \sum_{j=1}^{m(k)} \frac{R_{kj}(\lambda_k)}{(\lambda - \lambda_k)^j}\end{aligned}$$

$$= (-1)^{m(k)}\Delta(\lambda) \sum_{j=1}^{m(k)} R_{kj}(\lambda_k)(\lambda - \lambda_k)^{m(k)-j}$$
$$\triangleq (-1)^{m(k)}\Delta(\lambda)q(\lambda).$$

Applying the Cayley–Hamilton theorem,

$$\begin{aligned}(\boldsymbol{\lambda}_k \boldsymbol{I} - \boldsymbol{A})^{\mathbf{m(k)}}\mathbf{p}_{k1}(\boldsymbol{A}) &= (-1)^{m(k)}\boldsymbol{\Delta}(\boldsymbol{A})\mathbf{q}(\boldsymbol{A})\\ &= \boldsymbol{0}.\end{aligned}$$

It follows that $\mathbf{R}(\mathbf{p}_{k1}(\boldsymbol{A}))$ is a subspace of $\mathbf{N}\{(\boldsymbol{\lambda}_k\boldsymbol{I} - \boldsymbol{A})^{\mathbf{m(k)}}\}$ for $1 \leqslant k \leqslant q$. But since A is diagonalizable, from Exer. 5.2,

$$\begin{aligned}\mathbf{N}\{(\boldsymbol{\lambda}_k\boldsymbol{I} - \boldsymbol{A})^{\mathbf{m(k)}}\} &= \mathbf{H}(\boldsymbol{\lambda}_k)\\ &= \mathbf{G}(\boldsymbol{\lambda}_k)\\ &= \mathbf{N}\{(\boldsymbol{\lambda}_k\boldsymbol{I} - \boldsymbol{A})\}.\end{aligned}$$

Hence $(\boldsymbol{\lambda}_k\boldsymbol{I} - \boldsymbol{A})\mathbf{p}_{k1}(\boldsymbol{A}) = \boldsymbol{0}$ for $1 \leqslant k \leqslant q$. Thus from step 4 of Alg. 7.1,

$$\mathbf{p}_{kj}(\boldsymbol{A}) = \boldsymbol{0}, \qquad 2 \leqslant j \leqslant m(k).$$

Hence if f is analytic on $\sigma(A)$, then from Prop. 7.1

$$\mathbf{f}(\boldsymbol{A}) = \sum_{k=1}^{q} f(\lambda_k)\mathbf{p}_{k1}(\boldsymbol{A}). \qquad ■$$

Prop. 7.2 can be regarded as the finite-dimensional equivalent of Prop. 6.8.3, the spectral representation theorem for bounded self-adjoint operators. Note how the integral has been reduced to a finite sum in this case as the spectrum of A is reduced to the point spectrum $\{\lambda_1, \lambda_2, \ldots, \lambda_q\}$. Apart from the restriction on the dimension of A, Prop. 7.2 is more general than Prop. 6.8.3 since f does not have to be a polynomial and A does not have to be self-adjoint. The following exercises further explore the relationship between the finite- and infinite-dimensional versions of the spectral representation theorem.

Exercise 7.3. Let A be a finite-dimensional linear operator, and let λ_k be an eigenvalue of A of multiplicity $m(k)$ for $1 \leqslant k \leqslant q$. Let $\{p_{kj}\}$ be the spectral polynomials of A. Show that for $1 \leqslant k, j \leqslant q$

$$\mathbf{p}_{k1}(\boldsymbol{A})\mathbf{p}_{j1}(\boldsymbol{A}) = \boldsymbol{I}_{kj}\mathbf{p}_{k1}(\boldsymbol{A}). \qquad ■$$

Exercise 7.4 (Orthogonal Projections). Let A be a finite-dimensional diagonalizable linear operator with spectrum $\sigma(A) = \{\lambda_1, \lambda_2, \ldots, \lambda_q\}$ and spectral polynomials $\{p_{kj}\}$. Let $\mathbf{E}_k$ be the orthogonal projection operator induced by the eigenspace $\mathbf{G}(\lambda_k)$ for $1 \leqslant k \leqslant q$. Show that

$$\mathbf{p}_{k1}(A) = E_k, \qquad 1 \leqslant k \leqslant q. \qquad \blacksquare$$

Exercise 7.5 (Resolution of Identity). Let A be a self-adjoint finite-dimensional linear operator with $\sigma(A) = \{\lambda_1, \lambda_2, \ldots, \lambda_q\}$. Find the family of projections $E(\lambda)$ such that

$$A = \int_{\mathbf{R}} \lambda \, dE(\lambda). \qquad \blacksquare$$

Another way of representing a function of a finite-dimensional linear operator is through a power series. Suppose $f: \mathbf{C} \to \mathbf{C}$ is expressible as a power series expansion about a point λ_0. If the radius of convergence of this expansion is r, and if $\sigma(A)$ is contained in the ball $B(\lambda_0, r)$, then $\mathbf{f}(A)$ is well defined and can be represented by a power series. In particular, we have the following result.

Proposition 7.3 (Power Series). Let $f: \mathbf{C} \to \mathbf{C}$ be expressible by a power series expansion about λ_0 with coefficients $\{a_k\}$ and radius of convergence of r. If A is a finite-dimensional linear operator and $\sigma(A)$ is contained in the ball $B(\lambda_0, r)$, then $\mathbf{f}(A)$ exists and

$$\mathbf{f}(A) = \sum_{k=1}^{\infty} a_k (A - \lambda_0 I)^{k-1}.$$

Proof. Since f is expressible by a power series expansion about λ_0, valid in $B(\lambda_0, r)$, it follows that f is analytic on the open ball $B(\lambda_0, r)$. Because $\sigma(A)$ is a subset of $B(\lambda_0, r)$, f is analytic on $\sigma(A)$. Thus the operator $\mathbf{f}(A)$ is well defined by its interpolating polynomial. Let f_N be the Nth partial sum of the power series expansion of f. That is

$$f_N(\lambda) \triangleq \sum_{k=1}^{N} a_k (\lambda - \lambda_0)^{k-1}.$$

Next, let Δ be the characteristic polynomial of A, and suppose λ_k is an eigenvalue of A of multiplicity $m(k)$ for $1 \leqslant k \leqslant q$. If $\{p_{kj}\}$ are the spectral polynomials of A, then from Prop. 7.1

$$\begin{aligned}
\mathbf{f}(A) &= \sum_{k=1}^{q} \sum_{j=1}^{m(k)} f^{(j-1)}(\lambda_k) \mathbf{p}_{kj}(A) \\
&= \sum_{k=1}^{q} \sum_{j=1}^{m(k)} \left\{ \lim_{N \to \infty} f_N^{(j-1)}(\lambda_k) \right\} \mathbf{p}_{kj}(A) \\
&= \lim_{N \to \infty} \left\{ \sum_{k=1}^{q} \sum_{j=1}^{m(k)} f_N^{(j-1)}(\lambda_k) \mathbf{p}_{kj}(A) \right\}
\end{aligned}$$

$$= \lim_{N \to \infty} \{\mathbf{f_N}(A)\}$$

$$= \lim_{N \to \infty} \left\{ \sum_{k=1}^{N} a_k (A - \lambda_0 I)^{k-1} \right\}$$

$$= \sum_{k=1}^{\infty} a_k (A - \lambda_0 I)^{k-1}. \qquad \blacksquare$$

Thus the definition of $\mathbf{f}(A)$ originally proposed in Def. 6.3, although it may have appeared to be quite restrictive, is actually very general. It view of Prop. 7.3, we can easily obtain power series representations of functions of finite-dimensional linear operators. The following exercise illustrates some well-known examples.

Exercise 7.6 (Power Series of A). Let A be a finite-dimensional linear operator. Show that

1. $\sin(A) = \sum_{k=0}^{\infty} \frac{(-1)^k}{(2k+1)!} A^{2k+1}$,

2. $\cos(A) = \sum_{k=0}^{\infty} \frac{(-1)^k}{(2k)!} A^{2k}$,

3. $\exp(A) = \sum_{k=0}^{\infty} \frac{A^k}{k!}$. $\blacksquare$

Exercise 7.7. Show that the operator version of Euler's identity holds. That is, if A is a finite-dimensional linear operator, then

$$\exp(jA) = \cos(A) + j \sin(A). \qquad \blacksquare$$

Generally speaking, the most direct way to compute $\mathbf{f}(A)$ is to use the interpolating polynomial as in Alg. 6.1. The power series expansion technique is useful if an approximation to $\mathbf{f}(A)$ is desired. Alternately, if qualitative properties of $\mathbf{f}(A)$ are to be examined, the spectral representation constructed in Alg. 7.1 can be useful. Cauchy's residue theorem provides us with still another representation of $\mathbf{f}(A)$.

Property 7.4 (Cauchy). Let $A \in \mathbf{L}(\mathbf{V}, \mathbf{V})$ where $\mathbf{V}$ is finite-dimensional, and let C be a circle in the complex plane that encloses $\sigma(A)$. If f is analytic on and within C, then

$$\mathbf{f}(A) = \oint_C f(\lambda)(\lambda I - A)^{-1}\, d\lambda. \qquad \blacksquare$$

Here the integral is performed in a counterclockwise sense around the circle C. The circle C can be replaced by any simple rectifiable closed curve in $\mathbf{C}$ that encloses $\sigma(A)$. For a proof of Prop. 7.4, the reader is referred to Hille (1973). Cauchy's integral representation of $\mathbf{f}(A)$ is perhaps the most elegant of the four representations of $\mathbf{f}(A)$ that we have considered. As an example of how it might be used to establish a qualitative property of $\mathbf{f}(A)$, consider the following exercise that examines the relationship between a function of a matrix and a function of its transpose.

Exercise 7.8 (Transpose). Let $A \in \mathbf{C}^{\mathbf{n}\times\mathbf{n}}$, and let f be analytic on $\sigma(A)$. Show that $f(A^{\mathrm{T}}) = f(A)^{\mathrm{T}}$. ■

PROBLEMS

1. Let $\mathbf{V}$ be the space spanned by $\mathbf{B} = \{b^1, b^2, \ldots, b^n\}$ where $b^k(t) = \cos(kt)$ for $1 \leqslant k \leqslant n$ and $0 \leqslant t \leqslant 2\pi$. Next, let A be the following operator on $\mathbf{V}$. For each $x \in \mathbf{V}$

$$(Ax)(t) = \frac{d^2}{dt^2}\, x(t), \qquad 0 \leqslant t \leqslant 2\pi.$$

   a. Show that $A \in \mathbf{L}(\mathbf{V}, \mathbf{V})$.
   b. Show that $\mathbf{B}$ is a basis for $\mathbf{V}$.
   c. Find $[A]^{\mathbf{B},\mathbf{B}}$.
   d. Find $\sigma(A)$.

2. Let $A \in \mathbf{L}(\mathbf{V}, \mathbf{V})$ where $\mathbf{V}$ is finite-dimensional. Show that A is invertible $\Leftrightarrow 0 \notin \sigma(A)$.

3. Let $A \in \mathbf{F}^{\mathbf{n}\times\mathbf{n}}$ be upper triangular. That is, $A_{kj} = 0$ for $k > j$. Show that

$$\sigma(A) = \{A_{11}, A_{22}, \ldots, A_{nn}\}.$$

4. Show that every finite-dimensional linear operator can be represented by an upper triangular matrix.

5. Let $\mathbf{V} = \mathbf{P}^2(\mathbf{C})$, and let A be the following linear operator on $\mathbf{V}$. For each $x \in \mathbf{V}$

$$(Ax)(\lambda) = \frac{d^2}{d\lambda^2}\, x(\lambda) + x(\lambda), \qquad \lambda \in \mathbf{C}.$$

   a. Show that $\det(A) \neq 0$.
   b. Find an expression for A^{-1}.

6. Let $A \in \mathbf{L}(\mathbf{V}, \mathbf{V})$ where $\mathbf{V}$ is finite-dimensional. Show that A is diagonalizable $\Leftrightarrow$ the eigenvectors of A form a basis for $\mathbf{V}$.

7. Find a block diagonal representation of the following matrix A and the basis $\mathbf{B}$ relative to which it is expressed:

$$A = \begin{bmatrix} 1 & 0 & -2 \\ 0 & 2 & -3 \\ 0 & 0 & -1 \end{bmatrix}.$$

8. Let $A \in \mathbf{L}(\mathbf{V}, \mathbf{V})$ where $\mathbf{V}$ is finite-dimensional, and let f be analytic on $\sigma(A)$. Show that

$$\lambda \in \sigma(A) \Leftrightarrow f(\lambda) \in \sigma(\mathbf{f}(A)).$$

9. Find a block diagonal representation of the following matrix A and the basis $\mathbf{B}$ relative to which it is expressed:

$$A = \begin{bmatrix} 1 & 1 & 2 \\ 0 & 1 & 3 \\ 0 & 0 & 2 \end{bmatrix}.$$

10. Let $A \in \mathbf{L}(\mathbf{V}, \mathbf{V})$ where $\mathbf{V}$ is a finite-dimensional normed vector space. Show that if $\|A - I\| < 1$, then A is invertible and

$$\frac{1}{1 + \|A - I\|} \leqslant \|A^{-1}\| \leqslant \frac{1}{1 - \|A - I\|}.$$

11. Let $A \in \mathbf{C}^{\mathbf{n} \times \mathbf{n}}$. Show that A is invertible if A is *diagonally dominant*. That is,

$$|A_{kk}| > \sum_{j \neq k} |A_{kj}|, \qquad 1 \leqslant k, j \leqslant n.$$

12. Let $\mathbf{V} = \mathbf{P}^{\mathbf{m}}(\mathbf{C})$, and let $h \in \mathbf{C}$. Show that for each $x \in \mathbf{V}$

$$\left(\exp\left\{h \frac{d}{d\lambda}\right\} x\right)(\lambda) = x(\lambda + h), \qquad \lambda \in \mathbf{C}.$$

13. Let $\mathbf{V} = \mathbf{P}^{\mathbf{m}}(\mathbf{C})$, and let $h \in \mathbf{C}$. Next, let Δ_h denote the *backwards difference* operator on $\mathbf{V}$. That is, for each $x \in \mathbf{V}$

$$(\Delta_h x)(\lambda) = x(\lambda) - x(\lambda - h), \qquad \lambda \in \mathbf{C}.$$

Show that the derivative operator can be expressed in terms of backwards differences as follows (Dahlquist et al., 1974)

$$\frac{d}{d\lambda} = \frac{1}{h} \sum_{k=1}^{\infty} \frac{(-1)^{k-1}}{k} \Delta_h^k.$$

14. Consider the following linear operator on $\mathbf{R}^3$:

$$A = \begin{bmatrix} 1 & -3 & 3 \\ 3 & -5 & 3 \\ 6 & -6 & 4 \end{bmatrix}.$$

a. Find $\sigma(A)$.
b. Find the interpolating polynomial of $f(\lambda) = \lambda^{1/2}$.
c. Find $A^{1/2}$.

15. Consider the following linear operator on $\mathbf{R}^3$. For each $x \in \mathbf{R}^3$

$$Ax = [2x_1 + x_2, x_2 - x_3, 2x_2 + x_3]^T.$$

a. Find the representation of A induced by the standard basis $\mathbf{I}$.
b. Find $\sigma(A)$.
c. Find the spectral polynomials of A.
d. Find the spectral representation of $\cos(A)$.

16. Let $A(t) \in \mathbf{C}^{n \times n}$ be invertible and differentiable for $t \in \mathbf{R}$. Show that for each $t \in \mathbf{R}$

$$\frac{d}{dt} A^{-1}(t) = -A^{-1}(t) \left\{ \frac{d}{dt} A(t) \right\} A^{-1}(t).$$

17. Let $\mathbf{V} = \mathbf{P}^2(\mathbf{C})$, and let A be the following linear operator on $\mathbf{V}$. For each $x \in \mathbf{V}$

$$(Ax)(\lambda) = \frac{d}{d\lambda} x(\lambda) + x(\lambda), \qquad \lambda \in C.$$

a. Find the characteristic polynomial of A.
b. Find the spectral polynomials of A.
c. Find the spectral representation of $A^{1/3}$.

18. Let $A \in \mathbf{L}(\mathbf{V}, \mathbf{V})$ where $\mathbf{V}$ is finite-dimensional and suppose $\sigma(A)$ is contained in the ball $B(0, 1)$. Show that

$$(A - I)^{-1} = \sum_{k=0}^{\infty} A^k.$$

19. Let $A \in \mathbf{C}^{\mathbf{n}\times\mathbf{n}}$, and suppose the eigenvalues of A all have negative real parts. Show that

$$A^{-1} = -\int_0^\infty \exp(At)\, dt.$$

20. Let $A \in \mathbf{C}^{\mathbf{n}\times\mathbf{n}}$ be invertible. Consider the following second-order differential equation,

$$\frac{d^2}{dt^2}\, x(t) + Ax(t) = 0, \qquad 0 \leqslant t < \infty.$$

Show that if $x(0) = a$ and $x^{(1)}(0) = b$, then the solution is

$$x(t) = \cos(A^{1/2}t)a + A^{-1/2}\sin(A^{1/2}t)b.$$

21. Let $A \in \mathbf{C}^{\mathbf{n}\times\mathbf{n}}$, $B \in \mathbf{C}^{\mathbf{n}\times\mathbf{n}}$, and consider the following first-order differential equation:

$$d/dt\; Z(t) = AZ(t) + Z(t)B, \qquad 0 \leqslant t < \infty.$$

Find a solution when $Z(0) = C$ by looking for one of the form $Z(t) = \exp(At)Y(t)$.

22. Let A be a finite-dimensional linear operator and suppose that $\sigma(A)$ is contained in the ball $B(0, 1)$. Show that

$$\log(A + I) = \sum_{k=1}^{\infty} \frac{(-1)^{k-1}A^{\mathbf{k}}}{k}.$$

REFERENCES

Dahlquist, G., A. Bjorck, and N. Anderson, *Numerical Methods*, Prentice-Hall, Englewood Cliffs, NJ, 1974.

Hille, E., *Analytic Function Theory*, Vol. 1, Chelsea, New York, 1973.

Noble, B. and J. W. Deniel, *Applied Linear Algebra*, McGraw-Hill, New York, 1977.

8 LINEAR DIFFERENTIAL SYSTEMS

Since the time of Isaac Newton, differential equations have served as effective mathematical tools for modeling physical phenomena. Given the laws of physics, they arise repeatedly as mathematical models in a host of seemingly unrelated disciplines.

8.1. LINEAR DIFFERENTIAL SYSTEMS

A survey of mathematical models that appear in engineering applications reveals that many of the models can be cast as a coupled system of differential and algebraic equations of the following generic form:

$$\dot{x}(t) = f[x(t), u(t), t],$$

$$y(t) = g[x(t), u(t), t].$$

We will refer this system of equations as the *differential system* S. Here the independent variable $t \in \mathbf{R}$ represents *time*. The notation $\dot{x}(t)$ denotes the first derivative of $x(t)$ with respect to time, that is, $\dot{x} = x^{(1)}$. The vector $u(t) \in \mathbf{R}^{\mathbf{m}}$ represents the *input* or excitation applied to S at time t. In this case we say that S is an m-input system. The vector $x(t) \in \mathbf{R}^{\mathbf{n}}$ is called the *state of* S at time t, and we say that S is an *n-dimensional system.* Finally, the vector $y(t) \in \mathbf{R}^{\mathbf{p}}$ denotes the *output* or response of S at time t, and we call S a p-output system. A block diagram which illustrates the relationships between the input, state, and output of the system S is shown in Figure 8.1.

The differential equation in $x(t)$ represents the dynamic part of the system and is called the *state equation*, whereas the algebraic equation for $y(t)$ represents the static part of the system and is called the *output equation.* Note that the functions f and g characterize completely the *internal structure* of the system S. Before we proceed with a classification of systems based upon these functions, we examine a simple example.

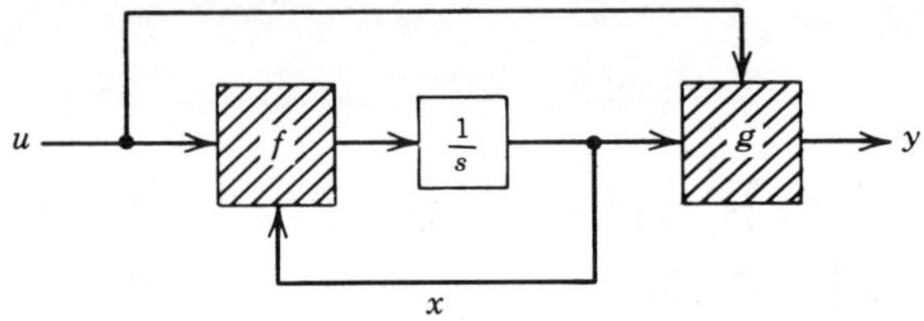

Figure 8.1. A differential system.

Example 1.1 (Population Growth). Consider the ecological problem of modeling the population growth of a single species in a limited environment. Let $x(t)$ denote the population (in appropriate units) at time t. The Verhulst–Pearl logistic equation for population growth (Pielou, 1969) can then be expressed in terms of $x(t)$ as

$$\dot{x}(t) = x(t)[a - bx(t)],$$

$$y(t) = x(t).$$

Here the parameters a and b are positive constants that depend upon the species and the environment. The right side of the state equation can be regarded as the first three terms of a Taylor series expansion of $f(x)$ about $x = 0$. Note that the constant term is zero, for there will be no population growth when $x = 0$. Furthermore, when $0 < x < a/b$, it is clear from the state equation that $\dot{x} > 0$. Similarly, when $x > a/b$, we find that $\dot{x} < 0$. Thus for each $x(0) > 0$

$$x(t) \to a/b \quad \text{as} \quad t \to \infty.$$

It is evident that the solution of the state equation approaches a stable equilibrium population, a/b. This point of equilibrium is the *carrying capacity* of the environment. ■

On the surface, it might appear that the restriction of S to a system of first-order differential equations is rather severe. Indeed, using the laws of physics, many mathematical models manifest themselves as second- or higher-order differential equations. However, we can often replace a single high-order differential equation by an equivalent system of first-order differential equations. That is, we can reduce the order of the system at the expense of increasing the dimension. The following result is an example of this.

Proposition 1.1 (High-Order Systems). Let S_q be a qth-order one-dimensional differential system of the following form:

$$y^{(q)}(t) = h[y(t), y^{(1)}(t), \ldots, y^{(q-1)}(t), u(t), t].$$

Define the state vector as $x = [y, y^{(1)}, \ldots, y^{(q-1)}]^T$. Then S_q is input/output equivalent to a first-order system S where

$$\dot{x}_k(t) = x_{k+1}(t), \qquad 1 \leqslant k \leqslant q-1,$$
$$\dot{x}_q(t) = h[x_1(t), x_2(t), \ldots, x_q(t), u(t), t],$$
$$y(t) = x_1(t).$$

■

The proof of Prop. 1.1 follows directly from the definition of the state vector since $\dot{x}_k = y^{(k)}$ for $1 \leqslant k \leqslant q$ and $x_1 = y$. The following example illustrates the use of Prop. 1.1 to convert a high-order physical model to the standard first-order form.

Example 1.2 (Pendulum). Consider the swinging pendulum shown in Figure 8.2. Here the angle of the pendulum with respect to vertical is y, the length of the pendulum is L, and the mass concentrated at the end of the pendulum is m.

If the acceleration due to gravity is g, the coefficient of viscous friction is η, and the applied torque is u, then from Newton's second law we get

$$mLy^{(2)}(t) + mg \sin[y(t)] + m\eta y^{(1)}(t) = u(t).$$

Using Prop. 1.1, we have $x = [y, y^{(1)}]^T$ and the equivalent first-order two-dimensional system is

$$\dot{x}_1(t) = x_2(t),$$
$$\dot{x}_2(t) = -\frac{g}{L}\sin[x_1(t)] - \frac{\eta}{L}x_2(t) + \frac{u(t)}{mL},$$
$$y(t) = x_1(t).$$

■

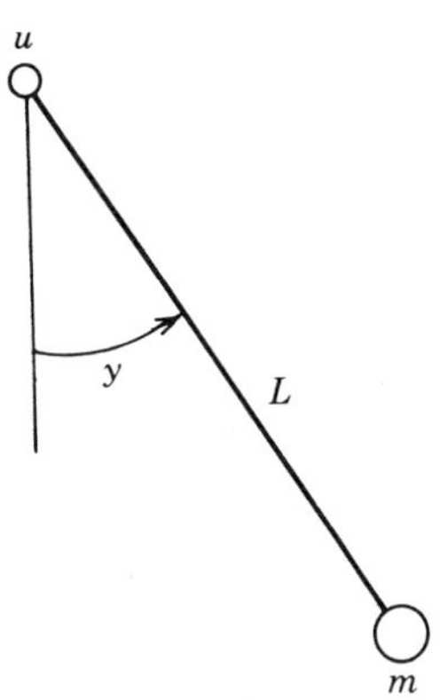

Figure 8.2. A swinging pendulum.

Differential systems can be classified into a number of useful categories. The classification schemes are based upon the nature of the *state function* f and *output function* g which characterize the structure of S.

Definition 1.1 (Time Invariance). A differential system S is *time-invariant* $\Leftrightarrow$

$$\dot{x}(t) = f[x(t), u(t)],$$
$$y(t) = g[x(t), u(t)]. \qquad \blacksquare$$

Thus S is time-invariant if and only if the functions f and g do not depend explicitly on t. The physical interpretation of time-invariance is that the results of input/output experiments do not depend upon the time at which those experiments are performed. It is in this sense that the behavior of the system S is invariant. A second classification of systems is based upon how the functions f and g depend on the state x and the input u.

Definition 1.2 (Linear System). A differential system S is *linear* $\Leftrightarrow$

$$\dot{x}(t) = A(t)x(t) + B(t)u(t),$$
$$y(t) = C(t)x(t) + D(t)u(t). \qquad \blacksquare$$

Thus a linear system is one whose state function f and output function g are linear in the augmented variable $[x(t), u(t)]$. For a system which is both linear and time-invariant, the *matrix parameters* $\{A, B, C, D\}$ are constant. Refer to Figure 8.3 for a block diagram that illustrates the structure of a linear differential system.

Note that neither Exam. 1.1 nor 1.2 is a linear system. However, they are both time-invariant. The following example is representative of a large class of systems that are both linear and time-invariant.

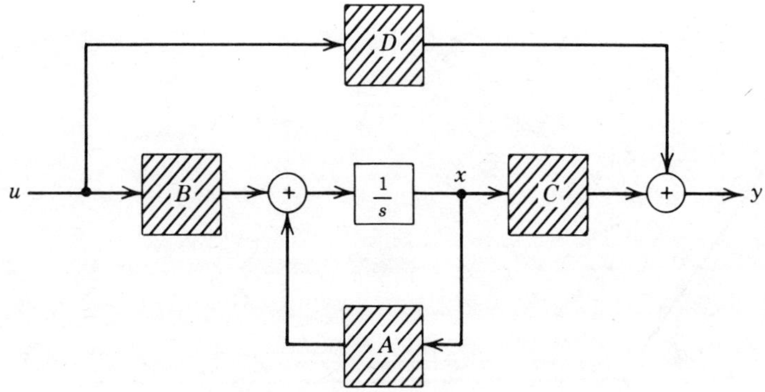

Figure 8.3. A linear differential system.

Example 1.3 (Electrical Network). Consider the two-port electrical network shown in Figure 8.4. Here R, L, and C denote resistance, inductance, and capacitance, respectively. The input u is the voltage applied at port 1, and the output y is the voltage measured at port 2.

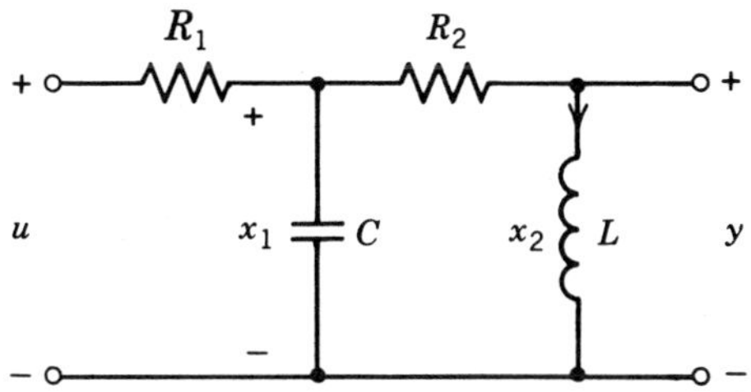

Figure 8.4. A two-port electrical network.

If we take the capacitor voltage and inductor current as the two state variables as indicated in Figure 8.4, then applying Kirchhoff's laws, the equations of motion of this network are

$$C\dot{x}_1(t) = \frac{u(t) - x_1(t)}{R_1} - x_2(t)$$

$$L\dot{x}_2(t) = x_1(t) - R_2 x_2(t),$$

$$y(t) = x_1(t) - R_2 x_2(t).$$

Dividing the first equation by C and the second equation by L transforms these equations into the standard form depicted in Def. 1.2. In this case the four matrix parameters are

$$A = \begin{bmatrix} \dfrac{1}{R_1 C} & \dfrac{1}{C} \\ \dfrac{1}{L} & -\dfrac{R_2}{L} \end{bmatrix}, \qquad B = \begin{bmatrix} \dfrac{1}{R_1 C} \\ 0 \end{bmatrix}, \qquad C = [1 \quad -R_2], \qquad D = [0].$$

■

Recall that Prop. 1.1 provided us with a means of recasting a qth order one-dimensional system as an equivalent first-order system of dimension q. Note, however, that the original qth-order system was constrained in the sense that no derivatives of the input u appeared. Within the class of linear time-invariant systems, this restriction on u can be relaxed as shown in the following proposition.

Proposition 1.2 (Linear High-Order Systems). Let S_q be a qth-order one-dimensional linear time-invariant differential system of the following form where

$a_0 = 1$:

$$\sum_{k=0}^{q} a_k y^{(q-k)}(t) = \sum_{k=0}^{q} b_k u^{(q-k)}(t).$$

Let the state vector $x(t) \in \mathbf{R}^q$ be defined as follows:

$$x^k = y^{(k-1)} - \sum_{j=0}^{k-1} \beta_j u^{(k-1-j)}, \qquad 1 \leqslant k \leqslant q.$$

Here $\beta_0 = b_0$ and

$$\beta_j = b_j - \sum_{k=1}^{j} a_k \beta_{j-k}, \qquad 1 \leqslant j \leqslant q.$$

Then S_q is input/output equivalent to a linear time-invariant system S with the following matrix parameters:

$$A = \begin{bmatrix} 0 & 1 & 0 & \cdots & 0 & 0 \\ 0 & 0 & 1 & \cdots & 0 & 0 \\ \vdots & \vdots & \vdots & & \vdots & \vdots \\ 0 & 0 & 0 & \cdots & 0 & 1 \\ -a_q & -a_{q-1} & -a_{q-2} & \cdots & -a_2 & -a_1 \end{bmatrix}, \qquad B = \begin{bmatrix} \beta_1 \\ \beta_2 \\ \vdots \\ \beta_{q-1} \\ \beta_q \end{bmatrix},$$

$$C = [1 \ \ 0 \ \ 0 \ \ \cdots \ \ 0 \ \ 0], \qquad D = [\beta_0]. \qquad \blacksquare$$

The interested reader can refer to Ogata (1970) for a proof of Prop. 1.2. When the system S is linear, Prop. 1.1 is actually a special case of Prop. 1.2, as can be seen from the following exercise.

Exercise 1.1. Show that the state assignment of Prop. 1.2 reduces to that of Prop. 1.1 when S is linear and $b_k = 0$ for $0 \leqslant k \leqslant q - 1$. $\blacksquare$

The following example illustrates the use of Prop. 1.2 to derive state and output equations for a high-order linear system in which derivatives of the input appear.

Exam. 1.4 [Seismograph (Ogata, 1970)]. Consider the seismograph pictured in Figure 8.5. Here u and y denote the position, in inertial space, of the case and the mass m, respectively. The spring constant is γ, while the coefficient of friction for the dashpot is η.

Applying Newton's second law, the equation of motion of this system is

$$my^{(2)}(t) + \eta[y^{(1)}(t) - u^{(1)}(t)] + \gamma[y(t) - u(t)] = 0.$$

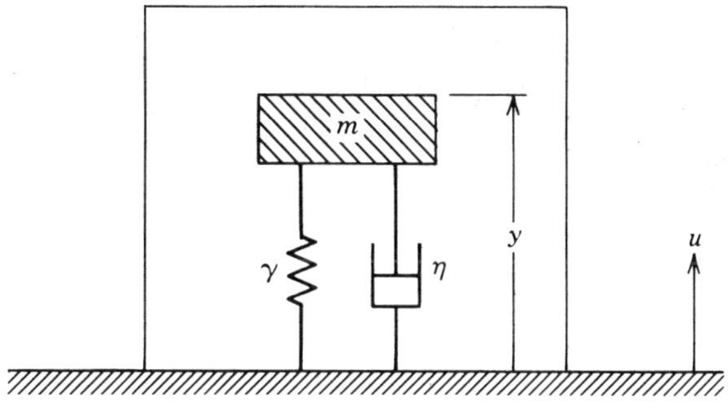

Figure 8.5. A seismograph system.

Thus we have a second-order one-dimensional system. Using Prop. 1.2, and identifying coefficients, we get $q = 2$, $b = \{0, \eta/m, \gamma/m\}$ and $a = \{1, \eta/m, \gamma/m\}$. Thus the state vector is

$$x_1 = y,$$
$$x_2 = y^{(1)} - (\eta/m)u.$$

The state and output equations of the seismograph system in this case reduce to the following:

$$\dot{x}(t) = \begin{bmatrix} 0 & 1 \\ -\gamma/m & -\eta/m \end{bmatrix} x(t) + \begin{bmatrix} \eta/m \\ (\gamma m - \eta^2)/(m^2) \end{bmatrix} u(t),$$
$$y(t) = [1 \quad 0]\, x(t).$$ ■

8.2. EXISTENCE AND UNIQUENESS

Next we consider conditions under which the existence of a unique solution to the state equation is guaranteed. Let $T = [t0, t1]$ be a subset of **R** denoting the *interval of time* over which the behavior of S is being investigated. We refer to $t0$ as the *initial time* and $t1$ as the *final time*. For each time $t \in T$ the vector $x(t) \in \mathbf{R}^n$ is then the state of S at time t. In particular, the state of S at the initial time, $x(t0)$, is called the *initial state* and the state $x(t1)$ is called the *final state*. In most practical cases we have $t1 < \infty$. However, in the event that the asymptotic behavior of $x(t)$ is of interest, we can take $t1$ to be infinite.

Given the initial time and initial state, the solution of the state equation is then determined by the nature of the input waveform u. We restrict our consideration to input signals that can be generated physically. In particular, let the *input space* U be the vector space of piecewise-continuous functions mapping

T into $\mathbf{R^m}$. That is,

$$\mathbf{U} \triangleq \{u \in \mathbf{F}(\mathbf{T}, \mathbf{R^m}) : u \text{ is piecewise-continuous}\}.$$

Here a function $u(t)$ is *piecewise-continuous* if and only if it is continuous except possibly on a set T_d of isolated points, a finite number per unit interval, where jump discontinuities can occur. Every continuous function is also piecewise-continuous since piecewise continuity is a weaker form of continuity. Examples of piecewise-continuous functions that are not continuous include square waves, rectangular pulse trains, and sawtooth waves.

Next let $x^0 \in \mathbf{R^n}$ be an arbitrary initial state. We then refer to the constraint, $x(t0) = x^0$, as the *initial condition* of S. To establish sufficient conditions for the existence of a unique solution to the state equation satisfying an initial condition, we must place additional continuity constraints on the state function f. In particular we require that f be Lipschitz-continuous in x, continuous in u, and piecewise-continuous in t. The notion of Lipschitz continuity is defined as follows.

Definition 2.1 (Lipschitz Continuity). Let $h: \mathbf{R^n} \times T \to \mathbf{R^m}$ where $h(x, t)$ is piecewise-continuous in t. Then h is *Lipschitz-continuous* in $x \Leftrightarrow$ there exists a piecewise-continuous $M : T \to [0, \infty)$ called a *Lipschitz function* such that for each $(t, x, y) \in T \times \mathbf{R^n} \times \mathbf{R^n}$:

$$\|h(x, t) - h(y, t)\| \leqslant M(t)\|x - y\|. \qquad \blacksquare$$

Note that if h is Lipschitz-continuous, then it is clearly also continuous. That is, Lipschitz continuity is a stronger form of continuity in the sense that the slope or derivative is bounded. Given the notion of Lipschitz continuity, we can now state some simple sufficient conditions that guarantee the existence of a unique solution over T satisfying the initial condition, $x(t0) = x^0$.

Proposition 2.1 (Existence and Uniqueness). Suppose $f(x, u, t)$ is Lipschitz-continuous in x, continuous in u, and piecewise-continuous in t with T_d denoting the points of discontinuity. Then for each $(x^0, u) \in \mathbf{R^n} \times \mathbf{U}$, there exists a unique continuous $x(t)$ such that

1. $\dot{x}(t) = f[x(t), u(t), t], \; t \in T - T_d$,
2. $x(t0) = x^0$. $\blacksquare$

We call $x(t)$ the *solution* of the state equation satisfying the initial condition. The interested reader is referred to Desoer (1970) for a proof of Prop. 2.1. Note that T_d represents the points in T at which $x(t)$ may not be differentiable even though it is continuous. T_d will be a set of measure zero and, in many instances, T_d is the empty set.

Exercise 2.1 (Linear Case). Show that a linear differential system has a unique solution for each $(x^0, u) \in \mathbf{R^n} \times \mathbf{U}$ if the matrices $A(t)$ and $B(t)$ are piecewise-continuous on T. ■

In light of Exer. 2.1, we will assume that the matrix parameters $A(t)$ and $B(t)$ are at least piecewise-continuous on T in which case a unique solution exists for each $(x^0, u) \in \mathbf{R^n} \times \mathbf{U}$. Once the existence of a unique solution has been established, an iterative technique can be used to construct an arbitrarily accurate approximation to that solution. Consider, in particular, the following successive approximation algorithm. Here a sequence of vectors is constructed in the Banach space $\mathbf{C^n_\infty(T)}$. Recall that $\mathbf{C^n_\infty(T)}$ represents the normed vector space of continuous functions mapping T into $\mathbf{R^n}$ which are bounded respect to the supremum or sup norm.

Algorithm 2.1 (Picard)

1. Set $k = 0$ and $x^0(t) = x^0$.
2. Compute $x^{k+1} = Ax^k$ using,

$$(Ax)(t) = x^0 + \int_{t0}^{t} f[x(\tau), u(\tau), \tau]\, d\tau.$$

3. Set $k = k + 1$, and go to 2. ■

Proposition 2.2 (Picard). Let $(x^0, u) \in \mathbf{R^n} \times \mathbf{U}$, and let $f(x, u, t)$ be Lipschitz-continuous in x with Lipschitz function $M(t)$, continuous in u, and piecewise-continuous in t where T_d denotes the points of discontinuity. If $|t1 - t0| < 1/M(t)$ for $t \in T$, then the sequence of vectors $\{x^k\}$ constructed by Alg. 2.1 converges in $\mathbf{C^n_\infty(T)}$ to a continuous $x(t)$ such that

1. $\dot{x}(t) = f[x(t), u(t), t]$, $t \in T - T_d$,
2. $x(t0) = x^0$.

Proof. Let $\mathbf{V} = \mathbf{C^n_\infty(T)}$, the space of continuous functions mapping $T = [t0, t1]$ into $\mathbf{R^n}$, where $t1 < \infty$. For each $x \in \mathbf{V}$

$$\|x\|_\infty = \sup_{t \in T}\{\|x(t)\|\}.$$

Here $\|x(t)\|$ is any norm in $\mathbf{R^n}$. With the ∞-norm on $\mathbf{V}$, the space $\mathbf{V}$ is complete and therefore a Banach space. Next, consider the following nonlinear operator A on $\mathbf{V}$. For each $x \in \mathbf{V}$

$$(Ax)(t) = x^0 + \int_{t0}^{t} f[x(\tau), u(\tau), \tau]\, d\tau, \qquad t \in T.$$

Since f is Lipschitz-continuous in x with Lipschitz function $M(t)$, for each $(y, z) \in \mathbf{V}^2$ we have

$$\begin{aligned}
\|(Ay)(t) - (Az)(t)\| &= \left\| \int_{t0}^{t} \{f[y(\tau), u(\tau), \tau] - f[z(\tau), u(\tau), \tau]\, d\tau \right\| \\
&\leqslant \int_{t0}^{t} \|f[y(\tau), u(\tau), \tau] - f[z(\tau), u(\tau), \tau]\|\, d\tau \\
&\leqslant \int_{t0}^{t} M(\tau)\| y(\tau) - z(\tau)\|\, d\tau \\
&\leqslant M_0 \int_{t0}^{t} \| y(\tau) - z(\tau)\|\, d\tau.
\end{aligned}$$

Here $M_0 \triangleq \sup_{t \in T}\{M(t)\}$. Since the interval T is finite, M_0 exists. If the final time $t1$ is selected such that $|t1 - t0| < 1/M_0$, it then follows that

$$\begin{aligned}
\|Ay - Az\|_\infty &= \sup_{t \in T}\{\|(Ay)(t) - (Az)(t)\| \\
&\leqslant M_0 \sup_{t \in T}\left\{\int_{t0}^{t} \| y(\tau) - z(\tau)\|\, d\tau\right\} \\
&\leqslant M_0 \sup_{t \in T}\left\{\int_{t0}^{t1} \| y(\tau) - z(\tau)\|\, d\tau\right\} \\
&= M_0 \int_{t0}^{t1} \| y(\tau) - z(\tau)\|\, d\tau \\
&= M_0 |t1 - t0| \sup_{t \in T}\| y(t) - z(t)\| \\
&= M_0(|t1 - t0|)\| y - z\|_\infty \\
&< \| y - z\|_\infty.
\end{aligned}$$

Thus when $|t1 - t0| < 1/M_0$, the operator A is a contraction mapping on the Banach space $\mathbf{V}$. From step 2 of Alg. 2.1, $x^{k+1} = Ax^k$ for $k \geqslant 1$. Thus from Prop. 2.8.2, there exists a unique $x \in \mathbf{V}$ called the *fixed point* of A such that

$$x^k \to x \quad \text{as} \quad k \to \infty.$$

Since $x \in \mathbf{V}$, it follows that $x(t)$ is continuous. Furthermore, since x is the fixed point of A, we have $Ax = x$. Hence for each $t \in T - T_d$

$$\begin{aligned}
\dot{x}(t) &= \frac{d}{dt}\, x(t) \\
&= \frac{d}{dt}\, (Ax)(t)
\end{aligned}$$

$$= \frac{d}{dt}\left(x^0 + \int_{t0}^{t} f[x(\tau), u(\tau), \tau]\, d\tau\right)$$

$$= \frac{d}{dt}\int_{t0}^{t} f[x(\tau), u(\tau), \tau]\, d\tau$$

$$= f[x(t), u(t), t].$$

Finally, if we evaluate $x^k(t)$ at the initial time we get $x^k(t0) = x^0$ for $k \geqslant 1$. Since $x^k \to x$ as $k \to \infty$, it follows that $x(t0) = x^0$. ■

Thus the iterative procedure in Alg. 2.1 specifies a contraction mapping on the Banach space $\mathbf{C}^n_\infty(\mathbf{T})$ if the interval T is short enough. The fixed point of this contraction mapping is $x(t)$, the unique solution of the nonlinear time-varying differential system S. The upper bound on $|t1 - t0|$ is not a serious limitation. Indeed, T can be shortened as necessary to satisfy the hypotheses of Prop. 2.2. Then the reduced $t1$ can be taken as a new initial time, and Alg. 2.1 can be repeated. In this way a solution over an arbitrarily long interval T can be constructed piece-by-piece.

Exercise 2.2 (Successive Approximation). Apply Alg. 2.1 to the following state equation with $T = [0, 1]$ and $x(0) = 1$. Compute and sketch $x^k(t)$ for $0 \leqslant k \leqslant 3$. What is $x(t)$ in this case?

$$\dot{x}(t) = -x(t)/2. \qquad ■$$

Exercise 2.3 (Convergence Rate). Let $\dot{x}(t) = f[x(t)]$ with Lipschitz constant M, and let $x^k(t)$ be the sequence constructed by Alg. 2.1. Show that $\{x^k\}$ converges to x with the following upper bound on the rate of convergence:

$$\|x^k - x\|_\infty < \frac{M^k(t1 - t0)^{k+1} f(x^0)}{1 - M(t1 - t0)}. \qquad ■$$

8.3. THE STATE SPACE REPRESENTATION

Suppose the differential system S is linear with $A(t)$ and $B(t)$ piecewise-continuous on T. To obtain the unique solution to the state equation satisfying the initial condition $x(t0) = x^0$, we begin by considering the *homogeneous* or undriven case when $u = 0$. Consider, in particular, a matrix $Q(t, t0)$ whose kth column $q^k(t, t0)$ is the solution of the homogeneous state equation given the initial condition $q^k(t0) = i^k$ where $I = [i^1, i^2, \ldots, i^n]$. The matrix $Q(t, t0)$ provides the key to the solution of the state equation in general. We call $Q(t, t0)$ the state transition matrix and define it formally as follows.

Definition 3.1 (State Transition Matrix). The *state transition matrix* $Q(t, t0)$ of a linear differential system S is the solution to the following matrix differential equation:

$$\dot{Q}(t, t0) = A(t)Q(t, t0), \qquad Q(t0, t0) = I. \qquad \blacksquare$$

The term *state transition* arises from the fact that when $u = 0$, the matrix $Q(t, t0)$ maps the state at time $t0$ into the state at time t, giving us the following result.

Proposition 3.1 (Homogeneous Solution: $u = 0$). Let S be a linear differential system with state transition matrix $Q(t, t0)$. If $u = 0$, then for each $x^0 \in \mathbf{R}^n$ the solution to the homogeneous state equation satisfying the initial condition $x(t0) = x^0$ is

$$x(t) = Q(t, t0)x^0, \qquad t \in T.$$

Proof. Let $z(t) = Q(t, t0)x^0$. Then, using Def. 3.1, for each $t \in T$

$$\begin{aligned} \frac{d}{dt} z(t) &= \frac{d}{dt} Q(t, t0)x^0 \\ &= A(t)Q(t, t0)x^0 \\ &= A(t)z(t). \end{aligned}$$

Again from Def. 3.1 we have

$$\begin{aligned} z(t0) &= Q(t0, t0)x^0 \\ &= Ix^0 \\ &= x^0. \end{aligned}$$

Since $A(t)$ is piecewise-continuous, the solution to the state equation satisfying the initial condition is unique. Hence

$$\begin{aligned} x(t) &= z(t) \\ &= Q(t, t0)x^0. \end{aligned} \qquad \blacksquare$$

Thus every solution to the homogeneous state equation can be expressed as a linear combination of the columns of $Q(t, t0)$. The coefficients of that linear combination are simply the components of the initial state vector, x^0. Refer to Figure 8.6 for a pictorial summary of Prop. 3.1.

In light of Prop. 3.1, it is worthwhile to examine the nature of the state transition matrix in more detail. Note from Def. 3.1 that $Q(t, t0)$ depends upon

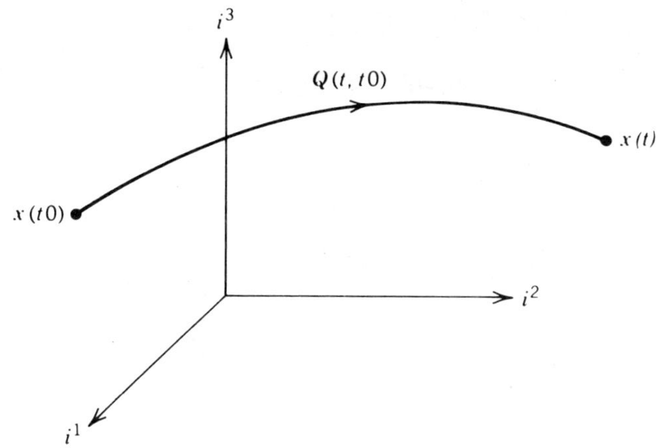

Figure 8.6. The state transition property of $Q(t, t0)$.

the matrix $A(t)$, which we assume is piecewise-continuous. Every state transition matrix $Q(t, t0)$ possesses a number of fundamental properties as summarized in the following result.

Proposition 3.2 (Properties of $Q(t, t0)$). Let $Q(t, t0)$ be the state transition matrix of a linear differential system S. Then for each $t \in T$

1. $Q(t, t) = I$,
2. $Q(t, t0) = Q(t, \tau)Q(\tau, t0)$, $\tau \in T$,
3. $Q^{-1}(t, t0) = Q(t0, t)$.

Proof. Since $t0 \in \mathbf{R}$ was arbitrary, from Def. 3.1 we have $Q(t, t) = I$. Let $P(t, t0) = Q(t, \tau)Q(\tau, t0)$. Then from Def. 3.1

$$\begin{aligned} \dot{P}(t, t0) &= \dot{Q}(t, \tau)Q(\tau, t0) \\ &= A(t)Q(t, \tau)Q(\tau, t0) \\ &= A(t)P(t, t0). \end{aligned}$$

Furthermore, from Def. 3.1 and property 1,

$$\begin{aligned} P(\tau, t0) &= Q(\tau, \tau)Q(\tau, t0) \\ &= IQ(\tau, t0) \\ &= Q(\tau, t0). \end{aligned}$$

It follows that $P(t, t0)$ and $Q(t, t0)$ satisfy the same differential equation and they are equal at $t = \tau$. Thus

$$\begin{aligned} Q(t, t0) &= P(t, t0) \\ &= Q(t, \tau)Q(\tau, t0). \end{aligned}$$

Finally, to establish property 3, note from properties 1 and 2 that,

$$Q(t0, t)Q(t, t0) = Q(t0, t0) = I,$$

$$Q(t, t0)Q(t0, t) = Q(t, t) = I.$$

Thus $Q(t, t0)$ is invertible and $Q^{-1}(t, t0) = Q(t0, t)$. ■

In light of Prop. 3.2, the state transition matrix $Q(t, t0)$ is always invertible even when $A(t)$ is singular. Moreover, the inverse of $Q(t, t0)$ is obtained by merely interchanging its arguments. Given the additional properties of the state transition matrix, we are now in a position to extend Prop. 3.1 to obtain the solution for the general or *nonhomogeneous* case when $u \neq 0$. We have the following result.

Proposition 3.3 (Nonhomogeneous Solution: $u \neq 0$). Let S be a linear differential system with state transition matrix $Q(t, t0)$. Then for each $(x^0, u) \in \mathbf{R}^n \times \mathbf{U}$ the solution to the state equation satisfying the initial condition $x(t0) = x^0$ is

$$x(t) = Q(t, t0)x^0 + \int_{t0}^{t} Q(t, \tau)B(\tau)u(\tau)\, d\tau.$$

Proof. Let $z(t)$ be the candidate solution. That is

$$z(t) = Q(t, t0)x^0 + \int_{t0}^{t} Q(t, \tau)B(\tau)u(\tau)\, d\tau.$$

Then, using Def. 3.1 and Prop. 3.2,

$$\begin{aligned} \dot{z}(t) &= \dot{Q}(t, t0)x^0 + Q(t, t)B(t)u(t) + \int_{t0}^{t} \dot{Q}(t, \tau)B(\tau)u(\tau)\, d\tau \\ &= A(t)Q(t, t0)x^0 + B(t)u(t) + \int_{t0}^{t} A(t)Q(t, \tau)B(\tau)u(\tau)\, d\tau \\ &= A(t)\{Q(t, t0)x^0 + \int_{t0}^{t} Q(t, \tau)B(\tau)u(\tau)\, d\tau\} + B(t)u(t) \\ &= A(t)z(t) + B(t)u(t). \end{aligned}$$

Evaluating $z(t)$ at $t = t0$ using Prop. 3.2,

$$\begin{aligned} z(t0) &= Q(t0, t0)x^0 \\ &= Ix^0 \\ &= x^0. \end{aligned}$$

Thus $z(t)$ and $x(t)$ satisfy the same differential equation and the same initial condition. With the solution being unique, it follows that $x(t) = z(t)$. ■

Note that Prop. 3.3 reduces to Prop. 3.1 given the undriven special case, $u = 0$. In light of Prop. 3.3, once the state transition matrix $Q(t, t0)$ is known, the general solution to the state equation can be found directly.

Exercise 3.1 (Decomposition). Show that the solution to the state equation of a linear differential system S can be decomposed as follows:

$$x(t) = x(t, x^0) + x(t, u).$$

Here $x(t, x^0)$ is the *zero-input solution* that results when $u = 0$. Similarly, $x(t, u)$ is the *zero-state solution* that results when $x^0 = 0$. ■

The following example illustrates the use of Prop. 3.3 to find the solution to a special form of time-varying linear system.

Example 3.1 (Uncoupled system). Consider an *uncoupled* linear time-varying system characterized by the following state equation:

$$\dot{x}_k(t) = a_k(t)x_k(t) + b_k(t)u(t), \qquad 1 \leqslant k \leqslant n.$$

Because the A matrix is diagonal in this case, an explicit expression can be obtained for the state transition matrix $Q(t, t0)$. In particular, using Def. 3.1 we have

$$Q_{kj}(t, t0) = \exp\left(\int_{t0}^{t} a_k(\tau)\, d\tau\right) I_{kj}, \qquad 1 \leqslant k, j \leqslant n.$$

Thus from Prop. 3.1, the part of the solution due to the initial condition, the zero-input solution $x(t, x^0)$, is

$$x_k(t, x^0) = Q_{kk}(t, t0)x_k^0, \qquad 1 \leqslant k \leqslant n.$$

Next from Prop. 3.3, the part of the solution due to the input, the zero-state solution $x(t, u)$ is

$$x_k(t, u) = \int_{t0}^{t} Q_{kk}(t, \tau)b_k(\tau)u(\tau)\, d\tau, \qquad 1 \leqslant k \leqslant n.$$

Finally, using Exer. 3.1, the complete solution is the sum of $x(t, x^0)$ and $x(t, u)$. That is, for $1 \leqslant k \leqslant n$

$$x_k(t) = Q_{kk}(t, t0)x_k^0 + \int_{t0}^{t} Q_{kk}(t, \tau)b_k(\tau)u(\tau)\, d\tau. \qquad \blacksquare$$

We have now succeeded in expressing the general solution of the state equation in terms of a set of n special solutions. These special solutions form the columns of the state transition matrix $Q(t, t0)$. The kth column is obtained by solving the homogeneous state equation using the kth column of the identity matrix as the initial state. Since the state transition matrix provides the key to the solution in general, we now focus on the problem of computing $Q(t, t0)$. When the linear system S is time-varying, we must often resort to a numerical solution of the differential equation in Def. 3.1. However in the event that S is time-invariant, we can obtain a simple explicit expression for $Q(t, t0)$, as shown in the following result.

Proposition 3.4 (Time-Invariant Case). Let S be a linear time-invariant differential system with state transition matrix $Q(t, t0)$. Then

$$Q(t, t0) = \exp[A(t - t0)].$$

Proof. Let $P(t, t0) = \exp[A(t - t0)]$. Then $P(t, t0)$ is an analytic function of A, parameterized by $t - t0$. Using Cauchy's integral representation of an analytic function of a matrix, we have from Prop. 7.7.4,

$$P(t, t0) = \oint_C \exp[\lambda(t - t0)](\lambda I - A)^{-1}\, d\lambda.$$

Here C is any circle containing $\sigma(A)$ traversed in a counterclockwise sense. Differentiating this expression for $P(t, t0)$ with respect to t, we get

$$\begin{aligned} \frac{d}{dt} P(t, t0) &= \oint_C \lambda \exp[\lambda(t - t0)](\lambda I - A)^{-1}\, d\lambda \\ &= A \exp[A(t - t0)] \\ &= AP(t, t0). \end{aligned}$$

Evaluating $P(t, t0)$ at the initial time $t = t0$ yields

$$\begin{aligned} P(t0, t0) &= \exp[A(t0 - t0)] \\ &= \exp(0) \\ &= I. \end{aligned}$$

It follows from Def. 3.1 that

$$\begin{aligned} Q(t, t0) &= P(t, t0) \\ &= \exp[A(t - t0)]. \end{aligned}$$ ■

Thus for time-invariant linear systems, the state transition matrix $Q(t, t0)$ is simply a matrix exponential. The fact that $Q(t, t0)$ depends only on the A matrix is apparent from this explicit formulation of the state transition matrix.

Exercise 3.2 (Elapsed Time). Let $Q(t, t0)$ be the state transition matrix of a linear time-invariant differential system S. Show that

$$Q(t, t0) = Q(t - t0, 0), \qquad t \in T.$$ ■

Note from Exer. 3.2 that for time-invariant linear systems it is the *elapsed time* $t - t0$ that is significant, not the initial time $t0$. Consequently, for time-invariant linear systems we will often assume $t0 = 0$ as a matter of convenience. In light of Prop. 3.4, the computation of $Q(t, t0)$ reduces to the computation of an analytic function of a matrix. There are several approaches we can take. The first is based upon the notion of the interpolating polynomial.

Proposition 3.5 (Interpolating Polynomial). Let $Q(t, 0)$ be the state transition matrix of a linear time-invariant differential system S. Then there exist functions $\{a_1(t), a_2(t), \ldots, a_n(t)\}$ such that

$$Q(t, 0) = \sum_{k=1}^{n} a_k(t) A^{k-1}.$$

Proof. Let g be the interpolating polynomial of f on $\sigma(A)$ where

$$f(\lambda) = \exp(\lambda t).$$

Since f is parameterized by t, it follows that the coefficients of g will be functions of t. That is

$$g(\lambda) = \sum_{k=1}^{n} a_k(t) \lambda^{k-1}.$$

Using Prop. 3.4, we then have

$$\begin{aligned} Q(t, 0) &= \exp(At) \\ &= f(A) \\ &= g(A) \\ &= \sum_{k=1}^{n} a_k(t) A^{k-1}. \end{aligned}$$

■

Thus the state transition matrix of a linear time-invariant system of dimension n can always be expressed as a time-varying polynomial of A of degree $n - 1$. The following example illustrates the use of this technique for computing $Q(t, 0)$.

Example 3.2 (Interpolating Polynomial). Consider the seismograph system discussed in Exam. 1.4. To make the example specific, suppose the parameter values are $m = 1$, $\gamma = 2$, $\eta = 3$. The A matrix in this case then reduces to

$$A = \begin{bmatrix} 0 & 1 \\ -2 & -3 \end{bmatrix}.$$

The characteristic polynomial of this matrix is

$$\begin{aligned} \Delta(\lambda) &= \det(\lambda I - A) \\ &= \lambda(\lambda + 3) + 2 \\ &= \lambda^2 + 3\lambda + 2 \\ &= (\lambda + 1)(\lambda + 2). \end{aligned}$$

Thus the spectrum of A is $\sigma(A) = \{-1, -2\}$. Since $A \in \mathbf{R}^{2\times 2}$, the interpolating polynomial will be a vector in $\mathbf{P}^1(\mathbf{C})$. That is

$$g(\lambda) = a_1 + a_2\lambda.$$

The two eigenvalues of A are simple. Hence the coefficients of the interpolating polynomial must satisfy

$$\begin{aligned} a_1 - a_2 &= \exp(-t) \\ a_1 - 2a_2 &= \exp(-2t). \end{aligned}$$

Solving these equations for the coefficient vector a then yields

$$\begin{aligned} a_1(t) &= 2\exp(-t) - \exp(-2t), \\ a_2(t) &= \exp(-t) - \exp(-2t). \end{aligned}$$

Thus from Prop. 3.5 the state transition matrix of the seismograph system is

$$Q(t, 0) = [2\exp(-t) - \exp(-2t)]I + [\exp(-t) - \exp(-2t)]A$$
$$= \begin{bmatrix} 2\exp(-t) - \exp(-2t) & \exp(-t) - \exp(-2t) \\ 2\exp(-2t) - 2\exp(-t) & 2\exp(-2t) - \exp(-t) \end{bmatrix}. \qquad \blacksquare$$

Note that in light of Exer. 3.2, $Q(t, t0)$ can be obtained from $Q(t, 0)$ by simply replacing t with $t - t0$. A second approach to computing the state transition matrix is based upon the block diagonal form of A. Here we transform to a convenient coordinate system, compute $Q(t, 0)$ in this coordinate system, and then transform back to the original coordinates, Example 7.6.2 is in fact an illustration of the use of this technique to compute the state transition matrix of a 3×3 matrix A having two diagonal blocks. In the event that A is diagonalizable, the block diagonal technique is particularly effective as can be seen from the time-varying example illustrated in Exam. 2.1.

A third approach to computing $Q(t, 0)$ is based upon the spectral representation theorem. Besides serving as a computational technique, the spectral representation of $Q(t, 0)$ also provides valuable qualitative information about the asymptotic behavior of the solution of the state equation of S.

Proposition 3.6 (Spectral Representation). Let $Q(t, 0)$ be the state transition matrix of a linear time-invariant differential system S. If λ_k is an eigenvalue of A of multiplicity $m(k)$ for $1 \leqslant k \leqslant q$ and if $\{p_{kj}\}$ are the spectral polynomials of A, then

$$Q(t, 0) = \sum_{k=1}^{q} \sum_{j=1}^{m(k)} t^{j-1} \exp(\lambda_k t) p_{kj}(A), \qquad t \in T.$$

Proof. This follows directly from Prop. 3.4 and the spectral representation theorem, Prop. 7.7.1. $\blacksquare$

In light of Prop. 3.6, it is clear that the behavior of $Q(t, 0)$ for large values of t is determined entirely by the location of $\sigma(A)$. In particular, we have the following stability result which is left as an exercise.

Exercise 3.3 (Stability). Let S be a linear time-invariant differential system, and suppose the eigenvalues of A all have negative real parts. Show that S is stable in the following sense. For each $x^0 \in \mathbf{R}^n$ the homogeneous solution of the state equation satisfying the initial condition $x(0) = x^0$ has the property that

$$x(t) \to 0 \quad \text{as} \quad t \to \infty. \qquad \blacksquare$$

The following example illustrates the use of Prop. 3.6 to compute the spectral representation of a state transition matrix $Q(t, 0)$.

Example 3.3 (Spectral Representation). Consider a linear time-invariant system characterized by the following A matrix:

$$A = \begin{bmatrix} -1 & 0 & 2 \\ 0 & -1 & 0 \\ 0 & 0 & -1 \end{bmatrix}.$$

Since A is upper triangular, the characteristic polynomial of A is simply

$$\begin{aligned} \Delta(\lambda) &= (\lambda + 1)^3 \\ &= (\lambda - \lambda_1)^{m(1)}. \end{aligned}$$

In Exam. 7.7.1 the spectral polynomials of A were found to be

$$\begin{aligned} p_{11}(\lambda) &= 1. \\ p_{12}(\lambda) &= \lambda + 1, \\ p_{13}(\lambda) &= (\lambda + 1)^2/2. \end{aligned}$$

Thus from Prop. 3.6 we have

$$\begin{aligned} Q(t, 0) &= \sum_{k=1}^{q} \sum_{j=1}^{m(k)} t^{j-1} \exp(\lambda_k t) p_{kj}(A) \\ &= \exp(-t)I + t\exp(-t)(A + I) + t^2 \exp(-t)\frac{(A + I)^2}{2} \\ &= \exp(-t)\begin{bmatrix} 1 & 0 & 0 \\ 0 & 1 & 0 \\ 0 & 0 & 1 \end{bmatrix} + t\exp(-t)\begin{bmatrix} 0 & 0 & 2 \\ 0 & 0 & 0 \\ 0 & 0 & 0 \end{bmatrix} \\ &= \exp(-t)\begin{bmatrix} 1 & 0 & 2t \\ 0 & 1 & 0 \\ 0 & 0 & 1 \end{bmatrix}. \end{aligned}$$

■

Note that if an expression for $Q(t, 0)$ has been obtained, the spectral matrices $\{p_{kj}(A)\}$ can be extracted by inspection. Once $\{p_{kj}(A)\}$ and $\sigma(A)$ are known, the problem of computing $f(A)$ for any analytic f is a trivial one.

Exercise 3.4 (Spectral Representation). Use the final results of Exam. 3.2 to write, by inspection, the spectral representation of $Q(t, 0)$ for the seismograph system. ■

Exercise 3.5 (Simple Eigenvalues). Let $Q(t, 0)$ be the state transition matrix of a linear time-invariant differential system S whose A matrix has simple eigenvalues $\{\lambda_1, \lambda_2, \ldots, \lambda_n\}$. Show that if E_k is a representation of the orthogonal projection operator induced by the kth eigenspace, then

$$Q(t, 0) = \sum_{k=1}^{n} \exp(\lambda_k t) E_k. \qquad \blacksquare$$

Exercise 3.6 (Power Series). Let $Q(t, 0)$ be the state transition matrix of a linear time-invariant system S. Show that

$$Q(t, 0) = \sum_{k=0}^{\infty} \frac{t^k A^k}{k!}. \qquad \blacksquare$$

8.4. THE IMPULSE RESPONSE REPRESENTATION

Thus far we have focused on the state equation and its solution. As such, we have concentrated on what might be called an *internal* description of S. We now examine the system S from the viewpoint of its *external* variables, the input u and the output y, as shown in Figure 8.7.

Recall that the solution of the state equation can be decomposed into the sum of the zero-input solution plus the zero-state solution. If the linear system S is to be represented in terms of its external variables, then some assumption must be placed on the initial internal state of S. In particular, we assume that the system S is relaxed.

Definition 4.1 (Relaxed System). A linear system S is *relaxed* at time $t0 \Leftrightarrow x(t0) = 0$. $\blacksquare$

Thus a linear differential system is relaxed when it is in the zero state. In this case, the *complete response* $y(t)$ includes only the zero-state response $y(t, u)$. That is, the output depends uniquely on the input if the system starts out relaxed. Note from Exer. 3.3 that if a linear time-invariant system S is stable, then in the absence of an input the internal state of S will approach the zero-state as time approaches infinity. As the response depends continuously on the initial state, for a stable physical system all we have to do is wait a sufficiently long period of time before we apply an input to ensure that S is relaxed.

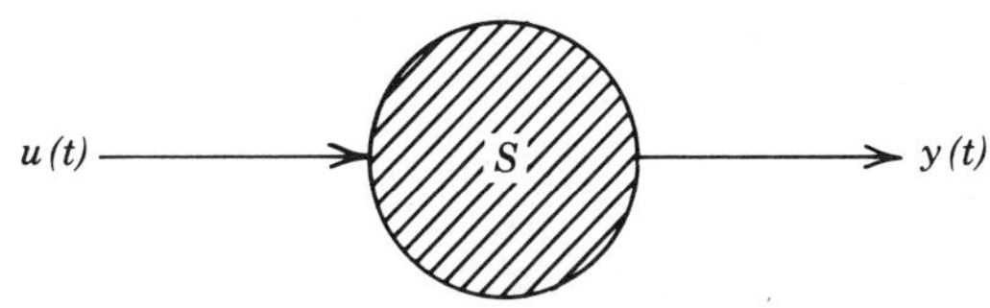

Figure 8.7. External representation of a system.

The mathematical relationship between the input and output of a relaxed linear system can be given a concise formulation if we introduce a second matrix associated with S called the impulse response matrix. We begin by examining the notion of a unit impulse.

Definition 4.2 (Impulse). The *unit impulse* is denoted $\delta(t)$ and is defined by the following two conditions:

1. $\delta(t) = 0,\ t \neq 0,$
2. $\int_R \delta(\tau)\, d\tau = 1.$ ■

Note that the unit impulse $\delta(t)$ is equal to zero everywhere except at $t = 0$ where it is in fact undefined. As such, $\delta(t)$ can not be regarded as a function mapping $\mathbf{R}$ into $\mathbf{R}$. However, upon integrating $\delta(t)$ once we do get a well defined function, the unit step function, $1(t)$. Therefore $\delta(t)$ is called a *generalized function* or *distribution* of order 1. The utility of the unit impulse lies in the following important property.

Proposition 4.1 (Sampling Property). Let $f:\mathbf{R} \to \mathbf{R}^n$ be continuous. Then for each $t0 \in \mathbf{R}$

$$\int_{\mathbf{R}} f(t)\delta(t - t0)\, dt = f(t0).$$

Proof. Using the two fundamental properties of $\delta(t)$ from Def. 4.2, and the continuity of f at $t0$, we have

$$\begin{aligned}\int_{\mathbf{R}} f(t)\delta(t - t0)\, dt &= \int_{t0-}^{t0+} f(t)\delta(t - t0)\, dt \\ &= \int_{t0-}^{t0+} f(t0)\delta(t - t0)\, dt \\ &= f(t0)\int_{t0-}^{t0+} \delta(t - t0)\, dt \\ &= f(t0)\int_{\mathbf{R}} \delta(t - t0)\, dt \\ &= f(t0)\int_{\mathbf{R}} \delta(t)\, dt \\ &= f(t0).\end{aligned}$$

■

Thus if a continuous function is modulated or multiplied by an impulse and the product is integrated, the effect is to *sample* or sift out the value of the

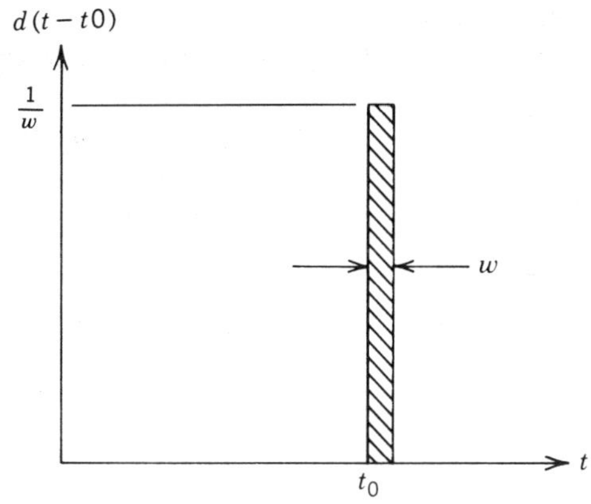

Figure 8.8. Approximation of unit impulse with brief pulse.

function at the time that the impulse occurs. From a practical standpoint, the unit impulse can be approximated, for example, by a brief pulse of unit area as shown in Figure 8.8.

Exercise 4.1 (General Sampling). Use integration by parts to show that Prop. 4.1 can be generalized as follows:

$$\int_{\mathbf{R}} f(t)\delta^{(k)}(t - t0)\, dt = (-1)^k f^{(k)}(t0), \qquad k \geqslant 0. \qquad \blacksquare$$

Next, we introduce a matrix called the impulse response matrix which is obtained by measuring the response of the system S to a unit impulse. More specifically, the impulse response matrix is a $p \times m$ matrix $G(t, t0)$ defined as follows where $I - [i^1, i^2, \ldots, i^m]^T$.

Definition 4.3 (Impulse Response Matrix). Let S be a linear differential system, and let $y^j(t)$ be the output of S assuming $x(t0) = 0$ and $u(t) = \delta(t - t0)i^j$. Then the *impulse response matrix* of S is denoted $G(t, t0)$ and is defined

$$G_{kj}(t, t0) \triangleq y_k^j(t), \qquad 1 \leqslant k \leqslant p, \quad 1 \leqslant j \leqslant m. \qquad \blacksquare$$

Thus the jth column of $G(t, t0)$ is the zero-state output of S at time t corresponding to an input $u^j(t)$ that is zero except for its jth component, which has an impulse at time $t0$. Note that the impulse response matrix $G(t, t0)$ can be obtained directly from input/output measurements of S. To derive the key property of the impulse response matrix, it is useful to first consider the relationship this matrix bears to the state transition matrix $Q(t, t0)$. In that regard we have the following result.

Proposition 4.2 (Impulse Response). Let S be a linear differential system with state transition matrix $Q(t, t0)$ and impulse response matrix $G(t, t0)$. Then

$$G(t, t0) = C(t)Q(t, t0)B(t0) + \delta(t - t0)D(t0).$$

Proof. Let $g^k(t, t0)$ be the kth column of $G(t, t0)$, and let $u^k(t) = \delta(t - t0)i^k$. Then, using Prop. 3.3 with $x(t0) = 0$ and the sampling property

$$\begin{aligned}
g^k(t, t0) &= C(t)x(t) + D(t)u^k(t) \\
&= C(t) \int_{t0}^{t} Q(t, \tau)B(\tau)u^k(\tau)\, d\tau + D(t)u^k(t) \\
&= C(t) \int_{t0}^{t} Q(t, \tau)B(\tau)\delta(\tau - t0)i^k\, d\tau + D(t)\delta(t - t0)i^k \\
&= \left\{C(t) \int_{t0}^{t} Q(t, \tau)B(\tau)\delta(\tau - t0)\, d\tau + \delta(t - t0)D(t0)\right\} i^k \\
&= \{C(t)Q(t, t0)B(t0) + \delta(t - t0)D(t0)\}\, i^k.
\end{aligned}$$

Since this holds for $1 \leqslant k \leqslant m$, it follows that

$$G(t, t0) = C(t)Q(t, t0)B(t0) + \delta(t - t0)D(t0). \qquad \blacksquare$$

The importance of the impulse response matrix $G(t, t0)$ lies in its usefulness in characterizing the input/output behavior of a relaxed linear system. In particular, we have the following fundamental result, which we refer to as the input/output representation of S.

Proposition 4.3 (Input/Output Representation). Let S be a linear differential system with impulse response matrix $G(t, t0)$. If $x(t0) = 0$, then for each $u \in \mathbf{U}$

$$y(t) = \int_{t0}^{t} G(t, \tau)u(\tau)\, d\tau, \qquad t \in T.$$

Proof. Applying Prop. 3.3 with $x(t0) = 0$, the sampling property, and Prop. 4.2, we have

$$\begin{aligned}
y(t) &= C(t)x(t) + D(t)u(t) \\
&= C(t) \int_{t0}^{t} Q(t, \tau)B(\tau)u(\tau)\, d\tau + D(t)u(t) \\
&= \int_{t0}^{t} C(t)Q(t, \tau)B(\tau)u(\tau)\, d\tau + D(t)u(t)
\end{aligned}$$

$$= \int_{t0}^{t} \{C(t)Q(t, \tau)B(\tau) + \delta(t - \tau)D(\tau)\}u(\tau)\, d\tau$$

$$= \int_{t0}^{t} G(t, \tau)u(\tau)\, d\tau. \qquad \blacksquare$$

Thus the impulse response matrix $G(t, t0)$ provides a complete characterization of the input/output behavior of a relaxed linear system. The following is an example of a linear system and its input/output representation based on the impulse response.

Example 4.1. Consider the seismograph system introduced in Exam. 1.4. Given the parameter values $m = 1, \gamma = 2, \eta = 3$, the state and output equations of S are

$$\dot{x}(t) = \begin{bmatrix} 0 & 1 \\ -2 & -3 \end{bmatrix} x(t) + \begin{bmatrix} 3 \\ -7 \end{bmatrix} u(t),$$

$$y(t) = [1 \quad 0]\ x(t).$$

Using the formulation of $Q(t, t0)$ obtained in Exam. 3.2, and applying Prop. 4.2, we have

$$\begin{aligned} G(t, t0) &= C(t)Q(t, t0)B(t0) + \delta(t - t0)D(t0) \\ &= [1, \quad 0]Q(t, t0)[3, \quad -7]^{\mathrm{T}} \\ &= 3Q_{11}(t, t0) - 7Q_{12}(t, t0) \\ &= 4 \exp[-2(t - t0)] - \exp[-(t - t0)]. \end{aligned}$$

It then follows from Prop. 4.3 that the relaxed or zero-state response of the seismograph system S for an arbitrary input $u \in \mathbf{U}$ is

$$y(t) = \int_{t0}^{t} \{4 \exp[-2(t - \tau)] - \exp[-(t - \tau)]\}u(\tau)\, d\tau. \qquad \blacksquare$$

Exercise 4.2 (Elapsed Time). Let S be a linear time-invariant differential system with impulse response matrix $G(t, t0)$. Show that

$$G(t, t0) = G(t - t0, 0), \qquad t \in T. \qquad \blacksquare$$

Note that Exer. 4.2 is analogous to Exer. 3.2, but for the impulse response matrix. That is, for time-invariant linear systems the initial time $t0$ is not important, only the elapsed time $t - t0$. The following result combines the impulse response matrix and the state transition matrix to yield a general characterization of the input/output behavior of a linear system S.

Proposition 4.4 (Complete Response). Let S be a linear differential system with state transition matrix $Q(t, t0)$ and impulse response matrix $G(t, t0)$. Then for each $(x^0, u) \in \mathbf{R}^n \times \mathbf{U}$, the output of S corresponding to initial condition $x(t0) = x^0$ is

$$y(t) = C(t)Q(t, t0)x^0 + \int_{t0}^{t} G(t, \tau)u(\tau)\, d\tau.$$

Proof. Let $(x^0, u) \in \mathbf{R}^n \times \mathbf{U}$. Then from Prop. 3.3 and Prop. 4.2

$$\begin{aligned} y(t) &= C(t)x(t) + D(t)u(t) \\ &= C(t)\left\{Q(t, t0)x^0 + \int_{t0}^{t} Q(t, \tau)B(\tau)u(\tau)\, d\tau\right\} + D(t)u(t) \\ &= C(t)Q(t, t0)x^0 + \int_{t0}^{t} C(t)Q(t, \tau)B(\tau)u(\tau)\, d\tau + D(t)u(t) \\ &= C(t)Q(t, t0)x^0 + \int_{t0}^{t} \{C(t)Q(t, \tau)B(\tau) + \delta(t-\tau)D(\tau)\}u(\tau)\, d\tau \\ &= C(t)Q(t, t0)x^0 + \int_{t0}^{t} G(t, \tau)u(\tau)\, d\tau. \end{aligned}$$

■

Exercise 4.3 (Decomposition). Show that the complete response of a linear differential system S can be decomposed as follows:

$$y(t) = y(t, x^0) + y(t, u).$$

Here $y(t, x^0)$ is the *zero-input response* that results when $u = 0$. Similarly, $y(t, u)$ is the *zero-state response* that results when $x^0 = 0$. ■

8.5. THE TRANSFER FUNCTION REPRESENTATION

The input/output representation of a relaxed linear system embodied in the impulse response matrix $G(t, t0)$ can be simplified considerably in the event that the system S is time-invariant. To see this, it is useful to first consider a linear transformation that allows us to recast a system of ordinary differential equations as an equivalent system of algebraic equations.

Definition 5.1 (Laplace Transform). Let $f: [0, \infty) \to \mathbf{R}^n$. Then the *Laplace transform* of f is denoted $F(s)$ or $L\{f(t)\}$ and is defined

$$F(s) \triangleq \int_{0}^{\infty} f(t) \exp(-st)\, dt, \qquad s \in W.$$

Here the subset **W** of **C** is called the *region of convergence* of $F(s)$. ■

Refer to Table 8.1 for a collection of commonly used Laplace transform pairs and their associated regions of convergence.

TABLE 8.1. Laplace Transform Pairs

$f(t)$	$F(s)$	W
$\delta^{(n)}(t)$	s^n	C
$t^n \exp(-\alpha t)$	$\dfrac{n!}{(s+\alpha)^{n+1}}$	$\mathrm{Re}(s) > -\alpha$
$\exp(-\alpha t)\cos(\omega t)$	$\dfrac{s+\alpha}{(s+\alpha)^2+\omega^2}$	$\mathrm{Re}(s) > -\alpha$
$\exp(-\alpha t)\sin(\omega t)$	$\dfrac{\omega}{(s+\alpha)^2+\omega^2}$	$\mathrm{Re}(s) > -\alpha$

It is evident from inspection of Table 8.1 that in each case $F(s)$ is the ratio of two polynomials in s. The zeros of the denominator polynomial are called the *poles* of $F(s)$. Note that the region of convergence, in each case, is that part of the complex plane to the right of the rightmost pole. That is, W is of the following form where α is the real part of the rightmost pole of $F(s)$:

$$W = \{s \in \mathbf{C} : \mathrm{Re}(s) > \alpha\}.$$

There are a number of useful properties that the Laplace transform satisfies. Some of the more fundamental ones are summarized in the following exercise.

Exercise 5.1 (Laplace Transform Properties). Let $f:[0,\infty) \to \mathbf{R}$ and $g:[0,\infty) \to \mathbf{R}$, and let $F(s)$ and $G(s)$ be the Laplace transforms of f and g, respectively. Show that for each $(\alpha, \tau) \in \mathbf{C} \times [0, \infty)$

1. $L\{\alpha f(t) + g(t)\} = \alpha F(s) + G(s)$,
2. $L\{f(t/\alpha)\} = \alpha F(\alpha s)$, $\alpha \neq 0$,
3. $L\{\exp(-\alpha t) f(t)\} = F(s+\alpha)$,
4. $L\{f(t-\tau)1(t-\tau)\} = \exp(-s\tau)F(s)$. ■

We refer to 1 of Exer. 5.1 as the linearity property, 2 as the scaling property, 3 as the shift property, and 4 as the delay property. Perhaps the most important property for system analysis purposes is the effect the Laplace transform has on the convolution of two functions. We define the convolution operation as follows.

Definition 5.2 (Convolution). Let $f:[0, \infty) \to \mathbf{R}$ and $g:[0, \infty) \to \mathbf{R}$. Then the *convolution* of f with g is denoted $f(t) * g(t)$ and defined as

$$f(t) * g(t) \triangleq \int_0^t f(t - \tau)g(\tau)\, d\tau. \qquad \blacksquare$$

The importance of the convolution operation lies in the fact that this is precisely the type of processing that a linear system performs when it is excited with input signal. In particular, the output of a relaxed linear time-invariant system S is simply the convolution of the input with the impulse response matrix as can be seen from the following exericse.

Exercise 5.2 (Convolution). Let S be a linear time-invariant differential system, and suppose $y(t)$ is the output of S associated with the input $u \in \mathbf{U}$ and the initial condition $x(0) = 0$. Show that

$$y(t) = G(t, 0) * u(t). \qquad \blacksquare$$

Although the convolution operation can involve substantial computational effort, the image of the convolution operation under the Laplace transformation is quite simple, as shown in the following result.

Proposition 5.1 (Convolution). Let f: $[0, \infty) \to \mathbf{R}$ and g: $[0, \infty) \to \mathbf{R}$, and let $F(s)$ and $G(s)$ denote the Laplace transforms of f and g, respectively. Then

$$L\{f(t) * g(t)\} = F(s)G(s).$$

Proof. Let $f_e(t)$ be an extension of the function f to the real line $\mathbf{R}$. Here $f_e(t) \triangleq 0$ for $t < 0$ and

$$f_e(t) \triangleq f(t), \qquad t \geqslant 0.$$

It then follows that $f_e(t - \tau)g(\tau) = f(t - \tau)g(\tau)$ for $\tau \in [0, t]$. Similarly, $f_e(t - \tau)g(\tau) = 0$ for $\tau \in (t, \infty)$. Thus from Def. 5.2

$$\begin{aligned} f(t) * g(t) &= \int_0^t f(t - \tau)g(\tau)\, d\tau \\ &= \int_0^t f_e(t - \tau)g(\tau)\, d\tau \\ &= \int_0^\infty f_e(t - \tau)g(\tau)\, d\tau. \end{aligned}$$

Using Def. 5.1, we then have

$$
\begin{aligned}
L\{f(t) * g(t)\} &= L\left\{\int_0^{\infty} f_e(t-\tau)g(\tau)\,d\tau\right\} \\
&= \int_0^{\infty}\int_0^{\infty} f_e(t-\tau)g(\tau)\,d\tau\,\exp(-st)\,dt \\
&= \int_0^{\infty}\int_0^{\infty} f_e(t-\tau)g(\tau)\exp(-st)\,d\tau\,dt \\
&= \int_0^{\infty}\int_0^{\infty} f_e(t-\tau)\exp[-s(t-\tau)]g(\tau)\exp(-s\tau)\,d\tau\,dt \\
&= \int_0^{\infty}\int_0^{\infty} f_e(t-\tau)\exp[-s(t-\tau)]g(\tau)\exp(-s\tau)\,dt\,d\tau \\
&= \int_0^{\infty}\int_0^{\infty} f_e(t-\tau)\exp[-s(t-\tau)]\,dt\,g(\tau)\,exp(-s\tau)\,d\tau \\
&= \int_0^{\infty}\int_{-\tau}^{\infty} f_e(\alpha)\exp(-s\alpha)\,d\alpha\,g(\tau)\exp(-s\tau)\,d\tau \\
&= \int_0^{\infty}\int_0^{\infty} f_e(\alpha)\exp(-s\alpha)\,d\alpha\,g(\tau)\exp(-s\tau)\,d\tau \\
&= \int_0^{\infty}\int_0^{\infty} f(\alpha)\exp(-s\alpha)\,d\alpha\,g(\tau)\exp(-s\tau)\,d\tau \\
&= \int_0^{\infty} F(s)g(\tau)\exp(-s\tau)\,d\tau \\
&= F(s)\int_0^{\infty} g(\tau)\exp(-s\tau)\,d\tau \\
&= F(s)G(s). \qquad \blacksquare
\end{aligned}
$$

Thus convolution in the time domain t maps into multiplication in the complex frequency domain s under the Laplace transformation. This observation, together with Exer. 5.2, provides the motivation for the following definition.

Definition 5.3 (Transfer Function Matrix). Let S be a linear time-invariant differential system with impulse response matrix $G(t, 0)$. Then the *transfer function matrix* $G(s)$ is defined as

$$G(s) \triangleq \mathrm{L}\,\{G(t, 0)\}. \qquad \blacksquare$$

Thus the transfer function matrix $G(s)$ is simply the Laplace transform of the impulse response matrix $G(t, t0)$ with $t0 = 0$. Note from Exer. 4.2 that the constraint on $t0$ is of no consequence as we are now restricting our attention to time-invariant systems.

Like the impulse response matrix it comes from, the transfer function matrix is an $p \times m$ matrix whose components can be obtained directly from input/output measurements of S. The transfer function matrix $G(s)$ provides us with a simple algebraic characterization of the input/output behavior of a relaxed linear time-invariant system, as shown in the following proposition.

Proposition 5.2 (Algebraic Representation). Let S be a linear time-invariant differential system with transfer function matrix $G(s)$, and let $y(t)$ be the output of S corresponding to $x(0) = 0$ and $u \in \mathbf{U}$. If $U(s)$ and $Y(s)$ denote the Laplace transforms of u and y, respectively, then

$$Y(s) = G(s)U(s).$$

Proof. Applying Exer. 5.2 and Prop. 5.1, we have

$$\begin{aligned} Y(s) &= L\{y(t)\} \\ &= L\{G(t, 0) * u(t)\} \\ &= L\{G(t, 0)\}U(s) \\ &= G(s)U(s). \end{aligned}$$ ■

Thus for a relaxed linear time-invariant system, the Laplace transform of the output is simply the transfer function times the Laplace transform of the input. Note that Prop. 5.2 effectively replaces a differential equation in $t \in \mathbf{R}$ with an algebraic equation in $s \in \mathbf{C}$.

Exercise 5.3. Let S be a linear time-invariant system with state transition matrix $Q(t, 0)$ and transfer function matrix $G(s)$. Show that

$$G(s) = CL\{Q(t, 0)\}B + D.$$ ■

As it now stands, the transfer function matrix is computed indirectly by first finding the impulse response matrix $G(t, 0)$ and then transforming it. Exercise 5.3 provides us with an analogous route based upon the state transition matrix $Q(t, 0)$. To develop a procedure for obtaining $G(s)$ directly from the system parameters $\{A, B, C, D\}$, it is useful to first consider the following additional fundamental property of the Laplace transform.

Proposition 5.3 (Derivative Property). Let $f:[0, \infty) \to \mathbf{R}$ be k times differentiable. Then

$$L\{f^{(k)}(t)\} = s^k F(s) - \sum_{j=0}^{k-1} f^{(j)}(0) s^{k-1-j}, \qquad k \geqslant 0.$$

Proof. We proceed by induction. The equation in Prop. 5.3 clearly holds when $k = 0$. Suppose it is true when $k = m$. Then, using Def. 5.1 and applying integration by parts,

$$\begin{aligned} L\{f^{(m+1)}(t)\} &= \int_0^\infty f^{(m+1)}(t) \exp(-st)\, dt \\ &= f^{(m)}(t) \exp(-st)|_0^\infty + s \int_0^\infty f^{(m)}(t) \exp(-st)\, dt \\ &= -f^{(m)}(0) + sL\{f^{(m)}(t)\} \\ &= sL\{f^{(m)}(t)\} - f^{(m)}(0) \\ &= s\left\{s^m F(s) - \sum_{j=0}^{m-1} f^{(j)}(0) s^{m-1-j}\right\} - f^{(m)}(0) \\ &= s^{m+1} F(s) - \sum_{j=0}^{m} f^{(j)}(0) s^{m-j}. \end{aligned}$$ ■

Thus when $f^{(k)}(0) = 0$ for $k \geqslant 0$, differentiation in the time domain corresponds to scaling by s in the Laplace transform domain. This in turn allows one to transform differential equations in t into algebraic equations in the complex variable s.

Exercise 5.4 (Transfer Function). Consider the following qth-order linear time-invariant differential equation where $a_0 = 1$:

$$\sum_{j=0}^{q} a_j y^{(j)}(t) = \sum_{j=0}^{q} b_j u^{(j)}(t).$$

Use the derivative property and Prop. 5.2 to show that the transfer function of this system is

$$G(s) = \frac{b_0 s^q + b_1 s^{q-1} + \cdots + b_q}{s^q + a_1 s^{q-1} + \cdots + a_q}.$$ ■

The derivative property of the Laplace transformation provides us with an alternative method for computing the state transition matrix $Q(t, 0)$, as can be seen from the following exercise.

Exercise 5.5 (State Transition Matrix). Let S be a linear time-invariant system with state transition matrix $Q(t, 0)$. Show that

$$L\{Q(t, 0)\} = (sI - A)^{-1}. \qquad \blacksquare$$

The results of Exers. 5.3 and 5.5 can now be combined. In doing so we generate the following direct formulation of the transfer function matrix $G(s)$ in terms of the matrix parameters of the system S.

Proposition 5.4 (Transfer Function Matrix). Let S be a linear time-invariant system with transfer function matrix $G(s)$. Then

$$G(s) = C(sI - A)^{-1}B + D. \qquad \blacksquare$$

The proof of Prop. 5.4 follows directly from Exer. 5.3 and Exer. 5.5. Typically, the transfer function $G(s)$ will be expressible as the ratio of two polynomials in s. This feature allows us to define something called the poles of the system S.

Definition 5.4 (Poles of S). Let S be a linear time-invariant system with transfer function matrix $G(s)$. Then $p \in \mathbf{C}$ is a *pole* of $S \Leftrightarrow p$ is a zero of the denominator polynomial of at least one component of the matrix $G(s)$. ■

Thus the poles of S are those points in the complex plane where the denominators of the components of $G(s)$ vanish. In light of Props. 5.4 and 7.3.2, every pole of S is an eigenvalue of A. However, the converse is not always true, as can be seen from the following exercise.

Exercise 5.6 (Poles of S). Let S be a linear time-invariant differential system with transfer function matrix $G(s)$. Show that every pole of S is an eigenvalue of A. Show that the converse is not true by finding a counterexample. ■

The following example illustrates the use of Prop. 5.4 to compute the transfer function matrix of a linear time-invariant system.

Example 5.1 (Transfer Function Matrix). Consider the seismograph system introduced in Exam. 1.4. Given the parameter values $m = 1$, $\gamma = 2$, $\eta = 3$, the state and output equations of S are

$$\dot{x}(t) = \begin{bmatrix} 0 & 1 \\ -2 & -3 \end{bmatrix} x(t) + \begin{bmatrix} 3 \\ -7 \end{bmatrix} u(t),$$

$$y(t) = [1 \quad 0]\, x(t).$$

In this case we have

$$(sI - A)^{-1} = \begin{bmatrix} s & -1 \\ 2 & s+3 \end{bmatrix}^{-1}$$
$$= \begin{bmatrix} s+3 & 1 \\ -2 & s \end{bmatrix} \Big/ \Delta(s).$$

Here $\Delta(s)$ is the characteristic polynomial of A. That is

$$\Delta(s) = \det(sI - A)$$
$$= s(s+3) + 2$$
$$= s^2 + 3s + 2.$$

Applying Prop. 5.4 to compute $G(s)$, we have

$$G(s) = C(sI - A)^{-1}B + D$$
$$= [1 \quad 0]\begin{bmatrix} s+3 & 1 \\ -2 & s \end{bmatrix}\begin{bmatrix} 3 \\ -7 \end{bmatrix} \Big/ \Delta(s)$$
$$= [s+3 \quad 1]\begin{bmatrix} 3 \\ -7 \end{bmatrix} \Big/ \Delta(s)$$
$$= \frac{3(s+3) - 7}{\Delta(s)}$$
$$= \frac{3s+2}{s^2 + 3s + 2}.$$

As a check, we know from Def. 5.3 that $G(s) = L\{G(t, 0)\}$. Using the expression for $G(t, 0)$ from Exam. 4.1, and Table 8.1,

$$G(s) = L\{G(t, 0)\}$$
$$= L\{4\exp(-2t) - \exp(-t)\}$$
$$= \frac{4}{s+2} - \frac{1}{s+1}$$
$$= \frac{4s + 4 - (s+2)}{(s+2)(s+1)}$$
$$= \frac{3s+2}{s^2 + 3s + 2}. \qquad \blacksquare$$

Proposition 5.4 provides us with a direct means of obtaining the transfer function matrix $G(s)$ from the system parameters $\{A, B, C, D\}$. However, it involves inverting the matrix $(sI - A)$, and this can be quite cumbersome for high-dimensional systems. The following algorithm provides an iterative procedure for obtaining a closed form expression for $(sI - A)^{-1}$. Here the notation *trace* (A) denotes the sum of the diagonal elements of A.

Algorithm 5.1 (Leverrier)

1. Set $k = 1$, $a_n = 1$, $P_n = I$.
2. Compute

$$a_{n-k} = \frac{1}{k}\,\text{trace}(AP_{n-k+1}),$$

$$R_{n-k} = a_{n-k}I + AR_{n-k+1}.$$

3. Set $k = k + 1$. If $k \leqslant n$, go to 2. ■

Proposition 5.5 (Leverrier). Let $\{a_k\}$ and $\{P_k\}$ be the scalars and matrices constructed by Alg. 5.1. Then

$$\det(sI - A) = \sum_{k=0}^{n} a_k s^k,$$

$$(sI - A)^{-1} = \frac{\sum_{k=1}^{n} R_k s^{k-1}}{\det(sI - A)}.$$ ■

The interested reader is referred to Zadeh and Desoer (1963) for a proof of Prop. 5.5. Note that in addition to producing $L\{Q(t, 0)\}$, Prop. 5.5 also provides us with an expression for the characteristic polynomial of an $n \times n$ matrix A. As a numerical check in applying Alg. 5.1, one should get $P_0 = 0$ on the last iteration.

To make the algebraic representation of S in Prop. 5.2 practical, we must have a means of inverting the Laplace transformation to recover $y(t)$ from $Y(s)$. This can be achieved in several ways, including the evaluation of an inversion integral in the complex plane. However, it is usually more convenient to take a less direct approach in which $F(s)$ is recast as a sum of simple expressions, each of which can be inverted by inspection with the help of Table 8.1. Then using the linearity property, the inverse transform of the sum is the sum of the inverse transforms.

The key to the table look-up method is to expand $F(s)$ into a sum of simple terms using the method of *partial fractions*. Suppose $F(s) = a(s)/b(s)$ where a and b are polynomials. In this case we say that $F(s)$ is a *rational polynomial* in s with numerator $a(s)$ and denominator $b(s)$. Furthermore, suppose that the degree of $a(s)$ is less than the degree of $b(s)$, in which case we say that $F(s)$ is a *strictly proper* rational polynomial. Note that if this is not so, we can always first process $F(s)$

with long division. The quotient will be a polynomial in s that will give rise to impulse terms in $f(t)$. The remainder will be a strictly proper rational polynomial with denominator $b(s)$. This remainder term can then be expanded into partial fraction terms using the following technique.

Proposition 5.6 (Partial Fractions). Let $F(s) = a(s)/b(s)$ be a strictly proper rational polynomial, and let p_k be a zero of $b(s)$ of multiplicity $m(k)$ for $1 \leqslant k \leqslant q$. Then $F(s)$ can be expanded into the following partial fraction representation:

$$F(s) = \sum_{k=1}^{q} \sum_{j=1}^{m(k)} \frac{R_{kj}(p_k)}{(s - p_k)^j}.$$

Here the coefficients of the partial fraction expansion $\{R_{kj}\}$ are computed as follows for $1 \leqslant k \leqslant q$ and $0 \leqslant j \leqslant m(k) - 1$:

$$R_{k,m(k)-j}(s) = \frac{1}{j!} \frac{d^j}{ds^j} \{(s - p_k)^{m(k)} F(s)\}. \qquad ■$$

A proof of Prop. 5.6 can be found in Zadeh and Desoer (1963). Note that each of terms in the partial fraction expansion of $F(s)$ appears in Table 8.1. The coefficients $\{R_{kj}\}$ are particularly simple to compute when the n zeros of $b(s)$ are all simple. In this case $R_{k1}(s) = (s - p_k)F(s)$ for $1 \leqslant k \leqslant n$.

Exercise 5.7 (Partial Fractions). Let $F(s)$ be expanded into partial fractions as in Prop. 5.5. Use Table 8.1 to show that

$$f(t) = \sum_{k=1}^{q} \sum_{j=1}^{m(k)} \frac{R_{kj}(p_k)t^{j-1} \exp(-p_k t)}{(j - 1)!}. \qquad ■$$

Example 5.2 (Relaxed Response). Consider a linear time-invariant system S with the following transfer function:

$$G(s) = \frac{s^2}{s^2 + 5s + 6}.$$

Suppose $x(0) = 0$ and $u(t) = 2\exp(-t)$ for $t \geqslant 0$. To find $y(t)$, first note from Prop. 5.2 and Table 8.1 that

$$\begin{aligned} Y(s) &= G(s)U(s) \\ &= \frac{s^2}{s^2 + 5s + 6} \frac{2}{s + 1} \\ &= \frac{2s^2}{(s + 2)(s + 3)(s + 1)} = \frac{-8}{s + 2} + \frac{9}{s + 3} + \frac{1}{s + 1}. \end{aligned}$$

Using Table 8.1 we can invert $Y(s)$ to get

$$y(t) = -8\exp(-2t) + 9\exp(-3t) + \exp(-t). \qquad \blacksquare$$

The three representations of linear differential systems based upon the state transition matrix $Q(t, t0)$, the impulse response matrix $G(t, t0)$, and the transfer function matrix $G(s)$ are summarized in Table 8.2. Note that the state space representation is a complete *internal* representation since it takes into account the effects of both the input u and the initial state x^0. The impulse response representation and the transfer function representation are less complete *external* representations that assume the system S is initially relaxed. Of these two input/output representations of S, the impulse response representation is the more general, for it applies to both time-varying and time-invariant linear systems, whereas the transfer function representation is applicable only to time-invariant linear systems.

TABLE 8.2. Three Representations of a Linear Differential System

Representation	Type	Relaxed	Time-Invariant
State space: $Q(t, t0)$	Internal	No	No
Impulse response: $G(t, t0)$	External	Yes	No
Transfer function: $G(s)$	External	Yes	Yes

PROBLEMS

1. Convert the following second-order differential equation to a first-order system S. Is S linear? Is S time-invariant?

$$y^{(2)}(t) + ty^{(1)}(t)\{1 - y^2(t)\} = u(t).$$

2. Convert the following third-order differential equation to a first-order system S.

$$y^{(3)}(t) + 2y^{(2)}(t) + 3y^{(1)}(t) + 4y(t) = 5u^{(2)}(t) + 6u^{(1)}(t) + 7u(t).$$

3. Consider the following linear time-invariant system S where $a > 0$ and $T = [0, t1]$. For what values of $t1$ will Alg. 2.1 converge to the solution?

$$\dot{x}(t) = -ax(t).$$

4. Consider the following nonlinear time-invariant system S where $x(0) = a$ and $T = [0, t1]$. Find $\{x^k(t)\}$ for $0 \leqslant k \leqslant 2$ using Alg. 2.1

$$\dot{x}(t) = -x^3(t).$$

5. Find an upper bound on the rate of convergence for the system in Exer. 2.3.

6. Consider a linear time-invariant system with the following state equation:

$$\dot{x} = \begin{bmatrix} -1 & 2 \\ 0 & -3 \end{bmatrix} x + \begin{bmatrix} 0 \\ 1 \end{bmatrix} u.$$

   a. Find the state transition matrix $Q(t, 0)$.
   b. Find the homogeneous solution when $x(0) = x^0$.

7. Let $P(t)$ be a *fundamental matrix* of $A(t)$. That is, let $P(t)$ be any solution of the following differential equation:

$$\dot{P}(t) = A(t)P(t), \qquad \det\{P(t0)\} \neq 0.$$

   a. Show that $\det\{P(t)\} \neq 0$ for $t \geqslant 0$.
   b. Show that $Q(t, t0) = P(t)P^{-1}(t0)$.

8. Find the zero-state solution $x(t, u)$ of the system in problem 6 when the input is $u(t) = \exp(-2t)$, $t \geqslant 0$.

9. Consider the following linear time-invariant state equation:

$$\dot{x} = \begin{bmatrix} -3 & -2 \\ 2 & 1 \end{bmatrix} x + \begin{bmatrix} 0 \\ 1 \end{bmatrix} u.$$

   a. Find the zero-input solution when $x(0) = [1, -1]^T$.
   b. Find the zero-state solution when $u(t) = 3$, $t \geqslant 0$.
   c. Use the results of a and b to write the complete solution when $x(0) = [2, -2]^T$ and $u(t) = 1$ for $t \geqslant 0$.

10. Compute $Q(t, 0)$ for the system in problem 9 using the method of interpolating polynomials.

11. Compute the spectral representation of $Q(t, 0)$ for the system in Problem 9.

12. Consider the following linear time-invariant system S.

$$\dot{x} = \begin{bmatrix} 1 & -3 \\ 2 & -4 \end{bmatrix} x + \begin{bmatrix} 0 \\ 1 \end{bmatrix} u,$$

$$y = [1 \quad 0]x + [3]u.$$

   a. Find the impulse response matrix $G(t, 0)$.
   b. Find the transfer function matrix $G(s)$.

13. Find the zero-input response $y(t, x^0)$ of the system in problem 12 when $x(0) = x^0$.

14. Find the zero-state response $y(t, u)$ of the system in problem 12 when $u(t) = \alpha$, $t \geqslant 0$.

15. Find the complete response of the system in problem 12 when $x(0) = [2, 0]^T$ and $u(t) = 10$, $t \geqslant 0$.

16. Consider the following linear time-invariant system with transfer function $G(s)$. Find the zero-state response to $u(t) = (\frac{1}{2})^t$, $t \geqslant 0$.

$$G(s) = \frac{s}{s^2 + 5s + 4}.$$

17. Let $f : [0, \infty) \rightarrow \mathbf{R}$, and let $F(s)$ be the Laplace transform of f. Show that the *initial value theorem* holds. That is, find conditions under which

$$f(0+) = \lim_{s \rightarrow \infty} \{sF(s)\},$$

18. Let $f : [0, \infty) \rightarrow \mathbf{R}$ and let $F(s)$ be the Laplace transform of f. Show that the *final value theorem* holds. That is, find conditions under which

$$f(\infty) = \lim_{s \rightarrow 0} \{sF(s)\}.$$

19. For a linear time-invariant system S consider the following matrix $W(t0, t1)$, which is called the *controllability Grammian* (Chen, 1984):

$$W(t0, t1) \triangleq \int_{t0}^{t1} Q(t0, t)BB^T Q^T(t0, t)\, dt.$$

Suppose $W(t0, t1)$ is nonsingular. Show that the following input drives the system S from the initial state x^0 at time $t0$ to the final state x^1 at time $t1$:

$$u(t) = -B^T Q^T(t0, t) W^{-1}(t0, t1)[x^0 - Q(t0, t1)x^1].$$

20. For a linear time-invariant system S consider the following matrix $V(t0, t1)$, which is called the *observability Grammian* (Chen, 1984):

$$V(t0, t1) \triangleq \int_{t0}^{t1} Q^T(t, t0)CC^T Q(t, t0)\, \mathrm{dt}.$$

Suppose $V(t0, t1)$ is nonsingular. Show that the initial state x^0 can be recovered from recordings of the input and output waveforms over $[t0, t1]$ as follows:

$$x^0 = V^{-1}(t0, t1) \int_{t0}^{t1} Q^T(t, t0)C\{y(t) - G(t, 0) * u(t)\}\, dt.$$

REFERENCES

Chen, C. T., *Linear System Theory and Design*, Holt, Rinehart and Winston, New York, 1984.

Desoer, C. A., *Notes for a Second Course on Linear Systems*, Van Nostrand Reinhold, New York, 1970.

Ogata, K., *Modern Control Engineering*, Prentice-Hall, Englewood Cliffs, NJ, 1970.

Pielou, E. C., *An Introduction to Mathematical Ecology*, Wiley-Interscience, New York, 1969.

Zadeh, L. A. and C. A. Desoer, *Linear System Theory—The State Space Approach*, McGraw-Hill, New York, 1963.

9 LINEAR DIFFERENCE SYSTEMS

A linear differential system can be regarded as a limiting special case of something called a linear difference system. Here the independent variable t is evaluated only at discrete instants of time $\{0, \Delta t, 2\Delta t, \dots\}$. Discrete-time or difference systems arise naturally in mathematical models of physical systems that employ computers.

9.1. LINEAR DIFFERENCE SYSTEMS

A survey of discrete-time models that arise in engineering applications reveals that many of the models can be cast as a coupled system of difference and algebraic equations of the following generic form:

$$x(k+1) = f[x(k), u(k), k],$$

$$y(k) = g[x(k), u(k), k]. \qquad \blacksquare$$

We will refer this system of equations as the *difference system* S. Here the independent integer variable $k \in \{0, 1, 2, \dots\}$ represents *time*. The vector $u(k) \in \mathbf{R}^m$ denotes the *input* or excitation applied to S at time k. In this case we say that S is an m-input system. The vector $x(k) \in \mathbf{R}^n$ is called the *state* of S at time k, and we say that S is an *n-dimensional system*. Finally, the vector $y(k) \in \mathbf{R}^p$ denotes the *output* or response of S at time k, and we call S a p-output system. A block diagram which illustrates the relationship between the input, state, and output of the difference system S is shown in Figure 9.1.

The difference equation in $x(k)$ represents the dynamic part of the system and is called the *state equation*, whereas the algebraic equation for $y(k)$ represents the static part of the system and is called the *output equation*. Note that the functions f and g completely characterize the *internal structure* of the system S. Before we proceed with a classification of systems based upon these functions, we examine a simple example.

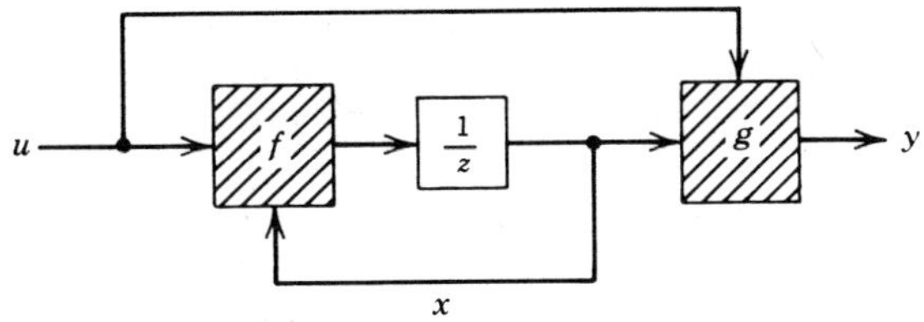

Figure 9.1. A difference system.

Example 1.1 (Interest Income). Consider the problem of modeling the balance in a savings account that earns ρ percent yearly interest compounded daily. Let $x(k)$ denote the balance at the beginning of day k, and let $u(k)$ denote the amount deposited (positive) or withdrawn (negative) during day k. Then the daily balance in the savings account can be modeled by the following difference system S:

$$x(k+1) = \left(1 + \frac{\rho}{365}\right) x(k) + u(k),$$

$$y(k) = x(k).$$

Suppose that the savings account is opened on day 0 with an initial balance of $x(0)$. If there are no subsequent deposits or withdrawals, then $u(k) = 0$ for $k \geqslant 0$. In this case the solution to the state equation is

$$x(k) = \left(1 + \frac{\rho}{365}\right)^k x(0).$$

This expression for $x(k)$ reveals the rate at which the balance will grow from accumulated interest. From this we can compute, for example, the effective annual interest or the number of days it takes for the opening deposit to double. ∎

On the surface, it might appear that the restriction of S to a system of first-order difference equations is rather severe. Many engineering models, for example, digital filters for speech processing, manifest themselves as higher-order difference equations. However, we can often replace a single high-order difference equation by an equivalent system of first-order difference equations. That is, we can reduce the order of the system at the expense of increasing the dimension. The following result demonstrates this.

Proposition 1.1 (High-order Systems). Let S_q be a qth-order one-dimensional difference system of the following form:

$$y(k+q) = h[y(k), y(k+1), \ldots, y(k+q-1), u(k), k].$$

Define the state as $x(k) = [y(k), y(k+1), \ldots, y(k+q-1)]^T$. Then S_q is input/output equivalent to a first-order system S where

$$
\begin{aligned}
x_j(k+1) &= x_{j+1}(k), \qquad 1 \leqslant j \leqslant q-1, \\
x_q(k+1) &= h[x_1(k), x_2(k), \ldots, x_q(k), u(k), k], \\
y(k) &= x_1(k).
\end{aligned}
$$

■

The proof of Prop. 1.1 follows directly from the definition of the state vector since $x_j(k) = y(k+j-1)$ for $1 \leqslant j \leqslant q$. The following example illustrates the use of Prop. 1.1 to convert a high-order model to the standard first-order form.

Example 1.2 (Fibonacci Numbers). Consider the problem of generating a sequence of numbers where the current number in the sequence is the sum of the previous two. If the process is initialized with the numbers $\{1, 1\}$, then the resulting numbers form something called the *Fibonacci sequence*, named after the mathematician who studied them. The Fibonacci numbers can be generated as the output of the following second-order difference system:

$$y(k+2) = y(k) + y(k+1) + u(k).$$

Here it is assumed that $y(k) = 0$ for $k < 0$. Suppose this difference equation is excited with the *unit pulse* input, $u(k) = \delta(k)$, which corresponds to a pulse of amplitude 1 at $k = 0$. Then the resulting output sequence $\{y(k)\}$ will be the Fibonacci sequence: $\{1, 1, 2, 3, 5, 8, 13, \ldots\}$. Using Prop. 1.1 to convert this second-order equation to the standard first-order form, we have $x(k) = [y(k), y(k+1)]^T$. The equivalent first-order *Fibonacci system* is then

$$
\begin{aligned}
x_1(k+1) &= x_2(k), \\
x_2(k+1) &= x_1(k) + x_2(k) + u(k), \\
y(k) &= x_1(k).
\end{aligned}
$$

■

Difference systems can be classified into a number of useful categories. The classification schemes are based upon the nature of the *state function* f and *output function* g, which characterize the structure of S.

Definition 1.1 (Time-Invariance). A difference system S is *time-invariant* $\Leftrightarrow$

$$
\begin{aligned}
x(k+1) &= f[x(k), u(k)], \\
y(k) &= g[x(k), u(k)].
\end{aligned}
$$

■

Thus S is time-invariant if and only if the functions f and g do not depend explicitly on the discrete-time variable k. The physical interpretation of time-invariance is that the results of input/output experiments do not depend upon the time at which those experiments are performed. It is in this sense that the behavior of the system S is invariant. A second classification of systems is based upon how the functions f and g depend on the state x and the input u.

Definition 1.2 (Linear System). A difference system S is *linear* $\Leftrightarrow$

$$x(k+1) = A(k)x(k) + B(k)u(k),$$
$$y(k) = C(k)x(k) + D(k)u(k). \qquad \blacksquare$$

Thus a linear system is one whose state function f and output function g are linear in the augmented variable $[x(k), u(k)]$. For a system which is both linear and time-invariant, the *matrix parameters* $\{A, B, C, D\}$ in Def. 1.2 are constants. Refer to Figure 9.2 for a block diagram which illustrates the structure of a linear difference system.

Note that both Exam. 1.1 and Exam. 1.2 are linear time-invariant difference systems. The following example illustrates how a linear difference system might be used to approximate a linear differential system.

Example 1.3 (Trapezoid Integrator). Suppose the sequence $\{u(k)\}$ denotes the samples of a continuous time signal $u(t)$ where the samples are taken every Δt seconds. Consider the problem of constructing a sequence $\{y(k)\}$ that approximates the integral of $u(t)$ over the interval $[0, k\Delta t]$. The curve $u(t)$ can be approximated by a sum of trapezoids with corners at $u(k)$ and $u(k+1)$ as shown in Figure 9.3.

Here if $y(k)$ is the approximate area under $u(t)$ up to time $k\,\Delta t$, then $y(k+1)$ is just $y(k)$ plus the area of the trapezoid starting at time k. Thus

$$y(k+1) = y(k) + \Delta t[u(k) + u(k+1)]/2.$$

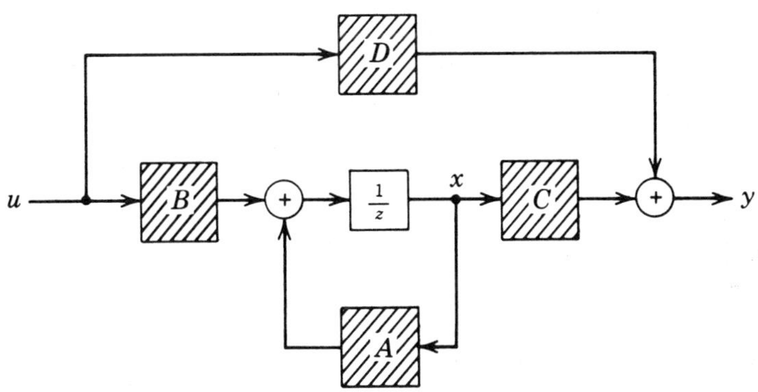

Figure 9.2. A linear difference system.

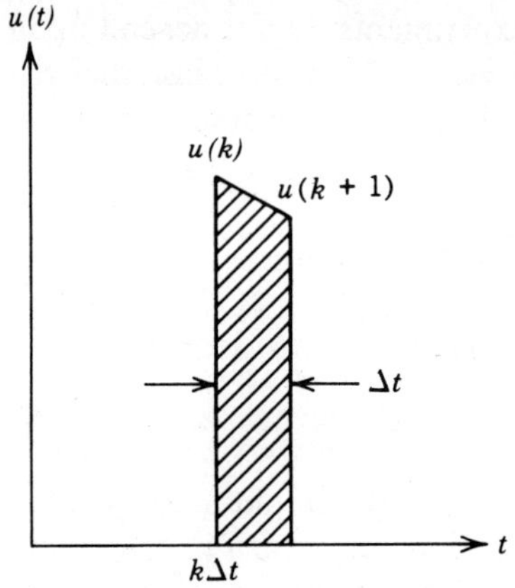

Figure 9.3. Approximation of integral with trapezoids.

To convert this to the standard first-order form, we use $x(k) = y(k) + \alpha u(k)$. Here the term $\alpha u(k)$ is included because Prop. 1.1 is not directly applicable in view of the presence of the $u(k + 1)$ term in the original equation. We determine a proper value for the coefficient α as follows:

$$\begin{aligned} x(k + 1) &= y(k + 1) + \alpha u(k + 1) \\ &= y(k) + \Delta t[u(k) + u(k + 1)]/2 + \alpha u(k + 1) \\ &= x(k) - \alpha u(k) + \Delta t[u(k) + u(k + 1)]/2 + \alpha u(k + 1) \\ &= x(k) + (\Delta t/2 - \alpha)u(k) + (\Delta t/2 + \alpha)u(k + 1). \end{aligned}$$

From Def. 1.2 it is clear that the $u(k + 1)$ term must be eliminated from the state equation. Thus we need $\alpha = -\Delta t/2$. Then the state and output equations for the *trapezoid integrator* reduce to

$$\begin{aligned} x(k + 1) &= x(k) + \Delta t u(k), \\ y(k) &= x(k) + \Delta t u(k)/2. \end{aligned}$$

In light of Def. 1.2, the trapezoid integrator is a linear time-invariant one-dimensional system with parameters $(A, B, C, D) = (1, \Delta t, 1, \Delta t/2)$. ■

Recall that Prop. 1.1 provided us with a means of recasting a qth-order one-dimensional system as an equivalent first-order system of dimension q. Note, however, that the original qth-order system was constrained in the sense that no input terms of the form $u(k + j)$, $j \geqslant 1$ appeared. Within the class of linear time-invariant systems, this restriction on $u(k)$ can be relaxed as the following result shows.

Proposition 1.2 (Linear High-Order Systems). Let S_q be a qth-order one-dimensional linear time-invariant difference system of the following form where $a_0 = 1$:

$$\sum_{j=0}^{q} a_j y(k+q-j) = \sum_{j=0}^{q} b_j u(k+q-j).$$

Let the state vector $x(k) \in \mathbf{R}^q$ be defined as follows:

$$x_j(k) = y(k+j-1) - \sum_{m=0}^{j-1} \beta_m u(k+j-m-1), \qquad 1 \leqslant j \leqslant q.$$

Here $\beta_0 = b_0$ and

$$\beta_j = b_j - \sum_{m=1}^{j} a_m \beta_{j-m}, \qquad 1 \leqslant j \leqslant q.$$

Then S_q is input/output equivalent to a linear time-invariant system S with the following matrix parameters:

$$A = \begin{bmatrix} 0 & 1 & 0 & \cdots & 0 & 0 \\ 0 & 0 & 1 & \cdots & 0 & 0 \\ \vdots & \vdots & \vdots & & \vdots & \vdots \\ 0 & 0 & 0 & \cdots & 0 & 1 \\ -a_q & -a_{q-1} & -a_{q-2} & \cdots & -a_2 & -a_1 \end{bmatrix}, \qquad B = \begin{bmatrix} \beta_1 \\ \beta_2 \\ \vdots \\ \beta_{q-1} \\ \beta_q \end{bmatrix},$$

$$c = [1 \quad 0 \quad 0 \quad \cdots \quad 0 \quad 0], \qquad D = [\beta_0].$$ ■

The interested reader can refer to Ogata (1970) for a proof of Prop. 1.2. When the system S is linear, Prop. 1.1 is actually a special case of Prop. 1.2, as can be seen from the following exercise.

Exercise 1.1. Show that the state assignment of Prop. 1.2 reduces to that of Prop. 1.1 when S is linear and $b_k = 0$ for $0 \leqslant k \leqslant q-1$. ■

The following example illustrates the use of Prop. 1.2 to derive state and output equations for a high-order linear system in which terms of the form $u(k+j)$ appear.

Example 1.4 (Moving-Average Filter). Consider the problem of constructing a difference equation which computes the moving-average of a sequence $\{u(k)\}$. If the number of terms in the average is $q+1$, then the difference equation for the *moving-average filter* is

$$y(k+q) = \frac{1}{q+1} \sum_{j=0}^{q} u(k+j).$$

Thus we have a qth-order one-dimensional system. Using Prop. 1.2, and identifying coefficients, yields $a = \{1, 0, \ldots, 0\}$, $b = \{1/(q+1), 1/(q+1), \ldots, 1/(q+1)\}$. Thus $\beta_j = b_j$ for $0 \leqslant j \leqslant q$. For the case $q = 3$, the state and output equations of the moving-average filter reduce to

$$x(k+1) = \begin{bmatrix} 0 & 1 & 0 \\ 0 & 0 & 1 \\ 0 & 0 & 0 \end{bmatrix} x(k) + \begin{bmatrix} \frac{1}{4} \\ \frac{1}{4} \\ \frac{1}{4} \end{bmatrix} u(k),$$

$$y(k) = [1 \quad 0 \quad 0]x(k) + [\tfrac{1}{4}]u(k).$$ ■

9.2. DISCRETE EQUIVALENT SYSTEMS

In certain engineering applications, discrete-time signals are generated from continuous-time signals by sampling at regular intervals of time. Consider, for example, the digital feedback control system (Franklin and Powell, 1980) shown in Figure 9.4.

Here the plant or system to be controlled is a linear differential system characterized by a transfer function matrix $G(s)$. The plant output $y_c(t)$ is sampled by the analog-to-digital (A/D) converter to produce a discrete-time signal $y(k)$ where $y(k) = y_c(k\,\Delta t)$. Here we refer to $\Delta t > 0$ as the *sampling interval.* A control algorithm is implemented in the digital computer. It takes the form of a linear difference equation of the following form where $a_0 = 1$:

$$\sum_{j=0}^{q} a_j u(q+j) = \sum_{j=0}^{q} b_j e(q+j).$$

Here $e(k) = y(k) - r(k)$ is the *error* signal, while $r(k)$ is the *reference input* or trajectory that we would like $y(k)$ to follow. The discrete-time control signal $u(k)$ from the computer is then converted to a continuous-time signal with a digital-to-analog (D/A) converter. The resulting piecewise-constant control signal $u_c(t)$ is then used to drive the plant.

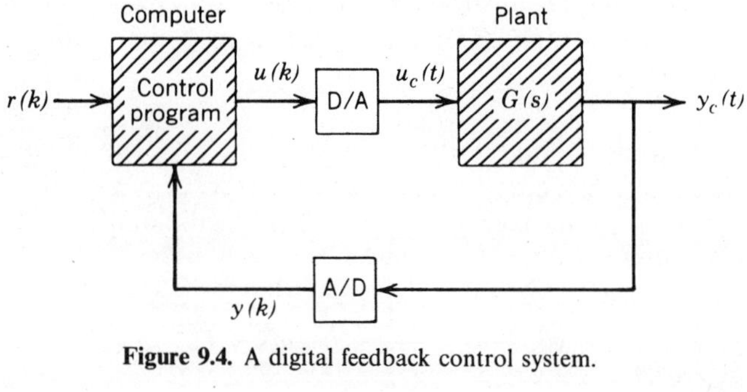

Figure 9.4. A digital feedback control system.

The feedback system in Figure 9.4 is a *sampled-data* system since it has a mixture of continuous-time variables and discrete-time variables. Thus the equations that govern the overall input/output behavior of this system are a combination of differential and difference equations. One way to analyze such a system is to replace the differential system part (the plant) by a difference system that is equivalent in the sense that the output is the same at the sampling instants. For a linear time-invariant differential system, an equivalent difference system can be obtained directly from the original system parameters, as can be seen from the following result.

Proposition 2.1 (Discrete Equivalent). Consider a linear time-invariant differential system S_c with matrix parameters $\{A, B, C, D\}$ and variables $\{u_c(t), x_c(t), y_c(t)\}$. Suppose the input $u_c(t)$ is piecewise-constant with $u_c(t) = u(k)$ for $k\,\Delta t \leqslant t < (k+1)\,\Delta t$. Then $y(k) = y_c(k\,\Delta t)$ where $y(k)$ is the output of the following equivalent linear time-invariant difference system:

$$\begin{aligned} x(k+1) &= Ex(k) + Fu(k), \\ y(k) &= Cx(k) + Du(k). \end{aligned}$$

Here the matrix parameters for the equivalent discrete-time system are

$$\begin{aligned} E &= \exp(A\,\Delta t), \\ F &= \left(\int_0^{\Delta t} \exp(At)\,dt\right) B. \end{aligned}$$

Proof. Since $u_c(t) = u(k)$ for $k\,\Delta t \leqslant t < (k+1)\,\Delta t$, using Prop. 8.3.3 and letting $x_c^{k+1} = x_c[(k+1)\,\Delta t]$, we have

$$\begin{aligned} x_c^{k+1} &= \exp[A(k+1)\,\Delta t]x_c(0) + \int_0^{(k+1)\Delta t} \exp\{A[(k+1)\,\Delta t - \tau]\}Bu(\tau)\,d\tau \\ &= \exp(A\,\Delta t)\left\{\exp(Ak\,\Delta t)x_c(0) + \int_0^{(k+1)\Delta t} \exp[A(k\,\Delta t - \tau)]Bu(\tau)\,d\tau\right\} \\ &= \exp(A\,\Delta t)\left\{x_c(k\,\Delta t) + \int_{k\Delta t}^{(k+1)\Delta t} \exp[A(k\,\Delta t - \tau)]Bu(\tau)\,d\tau\right\} \\ &= \exp(A\,\Delta t)x_c(k\,\Delta t) + \exp(A\,\Delta t)\int_{k\Delta t}^{(k+1)\Delta t} \exp[A(k\,\Delta t - \tau)]Bu(k)\,d\tau \\ &= \exp(A\,\Delta t)x_c(k\,\Delta t) + \exp(A\,\Delta t)\int_{k\Delta t}^{(k+1)\Delta t} \exp[A(k\,\Delta t - \tau)]\,d\tau\;Bu(k) \\ &= \exp(A\,\Delta t)x_c(k\,\Delta t) + \int_{k\Delta t}^{(k+1)\Delta t} \exp\{A[(k+1)\,\Delta t - \tau]\}\,d\tau\;Bu(k) \end{aligned}$$

$$= \exp(A\,\Delta t)x_c(k\,\Delta t) - \int_{\Delta t}^{0} \exp(At)\,dt\,Bu(k)$$

$$= \exp(A\,\Delta t)x_c(k\,\Delta t) + \int_{0}^{\Delta t} \exp(At)\,dt\,Bu(k)$$

$$= Ex_c(k\,\Delta t) + Fu(k)$$

$$= Ex_c(k\,\Delta t) + Fu_c(k\,\Delta t).$$

Thus $x(k) = x_c(k\,\Delta t)$ and from the output equation we have

$$\begin{aligned} y(k\,\Delta t) &= Cx_c(k\,\Delta t) + Du(k\,\Delta t) \\ &= Cx(k) + Du(k) \\ &= y(k). \end{aligned}$$

■

We refer to the difference system S in Prop. 2.1 with parameters $\{E, F, C, D\}$ as the *discrete equivalent* of the differential system S_c. Note that as the output equation is purely algebraic, the output parameters $\{C, D\}$ of the two systems are identical.

Example 2.1 (Discrete Equivalent). Find the discrete equivalent of the following linear time-invariant differential system assuming a sampling interval of $\Delta t = 100$ msec.

$$\dot{x}(t) = \begin{bmatrix} -1 & 0 \\ 1 & -3 \end{bmatrix} x(t) + \begin{bmatrix} 1 \\ -2 \end{bmatrix} u(t),$$

$$y(t) = [1 \quad 0]x(t) + [4]u(t).$$

First, we must compute the state transition matrix $Q(t, 0) = \exp(At)$. Using the method of Laplace transforms in Exer. 7.5.4, we have

$$\begin{aligned} (sI - A)^{-1} &= \begin{bmatrix} s+1 & 0 \\ -1 & s+3 \end{bmatrix}^{-1} \\ &= \begin{bmatrix} s+3 & 0 \\ 1 & s+1 \end{bmatrix} \frac{1}{\Delta(s)}. \end{aligned}$$

The characteristic polynomial of A is $\Delta(s) = (s + 1)(s + 3)$. Using partial fractions and Table 8.1 to invert the components of $(sI - A)^{-1}$, we get

$$\exp(At) = L^{-1}\{(sI - A)^{-1}\}$$

$$= L^{-1}\begin{bmatrix} \dfrac{1}{s+1} & 0 \\ \dfrac{\frac{1}{2}}{s+1} - \dfrac{\frac{1}{2}}{s+3} & \dfrac{1}{s+3} \end{bmatrix}$$

$$= \begin{bmatrix} \exp(-t) & 0 \\ \frac{1}{2}\exp(-t) - \frac{1}{2}\exp(-3t) & \exp(-3t) \end{bmatrix}.$$

Thus from Prop. 2.1, the state equation parameters for the discrete equivalent system are

$$E = \exp(A\ \Delta t)$$
$$= \begin{bmatrix} 0.905 & 0 \\ 0.082 & 0.741 \end{bmatrix},$$
$$F = \left(\int_0^{\Delta t} \exp(At)\ dt\right) B$$
$$= \begin{bmatrix} 0.095 & 0 \\ 0.0043 & 0.086 \end{bmatrix}\begin{bmatrix} 1 \\ -2 \end{bmatrix}$$
$$= [0.095,\ -0.168]^{\mathrm{T}}.$$

Finally from Prop. 2.1, the discrete equivalent of the linear time-invariant differential system generated using a sampling interval of $\Delta t = 0.1$ is

$$x(k+1) = \begin{bmatrix} 0.905 & 0 \\ 0.082 & 0.741 \end{bmatrix} x(k) + \begin{bmatrix} 0.095 \\ -0.168 \end{bmatrix} u(k),$$
$$y(k) = [1 \quad 0]x(k) + [4]u(k).$$ ■

9.3. THE STATE SPACE REPRESENTATION

Suppose the difference system S is linear. Let $K \triangleq \{k0, k0+1, \ldots, k1\}$ be a sequence of nonnegative integers denoting the *interval of time* over which the behavior of S is being investigated. We refer to $k0$ as the *initial time* and $k1$ as the *final time.* For each discrete time $k \in K$ the vector $x(k) \in \mathbf{R}^n$ is then the state of S at time k. In particular, the state of S at the initial time, $x(k0)$, is called the *initial state* and the state $x(k1)$ is called the *final state.* In most practical cases we have $k1 < \infty$. However, in the event that the asymptotic behavior of $x(k)$ is of interest, we can take $k1$ to be infinite.

Next, let $x^0 \in \mathbf{R}^n$ be an arbitrary initial state. We then refer to the constraint, $x(k0) = x^0$, as the *initial condition* of S. In order to obtain the solution to the state equation satisfying the initial condition $x(k0) = x^0$, we begin by considering the *homogeneous* or undriven case when $u = 0$. Consider, in particular, a matrix $Q(k, k0)$ whose jth column $q^j(k, k0)$ is the solution of the homogeneous state equation given the initial condition $q^j(k0) = i^j$ where $I = [i^1, i^2, \ldots, i^n]$. The matrix $Q(k, k0)$ provides the key to the solution of the state equation in general. We call $Q(k, k0)$ the state transition matrix of the difference system and define it formally as follows.

Definition 3.1 (State Transition Matrix). The *state transition matrix* $Q(k, k0)$ of a linear difference system S is the solution to the following matrix difference equation:

$$Q(k + 1, k0) = A(k)Q(k, k0), \qquad Q(k0, k0) = I. \qquad \blacksquare$$

The term *state transition* arises from the fact that when $u = 0$, the matrix $Q(k1, k)$ maps the state at time $k0$ into the state at time k, giving us the following result.

Proposition 3.1 (Homogeneous Solution: $u = 0$). Let S be a linear difference system with state transition matrix $Q(k, k0)$. If $u = 0$, then for each $x^0 \in \mathbf{R}^n$ the solution to the homogeneous state equation satisfying the initial condition $x(k0) = x^0$ is

$$x(k) = Q(k, k0)x^0, \qquad k \in K.$$

Proof. Let $z(k) = Q(k, k0)x^0$. Then, using Def. 3.1, for each $k \in K$

$$\begin{aligned} z(k + 1) &= Q(k + 1, k0)x^0 \\ &= A(k)Q(k, k0)x^0 \\ &= A(k)z(k). \end{aligned}$$

Again from Def. 3.1 we have

$$\begin{aligned} z(k0) &= Q(k0, k0)x^0 \\ &= Ix^0 \\ &= x^0. \end{aligned}$$

However, the solution to the state equation satisfying the initial condition is unique. Hence

$$\begin{aligned} x(k) &= z(k) \\ &= Q(k, k0)x^0. \end{aligned} \qquad \blacksquare$$

Thus every solution to the homogeneous state equation can be expressed as a linear combination of the columns of $Q(k, k0)$. The coefficients of that linear combination are simply the components of the initial state vector, x^0. Refer to Figure 9.5 for a pictorial summary of Prop. 3.1.

In light of Prop. 3.1, it is worthwhile to examine the nature of the state transition matrix in more detail. Note from Def. 3.1 that $Q(k, k0)$ depends upon the matrix $A(k)$. Every state transition matrix $Q(k, k0)$ possesses a number of fundamental properties as can be seen from the following result.

Proposition 3.2 (Properties of $Q(k, k0)$). Let $Q(k, k0)$ be the state transition matrix of a linear difference system S. Then for each $k \in K$

1. $Q(k, k) = I$,
2. $Q(k, k0) = Q(k, j)Q(j, k0), j \in K$,
3. $Q(k, k0) = A(k)A(k-1) \cdots A(k0)$.

Proof. Since $k0 \in \{0, 1, 2, \ldots\}$ was arbitrary, from Def. 3.1 we have $Q(k, k) = I$. Let $P(k, k0) = Q(k, j)Q(j, k0)$. Then from Def. 3.1

$$\begin{aligned} P(k+1, k0) &= Q(k+1, j)Q(j, k0) \\ &= A(k)Q(k, j)Q(j, k0) \\ &= A(k)P(k, k0). \end{aligned}$$

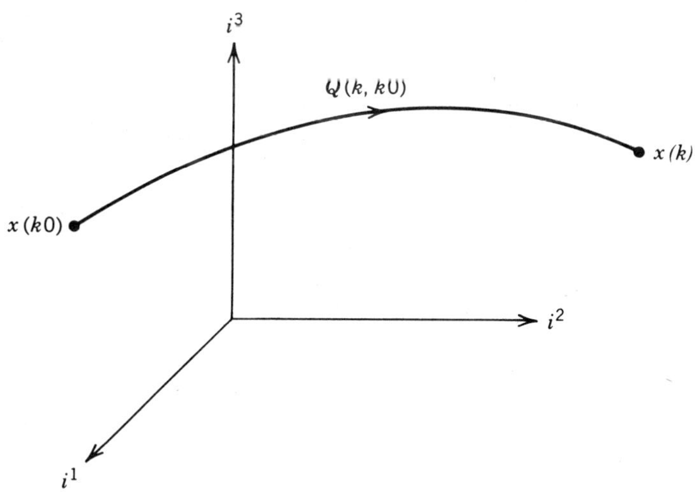

Figure 9.5. The state transition property of $Q(k, k0)$.

Furthermore, from Def. 3.1 and property 1

$$\begin{aligned}P(j, k0) &= Q(j, j)Q(j, k0)\\ &= IQ(j, k0)\\ &= Q(j, k0)\end{aligned}$$

It follows that $P(k, k0)$ and $Q(k, k0)$ satisfy the same difference equation and they are equal at $k = j$. Thus

$$\begin{aligned}Q(k, k0) &= P(k, k0)\\ &= Q(k, j)Q(j, k0).\end{aligned}$$

Finally, to establish property 3, note that if $u(k) = 0$ for $k \in K$, then by direct substitution into the state equation we have

$$\begin{aligned}&x(k0 + 1) = A(k0)x(k0),\\ &x(k0 + 2) = A(k0 + 1)x(k0 + 1) = A(k0 + 1)A(k0)x(k0),\\ &x(k0 + 3) = A(k0 + 2)x(k0 + 2) = A(k0 + 2)A(k0 + 1)A(k0)x(k0).\end{aligned}$$

In general, for $n \geqslant 1$ we have

$$x(k0 + n) = A(k0 + n)A(k0 + n - 1) \cdots A(k0)x(k0).$$

But from Prop. 3.1, $x(k) = Q(k, k0)x(k0)$ for $k \in K$. It follows that

$$Q(k, k0) = A(k)A(k - 1) \cdots A(k0). \qquad \blacksquare$$

Note that unlike the state transition matrix for a linear differential system, the state transition matrix for a linear difference system $Q(k, k0)$ is not always invertible. In light of Prop. 3.2, it is invertible if and only if the matrix $A(j)$ is invertible for $k0 \leqslant j \leqslant k$.

Exercise 3.1 (Inverse of $Q(k, k0)$). Let S be a linear time-invariant difference system with state transition matrix $Q(k, k0)$. Show that if $Q(k, k0)$ is invertible, then $Q^{-1}(k, k0) = Q(k0, k)$. ■

Given the additional properties of the state transition matrix, we are now in a position to extend Prop. 3.1 to obtain the solution to the general or *nonhomogeneous* case when $u \neq 0$. Here we let the *input space* $\mathbf{U}$ be the vector space of sequences mapping K into $\mathbf{R}^m$. That is, $\mathbf{U} \triangleq \{u: K \to \mathbf{R}^m\}$. We then have the following result.

Proposition 3.3 (Nonhomogeneous Solution: $u \neq 0$). Let S be a linear difference system with state transition matrix $Q(k, k0)$. Then for each $(x^0, u) \in \mathbf{R^n} \times \mathbf{U}$, the solution to the state equation satisfying the initial condition $x(k0) = x^0$ is

$$x(k) = Q(k, k0)x^0 + \sum_{j=k0}^{k-1} Q(k-1, j)B(j)u(j).$$

Proof. Let $z(k)$ be the candidate solution. That is,

$$z(k) = Q(k, k0)x^0 + \sum_{j=k0}^{k-1} Q(k-1, j)B(j)u(j).$$

Then, using Def. 3.1 and Prop. 3.2,

$$\begin{aligned}
z(k+1) &= Q(k+1, k0)x^0 + \sum_{j=k0}^{k} Q(k, j)B(j)u(j) \\
&= A(k)\left\{Q(k, k0)x^0 + \sum_{j=k0}^{k} Q(k-1, j)B(j)u(j)\right\} \\
&= A(k)z(k) + A(k)Q(k-1, k)B(k)u(k) \\
&= A(k)z(k) + Q(k, k)B(k)u(k) \\
&= A(k)z(k) + IB(k)u(k) \\
&= A(k)z(k) + B(k)u(k).
\end{aligned}$$

Evaluating $z(k)$ at $k = k0$ using Prop. 3.2,

$$\begin{aligned}
z(k0) &= Q(k0, k0)x^0 \\
&= Ix^0 \\
&= x^0.
\end{aligned}$$

Thus $z(k)$ and $x(k)$ satisfy the same difference equation and the same initial condition. Because the solution is unique, it follows that $x(k) = z(k)$. ■

Note that Prop. 3.3 reduces to Prop. 3,1, given the undriven special case, $u = 0$. In light of Prop. 3.3, once the state transition matrix $Q(k, k0)$ is known, the general solution to the state equation can be found directly.

Exercise 3.2 (Decomposition). Show that the solution to the state equation of a linear difference system S can be decomposed as follows:

$$x(k) = x(k, x^0) + x(k, u).$$

Here $x(k, x^0)$ is the *zero-input solution* that results when $u = 0$. Similarly, $x(k, u)$ is the *zero-state solution* that results when $x^0 = 0$. ■

The following example illustrates the use of Prop. 3.3 to find the solution to a special form of linear difference system.

Example 3.1 (Uncoupled System). Consider an *uncoupled* linear difference system characterized by the following state equation:

$$x_j(k+1) = a_j x_j(k) + b_j u(k), \qquad 1 \leqslant j \leqslant n.$$

Because the A matrix is diagonal in this case, an explicit expression can be obtained for the state transition matrix $Q(k, k0)$. In particular, using Def. 3.1, we have

$$Q_{ij}(k, k0) = a_i^{k-k0} I_{ij}, \qquad 1 \leqslant i, j \leqslant n.$$

Thus from Prop. 3.1, the part of the solution due to the initial condition, the zero-input solution $x(k, x^0)$, is

$$x_i(k, x^0) = a_i^{k-k0} x_i^0, \qquad 1 \leqslant i \leqslant n.$$

Next from Prop. 3.3, the part of the solution due to the input, the zero-state solution $x(k, u)$, is

$$x_i(k, u) = \sum_{j=k0}^{k-1} a_i^{k-1-j}\, b_i u(j), \qquad 1 \leqslant i \leqslant n.$$

Finally, using Exer. 3.2, the complete solution is the sum of $x(k, x^0)$ and $x(k, u)$. That is, for $1 \leqslant i \leqslant n$

$$x_i(k) = a_i^{k-k0} x_i^0 + \sum_{j=k0}^{k-1} a_i^{k-1-j} b_i u(j). \qquad ■$$

We have now succeeded in expressing the general solution of the state equation in terms of a set of n special solutions. These special solutions form the columns of the state transition matrix $Q(k, k0)$. The jth column is obtained by solving the homogeneous state equation using the jth column of the identity matrix as the initial state. Since the state transition matrix provides the key to the solution in general, we now focus on the problem of computing $Q(k, k0)$. When the linear system S is time-varying, we must often resort to a numerical solution of the difference equation in Def. 3.1. However, in the event that S is time-invariant, we can obtain a simple explicit expression for $Q(k, k0)$, as can be seen from the following result.

Proposition 3.4 (Time-Invariant Case). Let S be a linear time-invariant difference system with state transition matrix $Q(k, k0)$. Then

$$Q(k, k0) = A^{k-k0}.$$

Proof. Let $P(k, k0) = A^{k-k0}$. Then

$$\begin{aligned} P(k+1, k0) &= A^{k+1-k0} \\ &= A(A^{k-k0}) \\ &= AP(k, k0). \end{aligned}$$

Evaluating $P(k, k0)$ at the initial time $k = k0$ yields

$$\begin{aligned} P(k0, k0) &= A^{k0-k0} \\ &= A^0 \\ &= I. \end{aligned}$$

It follows from Def. 3.1 that

$$\begin{aligned} Q(k, k0) &= P(k, k0) \\ &= A^{k-k0}. \end{aligned}$$ ∎

Thus for time-invariant linear difference systems, the state transition matrix $Q(k, k0)$ is simply the constant matrix A raised to the $k - k0$ power. The fact that $Q(k, k0)$ depends only on the A matrix is apparent from this explicit formulation of the state transition matrix.

Exercise 3.3 (Elapsed Time). Let $Q(k, k0)$ be the state transition matrix of a linear time-invariant difference system S. Show that

$$Q(k, k0) = Q(k - k0, 0), \qquad k \in K.$$ ∎

Note from Exer. 3.3 that for time-invariant linear difference systems, it is the *elapsed time* $k - k0$ that is significant, not the initial time $k0$. Consequently, for time-invariant linear difference systems we will often assume $k0 = 0$ as a matter of convenience. In light of Prop. 3.4, the computation of $Q(k, k0)$ reduces to the computation of an analytic function of a matrix. There are several approaches we can take. The first is based upon the notion of the interpolating polynomial.

Proposition 3.5 (Interpolating Polynomial). Let $Q(k, 0)$ be the state transition matrix of a linear time-invariant difference system S. There exist discrete-time functions $\{a_1(k), a_2(k), \ldots, a_n(k)\}$ such that

$$Q(k, 0) = \sum_{j=1}^{n} a_j(k)A^{j-1}.$$

Proof. Let g be the interpolating polynomial of f on $\sigma(A)$ where

$$f(\lambda) = \lambda^k.$$

Because f is parameterized by k, it follows that the coefficients of g will be functions of k. That is

$$g(\lambda) = \sum_{j=1}^{n} a_j(k)\lambda^{j-1}.$$

Using Prop. 3.4, we then have

$$\begin{aligned} Q(k, 0) &= A^k \\ &= f(A) \\ &= g(A) \\ &= \sum_{j=1}^{n} a_j(k)A^{j-1}. \end{aligned}$$

■

Thus the state transition matrix of a time-invariant linear system of dimension n can always be expressed as a time-varying polynomial of A of degree $n - 1$. The following example illustrates the use of this technique for computing $Q(k, 0)$.

Example 3.2 (Interpolating Polynomial). Consider the difference system for generating the Fibonacci numbers introduced in Exam. 1.2. Here the A matrix is

$$A = \begin{bmatrix} 0 & 1 \\ 1 & 1 \end{bmatrix}.$$

The characteristic polynomial of this matrix is

$$\begin{aligned} \Delta(\lambda) &= \det(\lambda I - A) \\ &= \lambda(\lambda - 1) - 1 \\ &= \lambda^2 - \lambda - 1 \\ &= (\lambda - \alpha)(\lambda - \beta). \end{aligned}$$

Thus the spectrum of A is $\sigma(A) = \{\alpha, \beta\}$ where $\alpha = -0.618$ and $\beta = 1.618$. Since $A \in \mathbf{R}^{2\times 2}$, the interpolating polynomial will be a vector in $\mathbf{P}^1(\mathbf{C})$. That is,

$$g(\lambda) = a_1 + a_2\lambda.$$

The two eigenvalues of A are simple. Hence the coefficients of the interpolating polynomial must satisfy

$$a_1 + \alpha a_2 = \alpha^k,$$
$$a_1 + \beta a_2 = \beta^k.$$

Solving these equations for the coefficient vector a then yields

$$a_1(k) = \frac{\beta\alpha^k - \alpha\beta^k}{\beta - \alpha},$$

$$a_2(k) = \frac{\alpha^k - \beta^k}{\alpha - \beta}.$$

Thus from Prop. 3.5, the state transition matrix of the Fibonacci system is

$$Q(k, 0) = \frac{(\beta\alpha^k - \alpha\beta^k)I - (\alpha^k - \beta^k)A}{\beta - \alpha}$$

$$= \begin{bmatrix} \beta\alpha^k - \alpha\beta^k & \beta^k - \alpha^k \\ \beta^k - \alpha^k & (\beta - 1)\alpha^k - (\alpha - 1)\beta^k \end{bmatrix} \frac{1}{\beta - \alpha} \qquad \blacksquare$$

Note that in light of Exer. 3.3, $Q(k, k0)$ can be obtained from $Q(k, 0)$ by simply replacing k with $k - k0$. A second approach to computing the state transition matrix is based upon the block diagonal form of A. Here we transform to a convenient coordinate system, compute $Q(k, 0)$ in this coordinate system, and then transform back to the original coordinates. In the event that A is diagonalizable, the block diagonal technique can be particularly effective since $Q(k, 0)$ is also diagonalizable in this case.

A third approach to computing $Q(k, 0)$ is based upon the spectral representation theorem. Besides serving as a computational technique, the spectral representation of $Q(k, 0)$ also provides valuable qualitative information about the asymptotic behavior of the solution of the state equation of S.

Proposition 3.6 (Spectral Representation). Let $Q(k, 0)$ be the state transition matrix of a linear time-invariant difference system S. If λ_i is an eigenvalue of A of multiplicity $m(i)$ for $1 \leqslant i \leqslant q$ and if $\{p_{ij}\}$ are the spectral polynomials of A, then

$$Q(k, 0) = \sum_{i=1}^{q} \sum_{j=1}^{m(i)} \frac{k!}{(k+1-j)!} \lambda_i^{k+1-j} p_{ij}(A).$$

Proof. This follows directly from Prop. 3.4 and the spectral representation theorem, Prop. 7.7.1. ■

In light of Prop. 3.6, it is clear that the behavior of $Q(k, 0)$ for large values of k is determined entirely by the location of $\sigma(A)$. In particular, we have the following stability result, which is left as an exercise.

Exercise 3.4 (Stability). Let S be a linear time-invariant difference system, and suppose the eigenvalues of A all lie strictly inside the unit circle, $|\lambda| = 1$. Show that S is stable in the sense that if $u = 0$, then for each $x^0 \in \mathbf{R}^n$ the solution of the state equation satisfying the initial condition $x(0) = x^0$ has the property that

$$x(k) \to 0 \quad \text{as} \quad k \to \infty. \qquad ■$$

The following example illustrates the use of Prop. 3.6 to compute the spectral representation of a state transition matrix $Q(k, 0)$.

Example 3.3 (Spectral Representation). Consider a linear time-invariant difference system characterized by the following A matrix:

$$A = \begin{bmatrix} -1 & 0 & 2 \\ 0 & -1 & 0 \\ 0 & 0 & -1 \end{bmatrix}.$$

Since A is upper triangular, the characteristic polynomial of A is simply

$$\begin{aligned} \Delta(\lambda) &= (\lambda + 1)^3 \\ &= (\lambda - \lambda_1)^{m(1)}. \end{aligned}$$

In Exam. 7.7.1 the spectral polynomials of A were found to be

$$p_{11}(\lambda) = 1$$

$$p_{12}(\lambda) = \lambda + 1$$

$$p_{13}(\lambda) = (\lambda + 1)^2/2.$$

Thus from Prop. 3.6 we have

$$
\begin{aligned}
Q(k, 0) &= \sum_{i=1}^{q} \sum_{j=1}^{m(i)} \frac{k!}{(k+1-j)!} \lambda_i^{k+1-j} p_{kj}(A) \\
&= (-1)^k I + k(-1)^{k-1}(A+I) + k(k-1)(-1)^{k-2}\frac{(A+I)^2}{2} \\
&= (-1)^k \begin{bmatrix} 1 & 0 & 0 \\ 0 & 1 & 0 \\ 0 & 0 & 1 \end{bmatrix} + k(-1)^{k-1} \begin{bmatrix} 0 & 0 & 2 \\ 0 & 0 & 0 \\ 0 & 0 & 0 \end{bmatrix} \\
&= (-1)^k \begin{bmatrix} 1 & 0 & -2k \\ 0 & 1 & 0 \\ 0 & 0 & 1 \end{bmatrix}.
\end{aligned}
$$

■

Note that if an expression for $Q(k, 0)$ has been obtained, the spectral matrices $\{p_{kj}(A)\}$ can be extracted by inspection. Once $\{p_{kj}(A)\}$ and $\sigma(A)$ are known, the problem of computing $f(A)$ for *any* analytic f is a trivial one.

Exercise 3.5 (Spectral Representation). Use the results of Exam. 3.2 to write, by inspection, the spectral representation of $Q(k, 0)$ for the Fibonacci system. ■

Exercise 3.6 (Simple Eigenvalues). Let $Q(k, 0)$ be the state transition matrix of a linear time-invariant difference system S whose A matrix has simple eigenvalues $\{\lambda_1, \lambda_2, \ldots, \lambda_n\}$. Show that if E_i is a representation of the orthogonal projection operator induced by the ith eigenspace, then

$$Q(k, 0) = \sum_{i=1}^{n} \lambda_i^k E_i.$$

■

9.4. THE PULSE RESPONSE REPRESENTATION

Thus far we have focused on the state equation and its solution. As such, we have concentrated on what might be called an *internal* description of S. We now examine the system S from the point of view of its *external* variables, the input $u(k)$ and the output $y(k)$, as shown in Figure 9.6.

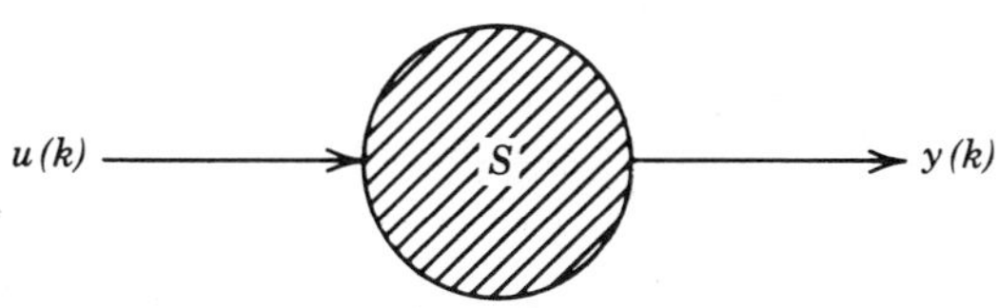

Figure 9.6. External representation of a system.

Recall that the solution of the state equation can be decomposed into the sum of the zero-input solution plus the zero-state solution. If the linear system S is to be represented in terms of its external variables, then some assumption must be placed on the internal state of S. In particular, we assume that the system S is relaxed.

Definition 4.1 (Relaxed System). A linear difference system S is *relaxed* at time $k0 \Leftrightarrow x(k0) = 0$. ■

Thus a linear difference system is relaxed when it is in the zero state. In this case, the *complete response* $y(k)$ includes only the zero-state response $y(k, u)$. That is, the output depends uniquely on the input if the system starts out relaxed. Note from Exer. 3.4 that if a linear time-invariant system S is stable, then in the absence of an input the internal state of S will approach the zero state as time approaches infinity. As the response depends continuously on the initial state, for a stable physical system all we have to do is wait a sufficiently long period of time before we apply an input to ensure that S is relaxed.

The mathematical relationship between the input and output of a relaxed linear system can be given a concise formulation if we introduce a second matrix associated with S called the pulse response matrix. We begin by examining the notion of a unit pulse.

Definition 4.2 (Unit Pulse). The *unit pulse* is denoted $\delta(k)$ and is defined by the following two conditions:

1. $\delta(k) = 0,\ k \neq 0,$
2. $\delta(0) = 1.$ ■

Note that the unit pulse $\delta(k)$ is similar to the unit impulse $\delta(t)$ in that it is zero everywhere except at $k = 0$. Furthermore, the area under the unit pulse is unity in the sense that the running sum of the unit pulse $\delta(k)$ is the *unit step* sequence $1(k)$. Like the unit impulse, the unit pulse satisfies a sampling property as shown in the following exercise.

Exercise 4.1 (Sampling Property). Let $f: K \to \mathbf{R}^n$. Show that for each $k0 \in K$

$$\sum_{k \in K} f(k)\delta(k - k0) = f(k0).$$ ■

Thus if a sequence is modulated or multiplied by a unit pulse and the product is summed, the effect is to *sample* or sift out the value of the sequence at the time that the pulse occurs. Next, we introduce a matrix called the pulse response matrix that is obtained by measuring the response of the system S to a unit pulse.

More specifically, the pulse response matrix is a $p \times m$ matrix $G(k, k0)$ defined as follows.

Definition 4.3 (Pulse Response Matrix). Let S be a linear difference system, and let $y^j(k)$ be the output of S assuming $x(k0) = 0$ and $u(k) = \delta(k - k0)i^j$. Then the *pulse response matrix* of S is denoted $G(k, k0)$ and is defined

$$G_{ij}(k, k0) \triangleq y_i^j(k), \qquad 1 \leqslant i \leqslant p, \quad 1 \leqslant j \leqslant m. \qquad \blacksquare$$

Thus the jth column of $G(k, k0)$ is the zero-state output of S at time k corresponding to an input $u^j(k)$ that is zero except for its jth component, which has a pulse at time $k0$. Note that the pulse response matrix $G(k, k0)$ can be obtained directly from input/output measurements of S. To derive the key property of the pulse response matrix, it is useful to first consider the relationship this matrix bears to the state transition matrix $Q(k, k0)$.

Proposition 4.1 (Pulse Response). Let S be a linear difference system with state transition matrix $Q(k, k0)$ and pulse response matrix $G(k, k0)$. Then

$$G(k, k0) = C(k)Q(k-1, k0)B(k0)1(k-1-k0) + \delta(k-k0)D(k0), \qquad k \in K.$$

Proof. Let $g^j(k, k0)$ be the jth column of $G(k, k0)$, and let $u^j(k) = \delta(k - k0)i^j$. Then, using Prop. 3.3 with $x(k0) = 0$ and the sampling property,

$$\begin{aligned}
g^j(k, k0) &= C(k)x(k) + D(k)u^j(k) \\
&= C(k) \sum_{i=k0}^{k-1} Q(k-1, i)B(i)u^j(i) + D(k)u^j(k) \\
&= C(k) \sum_{i=k0}^{k-1} Q(k-1, i)B(i)\delta(i-k0)i^j + D(k)\delta(k-k0)i^j \\
&= C(k)Q(k-1, k0)B(k0)1(k-k0)i^j + D(k0)\delta(k-k0)i^j \\
&= \{C(k)Q(k-1, k0)B(k0)1(k-1-k0) + D(k0)\delta(k-k0)\}i^j.
\end{aligned}$$

Here the step $1(k-1-k0)$ at $k = k0+1$ is included since $g^j(k0, k0) = D(k0)i^j$ because $x(k0) = 0$. Since the general expression for $g^j(k, k0)$ holds for $1 \leqslant j \leqslant m$, it follows that

$$G(k, k0) = C(k)Q(k-1, k0)B(k0)1(k-1-k0) + \delta(k-k0)D(k0). \qquad \blacksquare$$

The importance of the pulse response matrix $G(k, k0)$ lies in its usefulness in characterizing the input/output behavior of a relaxed linear difference system. In particular, we have the following fundamental result, which we refer to as the input/output representation of S.

Proposition 4.2 (Input/Output Representation). Let S be a linear difference system with pulse response matrix $G(k, k0)$. If $x(k0) = 0$, then for each $u \in \mathrm{U}$

$$y(k) = \sum_{j=k0}^{k} G(k, j)u(j), \quad k \in K.$$

Proof. Applying Prop. 3.3 with $x(k0) = 0$, the sampling property, and Prop. 4.1, we have

$$\begin{aligned}
y(k) &= C(k)x(k) + D(k)u(k) \\
&= C(k) \sum_{j=k0}^{k-1} Q(k-1, j)B(j)u(j) + D(k)u(k) \\
&= \sum_{j=k0}^{k-1} C(k)Q(k-1, j)B(j)u(j) + D(k)u(k) \\
&= \sum_{j=k0}^{k-1} C(k)Q(k-1, j)B(j)u(j)1(k-1-j) + D(k)u(k) \\
&= \sum_{j=k0}^{k} C(k)Q(k-1, j)B(j)u(j)1(k-1-j) + D(k)u(k) \\
&= \sum_{j=k0}^{k} \{C(k)Q(k-1, j)B(j)1(k-1-j) + \delta(k-j)D(j)\}u(j) \\
&= \sum_{j=k0}^{k} G(k, j)u(j).
\end{aligned}$$

■

Thus the pulse response matrix $G(k, k0)$ provides a complete characterization of the input/output behavior of a relaxed linear difference system.

Exercise 4.2 (Elapsed Time). Let S be a linear time-invariant difference system with pulse response matrix $G(k, k0)$. Show that

$$G(k, k0) = G(k - k0, 0), \qquad k \in K.$$

■

Note that Exer. 4.2 is analogous to Exer. 3.3, but for the pulse response matrix. That is, for time-invariant linear difference systems the initial time $k0$ is not important, only the elapsed time $k - k0$. The following is an example of a linear difference system and its input/output representation based on the pulse response.

Example 4.1 (Moving-Average Filter). Consider the moving-average filter introduced in Exam. 1.4. For a two-dimensional moving-average filter, the state and output equations are

$$x(k+1) = \begin{bmatrix} 0 & 1 \\ 0 & 0 \end{bmatrix} x(k) + \begin{bmatrix} \frac{1}{3} \\ \frac{1}{3} \end{bmatrix} u(k),$$

$$y(k) = [1 \quad 0]x(k) + [\tfrac{1}{3}]u(k).$$

From Prop. 3.4, the state transition matrix is $Q(k, 0) = A^k$. But in this case $A^2 = 0$, which implies $Q(k, 0) = 0$ for $k \geqslant 2$. Since $Q(0, 0) = I$, it follows that the state transition matrix of the moving-average filter is

$$Q(k, 0) = \begin{bmatrix} \delta(k) & \delta(k-1) \\ 0 & \delta(k) \end{bmatrix}.$$

Applying Prop. 4.1 to compute the pulse response matrix, we then have

$$\begin{aligned} G(k, 0) &= C(k)Q(k-1, 0)B(0)1(k-1) + \delta(k)D(0) \\ &= C(k)Q(k-1, 0)B(0)1(k-1) + \delta(k)/3 \\ &= [1, 0]Q(k-1, 0)[\tfrac{1}{3}, \tfrac{1}{3}]^{\mathrm{T}}1(k-1) + \delta(k)/3 \\ &= \{Q_{11}(k-1, 0)/3 + Q_{12}(k-1, 0)/3\}1(k-1) + \delta(k)/3 \\ &= \delta(k-1)/3 + \delta(k-2)/3 + \delta(k)/3. \end{aligned}$$

From Exer. 4.2 we have $G(k, j) = G(k-j, 0)$. It then follows from Prop. 4.2 that the relaxed or zero-state response of the moving-average filter for an arbitrary input $u \in \mathbf{U}$ is

$$\begin{aligned} y(k) &= \sum_{j=0}^{k} G(k, j)u(j) \\ &= \sum_{j=0}^{k} \{\delta(k-j) + \delta(k-1-j) + \delta(k-2-j)\} \frac{u(j)}{3} \\ &= \{u(k) + u(k-1)1(k-1) + u(k-2)1(k-2)\}/3. \end{aligned}$$

■

The following result combines the pulse response matrix and the state transition matrix to yield a general characterization of the input/output behavior of a linear difference system S.

Prop. 4.3 (Complete Response). Let S be a linear difference system with state transition matrix $Q(k, k0)$ and pulse response matrix $G(k, k0)$. Then for each $(x^0, u) \in \mathbf{R}^n \times \mathbf{U}$, the output of S corresponding to initial condition $x(k0) = x^0$ is

$$y(k) = C(k)Q(k, k0)x^0 + \sum_{j=k0}^{k} G(k, j)u(j)$$

Proof. Let $(x^0, u) \in \mathbf{R}^n \times \mathbf{U}$. Then from Exer. 3.2, Prop. 3.1, and Prop. 4.2

$$\begin{aligned} y(k) &= C(k)x(k) + D(k)u(k) \\ &= C(k)\{x(k, x^0) + x(k, u)\} + D(k)u(k) \\ &= C(k)x(k, x^0) + C(k)x(k, u) + D(k)u(k) \\ &= C(k)Q(k, k0)x^0 + C(k)x(k, u) + D(k)u(k) \\ &= C(k)Q(k, k0)x^0 + \sum_{j=k0}^{k} G(k, j)u(j). \end{aligned}$$ ■

Exercise 4.3 (Decomposition). Show that the complete response of a linear difference system S can be decomposed as follows:

$$y(k) = y(k, x^0) + y(k, u).$$

Here $y(k, x^0)$ is the *zero-input response* that results when $u = 0$. Similarly, $y(k, u)$ is the *zero-state response* that results when $x^0 = 0$. ■

9.5. THE TRANSFER FUNCTION REPRESENTATION

The input/output representation of a relaxed linear difference system embodied in the pulse response matrix $G(k, k0)$ can be simplified considerably in the event that the system S is time-invariant. To see this it is useful to first consider a linear transformation that allows us to recast a system of difference equations as an equivalent system of algebraic equations.

Definition 5.1 (Z-Transform). Let $f: \{0, 1, \ldots\} \to \mathbf{R}^n$. Then the *Z-transform* of f is denoted $F(z)$ or $Z\{f(k)\}$ and is defined as

$$F(z) \triangleq \sum_{k=0}^{\infty} f(k)z^{-k}, \qquad z \in W.$$

Here the subset W of C is called the *region of convergence* of $F(z)$. ■

Refer to Table 9.1 for a collection of commonly used Z-transform pairs and their associated regions of convergence. Note that in each case the region of convergence W is the area in the complex plane outside a closed ball of radius r where r is called the *radius of convergence.*

It is evident from inspection of Table 9.1 that in each case $F(z)$ is the ratio of two polynomials in z. The zeros of the denominator polynomials are called the *poles* of $F(z)$. Note that the radius of convergence in each case equals the radius of the outermost pole. Thus the region of convergence W is of the following form

TABLE 9.1. Z-Transform Pairs

$f(k)$	$F(z)$	W
$(k\,\Delta t)^j a^{k\Delta t}$	$(\Delta t)^j\left(-z\dfrac{d}{dz}\right)^j \dfrac{z}{(z-a)}$	$\lvert z\rvert > \lvert a\rvert$
$a^{k\Delta t}\cos(\omega k\,\Delta t)$	$\dfrac{z\{z - a\cos(\omega\,\Delta t)\}}{z^2 - 2a\cos(\omega\,\Delta t)z + a^2}$	$\lvert z\rvert > \lvert a\rvert$
$a^{k\Delta t}\sin(\omega k\,\Delta t)$	$\dfrac{a\sin(\omega\,\Delta t)z}{z^2 - 2a\cos(\omega\,\Delta t)z + a^2}$	$\lvert z\rvert > \lvert a\rvert$

where r is the radius of the outermost pole

$$W = \{z \in \mathbf{C} : |z| > r\}.$$

There are a number of useful properties that the Z-transform satisfies (Ahmed and Natarajan, 1983). Some of the more fundamental ones are summarized in the following exercise.

Exercise 5.1 (Z-Transform Properties). Let $f:\{0, 1, \ldots\} \to \mathbf{R}$ and $g:\{0, 1, \ldots\} \to \mathbf{R}$, and let $F(z)$ and $G(z)$ be the Z-transforms of f and g, respectively. Show that for each $(\alpha, \beta, i) \in \mathbf{C}^2 \times \{0, 1, \ldots\}$

1. $Z\{\alpha f(k) + \beta g(k)\} = \alpha F(z) + \beta G(z)$,
2. $Z\{\alpha^{-k} f(k)\} = F(\alpha z)$,
3. $Z\{kf(k)\} = -z\dfrac{dF(z)}{dz}$,
4. $Z\{f(k-i)1(k-i)\} = z^{-i}F(z)$. ■

We refer to 1 of Exer. 5.1 as the linearity property, 2 as the scaling property, 3 as the multiplication property, and 4 as the delay property. Perhaps the most important property for system analysis purposes is the effect the Z-transform has on the convolution of two sequences. We define the convolution operation for discrete-time signals as follows.

Definition 5.2 (Discrete Convolution). Let $f:\{0, 1, \ldots\} \to \mathbf{R}$ and $g:\{0, 1, \ldots\} \to \mathbf{R}$. Then the *discrete convolution* of f with g is denoted $f(k) * g(k)$ and is defined as

$$f(k) * g(k) \triangleq \sum_{j=0}^{k} f(k-j)g(j).$$ ■

The importance of the convolution operation lies in the fact that this is precisely the type of processing that a linear system performs when it is excited with input signal. In particular, the output of a relaxed linear time-invariant difference system S is simply the discrete convolution of the input sequence with the impulse response matrix as can be seen from the following exercise.

Exercise 5.2 (Convolution). Let S be a linear time-invariant difference system, and suppose $y(k)$ is the output of S associated with the input $u \in \mathbf{U}$ and the initial condition $x(0) = 0$. Show that

$$y(k) = G(k, 0) * u(k). \qquad \blacksquare$$

Although the convolution operation can involve substantial computational effort, the image of the convolution operation under the Z-transformation is quite simple, as can be seen from the following result.

Proposition 5.1 (Discrete Convolution). Let $f:\{0, 1, \ldots\} \to \mathbf{R}$ and $g:\{0, 1, \ldots\} \to \mathbf{R}$, and let $F(z)$ and $G(z)$ denote the Z-transforms of f and g, respectively. Then

$$Z\{f(k) * g(k)\} = F(z)G(z).$$

Proof. Let $f_e(k)$ be an extension of the function f to all integers. Here $f_e(k) \triangleq 0$ for $k < 0$, and

$$f_e(k) \triangleq f(k), \qquad k \geqslant 0.$$

It then follows that $f_e(k-j)g(j) = f(k-j)g(j)$ for $0 \leqslant j \leqslant k$. Similarly, $f_e(k-j)g(j) = 0$ for $j > k$. Thus from Def. 5.2

$$\begin{aligned} f(k) * g(k) &= \sum_{j=0}^{k} f(k-j)g(j) \\ &= \sum_{j=0}^{k} f_e(k-j)g(j) \\ &= \sum_{j=0}^{\infty} f_e(k-j)g(j). \end{aligned}$$

Using Def. 5.1 we then have

$$\begin{aligned} Z\{f(k) * g(k)\} &= Z\left\{\sum_{j=0}^{\infty} f_e(k-j)g(j)\right\} \\ &= \sum_{k=0}^{\infty}\sum_{j=0}^{\infty} f_e(k-j)g(j)z^{-k}. \end{aligned}$$

Let $r \geqslant 0$ be the radius of convergence. Since the convergence of the series is absolute, it is uniform over $|z| > r$. Thus we can interchange the order of summation. Using Exer. 5.1, we then have

$$\begin{aligned}
Z\{f(k) * g(k)\} &= \sum_{j=0}^{\infty} \sum_{k=0}^{\infty} f_e(k - j) g(j) z^{-k} \\
&= \sum_{j=0}^{\infty} \left\{ \sum_{k=0}^{\infty} f_e(k - j) z^{-k} \right\} g(j) \\
&= \sum_{j=0}^{\infty} \left\{ \sum_{k=0}^{\infty} f_e(k - j) z^{-(k-j)} \right\} z^{-j} g(j) \\
&= \sum_{j=0}^{\infty} \left\{ \sum_{i=-j}^{\infty} f_e(i) z^{-i} \right\} z^{-j} g(j) \\
&= \sum_{j=0}^{\infty} \left\{ \sum_{i=0}^{\infty} f_e(i) z^{-i} \right\} z^{-j} g(j) \\
&= \sum_{j=0}^{\infty} \left\{ \sum_{i=0}^{\infty} f(i) z^{-i} \right\} z^{-j} g(j) \\
&= \sum_{j=0}^{\infty} F(z) z^{-j} g(j) \\
&= F(z) \sum_{j=0}^{\infty} g(j) z^{-j} \\
&= F(z) G(z). \qquad \blacksquare
\end{aligned}$$

Thus discrete convolution in the time domain k maps into multiplication in the complex frequency domain z under the Z-transform. This observation, together with Exer. 5.2, provides the motivation for the following definition.

Definition 5.3 (Transfer Function Matrix). Let S be a linear time-invariant difference system with pulse response matrix $G(k, 0)$. Then the *transfer function matrix* $G(z)$ is defined as

$$G(z) \triangleq Z\{G(k, 0)\}. \qquad \blacksquare$$

Thus the transfer function matrix $G(z)$ is simply the Z-transform of the pulse response matrix $G(k, k0)$ with $k0 = 0$. Note from Exer. 4.2 that the constraint on $k0$ is of no consequence as we are now restricting our attention to time-invariant systems.

Like the pulse response matrix it comes from, the transfer function matrix is a $p \times m$ matrix whose components can be obtained directly from input/output measurements of S. The transfer function matrix $G(z)$ provides us with a simple algebraic characterization of the input/output behavior of a relaxed linear time-invariant system, as the following result demonstrates.

Proposition 5.2 (Algebraic Representation). Let S be a linear time-invariant difference system with transfer function matrix $G(z)$, and let $y(k)$ be the output of S corresponding to $x(0) = 0$ and $u \in \mathbf{U}$. If $U(z)$ and $Y(z)$ denote the Z-transforms of u and y, respectively, then

$$Y(z) = G(z)U(z).$$

Proof. Applying Exer. 5.2 and Prop. 5.1, we have

$$\begin{aligned} Y(z) &= Z\{y(k)\} \\ &= Z\{G(k, 0) * u(k)\} \\ &= Z\{G(k, 0)\}U(z) \\ &= G(z)U(z). \end{aligned}$$ ■

Thus for a relaxed linear time-invariant system, the Z-transform of the output is simply the transfer function times the Z-transform of the input. Note that Prop. 5.2 effectively replaces a difference equation in $k \in \mathbf{K}$ with an algebraic equation in $z \in \mathbf{C}$.

Exercise 5.3. Let S be a linear time-invariant system with state transition matrix $Q(k, 0)$ and transfer function matrix $G(z)$. Show that

$$G(z) = CZ\{Q(k-1, 0)1(k-1)\}B + D.$$ ■

As it now stands, the transfer function matrix is computed indirectly by first finding the pulse response matrix $G(k, 0)$ and then transforming it. Exercise 5.3 provides us with an analogous route based upon the state transition matrix $Q(k, 0)$. To develop a procedure for obtaining $G(z)$ directly from the system parameters $\{A, B, C, D\}$, it is useful to first consider the following additional fundamental property of the Z-transform.

Proposition 5.3 (Advance Property). Let $f: \{0, 1, \ldots\} \to \mathbf{R}$, and let $F(z)$ be the Z-transform of $f(k)$. Then

$$Z\{f(k+j)\} = z^j \left\{F(z) - \sum_{i=0}^{j-1} f(i)z^{-i}\right\}, \qquad j \geq 0.$$

Proof. Applying Def. 5.1,

$$\begin{aligned} Z\{f(k+j)\} &= \sum_{k=0}^{\infty} f(k+j)z^{-k} \\ &= z^j \sum_{k=0}^{\infty} f(k+j)z^{-(k+j)} \\ &= z^j \sum_{i=j}^{\infty} f(i)z^{-i} \\ &= z^j \left\{ \sum_{i=0}^{\infty} f(i)z^{-i} - \sum_{i=0}^{j-1} f(i)z^{-i} \right\} \\ &= z^j \left\{ F(z) - \sum_{i=0}^{j-1} f(i)z^{-i} \right\}. \end{aligned}$$ ■

Thus when $f(0) = 0$, an advance in the time domain corresponds to scaling by z in the Z-transform domain. It is this property that allows one to transform a difference system in k into an algebraic system in the complex variable z.

Exercise 5.4 (Transfer Function). Consider the following qth-order linear time-invariant difference equation where $a_0 = 1$:

$$\sum_{j=0}^{q} a_j y(k+q-j) = \sum_{j=0}^{q} b_j u(k+q-j).$$

Use the advance property and Prop. 5.2 to show that the transfer function of this system is

$$G(z) = \frac{b_0 z^q + b_1 z^{q-1} + \cdots + b_q}{z^q + a_1 z^{q-1} + \cdots + a_q}.$$ ■

The advance property of the Z-transform provides us with an alternative method for computing the state transition matrix $Q(k, 0)$, as can be seen from the following exercise.

Exercise 5.5 (State Transition Matrix). Let S be a linear time-invariant difference system with state transition matrix $Q(k, 0)$. Show that

$$Z\{Q(k, 0)\} = z(zI - A)^{-1}.$$ ■

The results of Exers. 5.3 and 5.5 can now be combined. In doing so we generate the following direct formulation of the transfer function matrix in terms of the matrix parameters of the system S.

Proposition 5.4 (Transfer Function Matrix). Let S be a linear time-invariant difference system with transfer function matrix $G(z)$. Then

$$G(z) = C(zI - A)^{-1}B + D. \qquad \blacksquare$$

The proof of Prop. 5.4 follows directly from Exers. 5.3 and 5.5 and the advance property. Typically, the transfer function $G(z)$ will be expressible as the ratio of two polynomials in z. This feature allows us to define something called the poles of S.

Definition 5.4 (Poles of S). Let S be a linear time-invariant difference system with transfer function matrix $G(z)$. Then $p \in \mathbf{C}$ is a *pole* of $S \Leftrightarrow p$ is a zero of the denominator polynomial of at least one component of the matrix $G(z)$. ■

Thus the poles of S are those points in the complex plane where the denominators of the components of $G(z)$ vanish. In light of Props. 5.4 and 7.3.2, every pole of S is an eigenvalue of A. However, the converse is not always true, as can be seen from the following exercise.

Exercise 5.6 (Poles of S). Let S be a linear time-invariant difference system with transfer function matrix $G(z)$. Show that every pole of S is an eigenvalue of A. Show that the converse is not true by finding a counterexample. ■

The following example illustrates the use of Prop. 5.4 to compute the transfer function matrix of a linear time-invariant difference system.

Example 5.1 (Transfer Function Matrix). Consider the moving-average filter introduced in Exam. 1.4. For the two-dimensional filter the state and output equations are

$$x(k+1) = \begin{bmatrix} 0 & 1 \\ 0 & 0 \end{bmatrix} x(k) + \begin{bmatrix} \frac{1}{3} \\ \frac{1}{3} \end{bmatrix} u(k),$$

$$y(k) = [1 \quad 0]x(k) + [\tfrac{1}{3}]u(k).$$

In this case we have

$$\begin{aligned}(zI - A)^{-1} &= \begin{bmatrix} z & -1 \\ 0 & z \end{bmatrix}^{-1} \\ &= \begin{bmatrix} z & 1 \\ 0 & z \end{bmatrix} \frac{1}{z^2}.\end{aligned}$$

Applying Prop. 5.4 to compute $G(z)$ yields

$$\begin{aligned}
G(z) &= C(zI - A)^{-1}B + D \\
&= [1 \quad 0]\begin{bmatrix} z & 1 \\ 0 & z \end{bmatrix}\begin{bmatrix} \frac{1}{3} \\ \frac{1}{3} \end{bmatrix}\frac{1}{z^2} + \frac{1}{3} \\
&= [z \quad 1]\begin{bmatrix} \frac{1}{3} \\ \frac{1}{3} \end{bmatrix}\frac{1}{z^2} + \frac{1}{3} \\
&= \frac{1}{3} + \frac{z^{-1}}{3} + \frac{z^{-2}}{3} \\
&= \frac{z^2/3 + z/3 + 1/3}{z^2}.
\end{aligned}$$

As a check, we know from Def. 5.3 that $G(z) = Z\{G(k, 0)\}$. Using the expression for $G(k, 0)$ from Exam. 4.1,

$$\begin{aligned}
G(z) &= Z\{G(k, 0)\} \\
&= Z\left(\frac{\delta(k) + \delta(k-1) + \delta(k-2)}{3}\right) \\
&= \frac{1}{3} + \frac{z^{-1}}{3} + \frac{z^{-2}}{3} \\
&= \frac{z^2/3 + z/3 + 1/3}{z^2}.
\end{aligned}$$

Thus a two-dimensional moving-average filter has a double pole located at $z = 0$. ■

Proposition 5.4 provides us with a direct means of obtaining the transfer function matrix $G(z)$ from the system parameters $\{A, B, C, D\}$. Note that it involves inverting the matrix $(zI - A)$. One way to compute $(zI - A)^{-1}$ is to apply Alg. 8.5.1, substituting the Z-transform variable z for the Laplace transform variable s.

To make the algebraic representation of S in Prop. 5.2 practical, we must have a means of inverting the Z-transform to recover $y(k)$ from $Y(z)$. This can be achieved in several ways. One technique is to recast $F(z)$ as a sum of simple expressions, each of which can be inverted by inspection with the help of Table 9.1. Then using the linearity property, the inverse transform of the sum is the sum of the inverse transforms.

The key to the table look-up method is to expand $F(z)/z$ into a sum of simple terms using the method of partial fractions. We then multiply both sides by z and use the table. Note that $F(z)/z$ is expanded rather than $F(z)$ since the entries in Table 9.1 each have a z in the numerator. Since $F(z)/z$ will be the ratio of two polynomials in z, Prop. 8.5.6 can be used to perform the required partial fraction expansion.

An alternative technique for inverting $F(z)$ to recover $f(k)$ is the residue method. This method, based on Cauchy's residue theorem, is more effective than partial fractions in the sense that for multiple poles it involves fewer computations. Furthermore, the residue method does not require the use of a table.

Proposition 5.5 (Residue Method). Let p_i be a pole of $F(z)z^{k-1}$ of multiplicity $m(i)$ for $1 \leqslant i \leqslant q$. Let $R_i(z)$ be the *residue* of $F(z)z^{k-1}$ at the pole p_i. That is

$$R_i(z) \triangleq \frac{1}{(m(i)-1)!} \frac{d^{m(i)-1}}{dz^{m(i)-1}} \{(z-p_i)^{m(i)} F(z) z^{k-1}\}.$$

Then for each $k \geqslant 0$ we have

$$f(k) = \sum_{i=1}^{q} R_i(p_i).$$

Proof. Let C represent a circle in the complex plane of radius R where $R > |p_i|$ for $1 \leqslant i \leqslant q$. Thus R encloses all of the poles of $F(z)z^{k-1}$. First, we show that $f(k)$ can be expressed by the following contour integral where $j = \sqrt{-1}$:

$$f(k) = \frac{1}{j2\pi} \oint_C F(z) z^{k-1}\, dz, \qquad k \geqslant 0.$$

Here the integral is around C in the counterclockwise sense. Let f_k denote the candidate expression for $f(k)$, and consider the change of variable $z = R\exp(j\theta)$. Then $dz = jz\, d\theta$ and

$$\begin{aligned}
f_k &\triangleq \frac{1}{j2\pi} \oint_C F(z) z^{k-1}\, dz \\
&= \frac{1}{j2\pi} \oint_C \left(\sum_{i=0}^{\infty} f(i) z^{-i} \right) z^{k-1}\, dz \\
&= \frac{1}{j2\pi} \sum_{i=0}^{\infty} f(i) \oint_C z^{k-i-1}\, dz \\
&= \frac{1}{j2\pi} \sum_{i=0}^{\infty} f(i) \oint_C z^{k-i} \frac{dz}{z} \\
&= \frac{1}{2\pi} \sum_{i=0}^{\infty} f(i) \int_0^{2\pi} R^{k-i} \exp[j(k-i)\theta]\, d\theta
\end{aligned}$$

$$= \frac{1}{2\pi} \sum_{i=0}^{\infty} f(i)R^{k-i} \int_{0}^{2\pi} \exp[j(k-i)\theta]\, d\theta$$

$$= \frac{1}{2\pi} \sum_{i=0}^{\infty} f(i)R^{k-1} 2\pi\delta(k-i)$$

$$= \sum_{i=0}^{\infty} f(i)R^{k-i}\delta(k-i)$$

$$= f(k).$$

Thus $f_k = f(k)$ for $k \geqslant 0$ where f_k is the contour integral of $F(z)z^{k-1}$. But from Cauchy's residue theorem (Hille, 1973) this contour integral equals the sum of the residues of the poles of $F(z)z^{k-1}$ enclosed by C. Thus

$$\begin{aligned} f(k) &= f_k \\ &= \sum_{i=1}^{q} R_i(p_i). \end{aligned}$$ ■

Note from Prop. 8.5.6 that the residues are in fact the partial fraction expansion coefficients $\{R_{i1}(z), 1 \leqslant i \leqslant q\}$ associated with the terms of the form $1/(z - p_i)$. For the case of simple poles, the residues are quite easy to compute, for in this case $R_i(z) = (z - p_i)F(z)z^{k-1}$.

Example 5.2 (Relaxed Response). Consider a linear time-invariant system S with the following transfer function:

$$G(z) = \frac{z}{z^2 + 3z/4 + 1/8}.$$

Suppose $x(0) = 0$ and $u(k) = (1/3)^k$ for $k \geqslant 0$. To find $y(k)$, first note from Prop. 5.2 and Table 9.1 that

$$\begin{aligned} Y(z) &= G(z)U(z) \\ &= \frac{z}{z^2 + 3z/4 + 1/8} \frac{z}{(z - 1/3)} \\ &= \frac{z^2}{(z + 1/2)(z + 1/4)(z - 1/3)}. \end{aligned}$$

Thus from Prop. 5.5, $y(k)$ is the sum of the residues of $Y(z)z^{k-1}$. That is,

$$y(k) = R_1(-\tfrac{1}{2}) + R_2(-\tfrac{1}{4}) + R_3(\tfrac{1}{3}).$$

Here

$$R_1\left(-\frac{1}{2}\right)=\frac{(-1/2)^{k+1}}{(-1/4)(-5/6)}=\left(\frac{24}{5}\right)\left(-\frac{1}{2}\right)^{k+1},$$

$$R_2\left(-\frac{1}{4}\right)=\frac{(-1/4)^{k+1}}{(1/4)(-7/12)}=-\left(\frac{48}{7}\right)\left(-\frac{1}{4}\right)^{k+1},$$

$$R_3\left(\frac{1}{3}\right)=\frac{(1/3)^{k+1}}{(5/6)(7/12)}=\left(\frac{72}{35}\right)\left(\frac{1}{3}\right)^{k+1}.$$

Thus the zero-state response of this system to the input $u(k) = (1/3)^k$ for $k \geqslant 0$ is

$$y(k) = -(\tfrac{12}{5})(-\tfrac{1}{2})^k + (\tfrac{12}{7})(-\tfrac{1}{4})^k + (\tfrac{24}{35})(\tfrac{1}{3})^k. \qquad \blacksquare$$

The three representations of linear difference systems based upon the state transition matrix $Q(k, k0)$, the pulse response matrix $G(k, k0)$, and the transfer function matrix $G(z)$ are summarized in Table 9.2. Note that the state space representation is a complete *internal* representation because it takes into account the effects of both the input u and the initial state x^0. The pulse response representation and the transfer function representation are less complete *external* representations that assume the system S is relaxed. Of these two input/output representations of S, the pulse response representation is the more general, for it applies to both time-varying and time-invariant linear systems, whereas the transfer function representation is applicable only to time-invariant linear systems.

TABLE 9.2. Three Representations of Linear Difference Systems

Representation	Type	Relaxed	Time-Invariant
State space: $Q(k, k0)$	Internal	No	No
Impulse response: $G(k, k0)$	External	Yes	No
Transfer function: $G(z)$	External	Yes	Yes

PROBLEMS

1. Convert the following second-order difference equation to a first-order system S. Is S linear? Is S time-invariant?

$$y(k + 2) + ky(k + 1)\{1 - y^2(k)\} = u(k).$$

2. Convert the following third-order difference equation to a first-order system S.

$$y(k+3) + 2y(k+2) + 3y(k+1) + 4y(k) = 5u(k+2) + 6u(k).$$

3. Consider a linear time-invariant difference system with the following state equation:

$$x(k+1) = \begin{bmatrix} \frac{2}{5} & 1 \\ 0 & \frac{4}{5} \end{bmatrix} x(k) + \begin{bmatrix} 0 \\ 1 \end{bmatrix} u(k).$$

a. Find the state transition matrix $Q(k, 0)$.

b. Find the homogeneous solution when $x(0) = x^0$.

4. Let $P(k)$ *be a fundamental matrix* of A. That is, let $P(k)$ be any solution of the following difference equation:

$$P(k+1) = AP(k), \qquad \det\{P(0)\} \neq 0.$$

a. Show that if $\det(A) \neq 0$, then $\det\{P(k)\} \neq 0$ for $k \geqslant 0$.

b. Show that $Q(k, k0) = P(k)P^{-1}(k0)$.

5. Find the zero-state solution $x(k, u)$ of the system in problem 3 when the input is $u(k) = (\frac{1}{2})^k$, $k \geqslant 0$.

6. Consider the following linear time-invariant state equation:

$$x(k+1) = \begin{bmatrix} 1 & \frac{5}{4} \\ -1 & -1 \end{bmatrix} x(k) + \begin{bmatrix} 0 \\ 1 \end{bmatrix} u(k).$$

a. Find the zero-input solution when $x(0) = [1, -1]^T$.

b. Find the zero-state solution when $u(k) = 3$, $k \geqslant 0$.

c. Use the results of a and b to write the complete solution when $x(0) = [2, -2]^T$ and $u(k) = 1$ for $k \geqslant 0$.

7. Compute $Q(k, 0)$ for the system in problem 6 using the method of interpolating polynomials.

8. Compute the spectral representation of $Q(k, 0)$ for the system in problem 6.

9. Consider the following linear time-invariant difference system S.

$$x(k+1) = \begin{bmatrix} 1 & -\frac{1}{2} \\ \frac{3}{2} & -1 \end{bmatrix} x(k) + \begin{bmatrix} 0 \\ 1 \end{bmatrix} u(k),$$

$$y(k) = [1 \quad 0]x(k) + [3]u(k).$$

a. Find the impulse response matrix $G(k, 0)$.

b. Find the transfer function matrix $G(z)$.

10. Find the zero-input response $y(k, x^0)$ of the system in problem 9 when $x(0) = x^0$.

11. Find the zero-state response $y(k, u)$ of the system in problem 9 when $u(k) = \alpha$, $k \geqslant 0$.

12. Find the complete response of the system in problem 9 when $x(0) = [2, 0]^T$ and $u(k) = 10$, $k \geqslant 0$.

13. Consider the following linear time-invariant difference system with transfer function $G(z)$. Find the zero-state response $y(k, u)$ when $u(k) = \exp(-2k)$, $k \geqslant 0$.

$$G(z) = \frac{z}{z^2 + z/2 - 3/16}.$$

14. Let $f:\{0, 1, \ldots\} \to \mathbf{R}$, and let $F(z)$ be the Z-transform of f. Show that the *initial value theorem* holds. That is, show that

$$f(0) = \lim_{z \to \infty} \{F(z)\}.$$

15. Let $f:\{0, 1, \ldots\} \to \mathbf{R}$, and let $F(z)$ be the Z-transform of f. Show that the *final value theorem* holds. That is, show that if the poles of $(z - 1)F(z)$ all lie strictly inside the unit circle, $|z| = 1$, then

$$f(\infty) = \lim_{z \to 1} \{(z - 1)F(z)\}.$$

16. Show that the following series, known as the *geometric series*, converges in the manner indicated. (*Hint*: let S_n be the nth partial sum, and examine the error in the limit as $n \to \infty$.)

$$\sum_{k=0}^{\infty} \{g(z)\}^k = \frac{1}{1 - g(z)}, \qquad |g(z)| < 1.$$

17. Use the results of problem 16 to generalize the geometric series as follows:

$$\sum_{k=m}^{\infty} \{g(z)\}^k = \frac{\{g(z)\}^m}{1 - g(z)}, \qquad |g(z)| < 1.$$

18. Use the results of problem 17 to generalize the geometric series to finite sums as follows:

$$\sum_{k=m}^{n} \{g(z)\}^k = \frac{\{g(z)\}^m - \{g(z)\}^{n+1}}{1 - g(z)}, \qquad n \geqslant m.$$

19. Use the geometric series to derive the following entry in Table 9.1:

$$Z\{a^k 1(k)\} = \frac{z}{z - a}, \qquad |z| > |a|.$$

20. For a linear time-invariant difference system S consider the following matrix $U_c(n)$, which is called the *controllability matrix*:

$$U_c(n) \triangleq [B, AB, A^2B, \ldots, A^{n-1}B].$$

Suppose $U_c(n)$ is nonsingular and S is a single-input system. Let $w = [u(n-1), u(n-2), \ldots, u(0)]^T$. Show that the input $\{w(k)\}$ drives the system S from the initial state x^0 at time 0 to the final state x^1 at time n if

$$w = U_c^{-1}(n)(x^1 - A^n x^0).$$

21. For a linear time-invariant difference system S consider the following matrix $U_o(n)$ called the *observability matrix*:

$$U_o(n) \triangleq [C^T, A^T C^T, \ldots, (A^T)^{n-1} C^T]^T.$$

Suppose $U_o(n)$ is nonsingular and S is a single-output system. Let $v = [y(0, x^0), y(1, x^0), \ldots y(n-1, x^0)]^T$ where $y(k, x^0)$ is the zero-input part of the response. Show that the initial state x^0 can be recovered from recordings of the input and output waveforms over $\{0, 1, \ldots n-1\}$ as follows:

$$x^0 = U_o^{-1}(n)v.$$

REFERENCES

Ahmed, N. and T. Natarajan, *Discrete-Time Signals and Systems*, Reston Publishing, Reston, VA, 1983.

Franklin, G. F. and J. D. Powell, *Digital Control of Dynamic Systems*, Addison-Wesley, Reading, MA, 1980.

Hille, E., *Analytic Function Theory, Vol. 1*, Chelsea, New York, 1973.

Ogata, K., *Modern Control Engineering*, Prentice-Hall, Englewood Cliffs, NJ, 1970.

10 ORTHOGONAL TRANSFORMATIONS AND AMPLITUDE ESTIMATION

The vector space approach is a very powerful tool for engineering analysis because it provides us with a conceptual framework within which many different problems can be formulated. In this chapter we investigate the potential of vector space methods by applying these techniques to a number of basic engineering problems. We examine a variety of orthogonal transformations that are used in engineering analysis. By employing a vector space model, we are able to provide a unified treatment based on a simple standard formulation. We also examine the problem of estimating the amplitude of a signal that is corrupted with noise. Here vector space techniques are used to formulate several different approaches to the solution

10.1. A VECTOR SPACE MODEL

Many orthogonal transformations are routinely used in engineering analysis. Generally, the introduction of an orthogonal transformation pair (the transformation and its inverse) is begun with a definition. This is followed by proofs of orthogonality and associated properties of the transform pair including Parseval's relation. This procedure has to be repeated for various transform relations. In the first part of this chapter, we present a generalized structure for an orthogonal transformation relation by using a vector space model (Lee, 1985a). This method enables us to visualize the similarities as well as the differences among the orthogonal transformations. Transform pairs, including the Fourier transform, Fourier series, discrete-time Fourier transform (DFT), Hankel transform, Hilbert transform, and the sampling theorem, are shown to be special cases of the generalized model. This approach can be used for extension or modification of existing orthogonal transformations. It can be also applied to the design of special-purpose orthogonal mapping techniques.

The structure of an orthogonal transform pair can be generalized by using a simple vector space model. The requirement for formulating a transform pair consists of three basic items: (1) two vector spaces, (2) a basis, and (3) two valid inner products. The two vector spaces, say **V** and **W**, are defined in domains, A

and B, respectively. We let α be the parameter associated with A, and β be the parameter associated with B. The domains, A and B, can be continuous or discrete, finite or infinite, and single or multidimensional. The vectors in $\mathbf{V}$ and $\mathbf{W}$ can be arrays, sequences, or functions of finite or infinite dimensions. For each vector space with its associated domain, one valid inner product is defined. We then select a basis set which spans both $\mathbf{V}$ and $\mathbf{W}$. The basis $\{\phi(\alpha, \beta)\}$ must be orthogonal in both vector spaces with respect to the corresponding inner products in the following sense:

$$\langle \phi(\alpha, \beta), \phi(\alpha, \beta')\rangle_A = \|\phi(\alpha, \beta)\|_A^2 \delta(\beta - \beta')$$
$$\langle \phi(\alpha, \beta), \phi(\alpha', \beta)\rangle_B = \|\phi(\alpha, \beta)\|_B^2 \delta(\alpha - \alpha').$$

Here $\delta(\,)$ is the *Kronecker delta* (unit pulse) or the *Dirac* delta (unit impulse), as appropriate, depending on whether the domains A and B are discrete or continuous.

The basis $\{\phi(\alpha, \beta)\}$ is also required to satisfy the following conditions in order to produce the uniqueness of the complete mapping operation:

$$\|\phi(\alpha, \beta)\|_A \|\delta(\beta - \beta')\|_B = 1,$$
$$\|\phi(\alpha, \beta)\|_B \|\delta(\alpha - \alpha')\|_A = 1.$$

The orthogonal transformation can now be described as a one-to-one mapping, which transforms an element $x(\alpha)$ in $\mathbf{V}$ uniquely into an element $X(\beta)$ in $\mathbf{W}$. The transform and its inverse can be formulated as

$$X(\beta) = \langle x(\alpha), \phi(\alpha, \beta)\rangle_A,$$
$$x(\alpha) = \langle X(\beta), \bar{\phi}(\alpha, \beta)\rangle_B.$$

The uniqueness of the mapping can be shown by utilizing the orthogonality. In particular,

$$\begin{aligned} X(\beta) &= \langle x(\alpha), \phi(\alpha, \beta)\rangle_A \\ &= \langle\langle X(\beta'), \bar{\phi}(\alpha, \beta')\rangle_B, \phi(\alpha, \beta)\rangle_A \\ &= \langle X(\beta'), \langle \phi(\alpha, \beta'), \phi(\alpha, \beta)\rangle_A\rangle_B \\ &= \langle X(\beta'), \|\phi(\alpha, \beta)\|_A^2 \delta(\beta' - \beta)\rangle_B = X(\beta). \end{aligned}$$

Parseval's relation can also be derived for this general vector space model of an orthogonal transform. Here

$$\begin{aligned} \langle x(\alpha), y(\alpha)\rangle_A &= \langle x(\alpha), \langle Y(\beta), \bar{\phi}(\alpha, \beta)\rangle_B\rangle_A \\ &= \langle\langle x(\alpha), \phi(\alpha, \beta)\rangle_A, Y(\beta)\rangle_B \\ &= \langle X(\beta), Y(\beta)\rangle_B, \end{aligned}$$

where $X(\beta)$ and $Y(\beta)$ are the transforms of $x(\alpha)$ and $y(\alpha)$, respectively. For the case when $x(\alpha) = y(\alpha)$, this reduces to

$$\begin{aligned}\|x(\alpha)\|_A^2 &= \langle x(\alpha), x(\alpha)\rangle_A \\ &= \langle X(\beta), X(\beta)\rangle_B \\ &= \|X(\beta)\|_B^2.\end{aligned}$$

It follows that the norm of a vector is invariant under an orthogonal transformation.

10.2. FOURIER TRANSFORM

The Fourier transform is probably the most widely used orthogonal transformation in engineering analysis for analog signals and systems (Gabel and Roberts, 1980; Frederick and Carlson, 1971). For the case of the Fourier transform, the vector spaces, **V** and **W**, are of infinite dimension. Both **V** and **W** are vector spaces of all energy and power signals. Time and frequency are commonly used to denote the associated parameters. Both parameters, t (time) and f (frequency), are defined in the interval $(-\infty, \infty)$. That is,

$$\begin{aligned}A &= \{t: -\infty < t < \infty\}, \\ B &= \{f: -\infty < f < \infty\}.\end{aligned}$$

The basis functions for the Fourier transform are

$$\phi(t, f) = \exp(j2\pi ft).$$

Let $x(t)$ and $y(t)$ be two arbitrary functions in **V**. The corresponding inner product is defined as

$$\langle x(t), y(t)\rangle_t = \int_{-\infty}^{\infty} x(t)\bar{y}(t)\, dt.$$

Similarly, the definition of the inner product in **W** can be formulated for any two frequency-domain functions, $X(f)$ and $Y(f)$, as

$$\langle X(f), Y(f)\rangle_f = \int_{-\infty}^{\infty} X(f)\bar{Y}(f)\, df.$$

The basis functions are orthonormal with respect to both inner products:

$$\begin{aligned}\langle \phi(t, f), \phi(t, f')\rangle_t &= \int_{-\infty}^{\infty} \exp[j2\pi(f - f')t]\, dt = \delta(f - f'), \\ \langle \phi(t, f), \phi(t', f)\rangle_f &= \int_{-\infty}^{\infty} \exp[j2\pi f(t - t')]\, df = \delta(t - t').\end{aligned}$$

These are Dirac delta or unit impulse functions because both **V** and **W** have continuous domains. The Fourier transform can be regarded as a mapping defined by the inner product in the vector space **V**. That is,

$$\begin{aligned} X(f) &= F\{x(t)\} \\ &= \langle x(t), \phi(t, f)\rangle_t \\ &= \int_{-\infty}^{\infty} x(t)\bar{\phi}(t, f)\, dt \\ &= \int_{-\infty}^{\infty} x(t) \exp(-j2\pi ft)\, dt. \end{aligned}$$

Similarly, the inverse Fourier transform can be defined as

$$\begin{aligned} x(t) &= F^{-1}\{X(f)\} \\ &= \langle X(f), \bar{\phi}(t, f)\rangle_f \\ &= \int_{-\infty}^{\infty} X(f)\phi(t, f)\, df \\ &= \int_{-\infty}^{\infty} X(f) \exp(j2\pi ft)\, df. \end{aligned}$$

Parseval's relation for Fourier transform in this case is as follows:

$$\begin{aligned} \int_{-\infty}^{\infty} |x(t)|^2\, dt &= \langle x(t), x(t)\rangle_t \\ &= \langle X(f), X(f)\rangle_f \\ &= \int_{-\infty}^{\infty} |X(f)|^2\, df, \end{aligned}$$

For the case of the Fourier transform, it should be pointed out that the mapping is formulated in the same vector space (**V** = **W**), same domain ($A = B$), with inner products of the same structure. There is no weighting in the inner products.

10.3. FOURIER SERIES EXPANSION

The Fourier series can be regarded as a degenerate special case of the Fourier transformation. It transforms a periodic function into a set of Fourier coefficients (Gabel and Roberts, 1980; Frederick and Carlson, 1971). In terms of the generalized model, **V** is the vector space containing all the periodic functions of period T, while **W** is the vector space of infinite sequences. The associated parameters are t (time) and n (an integer index) defined in the domains A and B, respectively. Here

$$A = \{t: 0 \leqslant t < T\},$$
$$B = \{n: -\infty < n < \infty,\}.$$

It can be seen from the corresponding parameters and domains that a Fourier series expansion represents a mapping from a continuous-time, finite-interval vector space **V** to a discrete, infinite-range vector space **W**. The dimension of the mapping remains infinite, and the basis functions are $\phi(t, n) = \exp(j2\pi nt/T)$ where T is the period.

Corresponding to the parameter and domain, the inner product in **V** is defined over a continuous and finite time interval $[0, T)$ and can be written as

$$\langle x(t), y(t)\rangle_t = \frac{1}{T}\int_0^T x(t)\bar{y}(t)\, dt.$$

Here $x(t)$ and $y(t)$ are two "arbitrary" periodic functions of period T. The inner product for the vector space **W** is then defined over a discrete infinite interval as follows:

$$\langle a(n), b(n)\rangle_n = \sum_{n=-\infty}^{\infty} a(n)\bar{b}(n).$$

Here $a(n)$ and $b(n)$ denote two sequences with the same sample spacing. It can be shown that the basis functions are elements in both **V** and **W** and orthogonal with respect to both inner products. In particular,

$$\langle \phi(t, n), \phi(t, n')\rangle_t = \frac{1}{T}\int_0^T \exp\left[\frac{j2\pi(n - n')t}{T}\right] dt = \delta(n - n').$$

$$\langle \phi(t, n), \phi(t', n)\rangle_n = \sum_{n=-\infty}^{\infty} \exp\left[\frac{j2\pi n(t - t')}{T}\right] = T\,\delta(t - t').$$

Here $\delta(n - n')$ is the Kronecker delta function or unit pulse, while $\delta(t - t')$ is the Dirac delta function or unit impulse. The Fourier series expansion can be obtained from a periodic function $x(t)$ by an inner product operation,

$$\begin{aligned} a(n) &= \langle x(t), \phi(t, n)\rangle_t \\ &= \frac{1}{T}\int_0^T x(t)\exp\left(\frac{-j2\pi nt}{T}\right) dt. \end{aligned}$$

Similarly, the reconstruction of the periodic signal from the Fourier coefficients can also be obtained using

$$x(t) = \langle a(n), \bar{\phi}(t, n) \rangle_n$$
$$= \sum_{n=-\infty}^{\infty} a(n) \exp\left(\frac{j2\pi nt}{T}\right).$$

The associated Parseval's relation in this case can be formulated as

$$\frac{1}{T} \int_0^T |x(t)|^2 \, dt = \langle x(t), x(t) \rangle_t$$
$$= \langle a(n), a(n) \rangle_n$$
$$= \sum_{n=-\infty}^{\infty} |a(n)|^2.$$

10.4. DISCRETE-TIME FOURIER TRANSFORM

The discrete-time Fourier transform is also a degenerate special case of the Fourier transformation. It is frequently used in digital system and signal analysis (Oppenheim and Schafer, 1975). The structure of the discrete-time Fourier transform pair represents the dual operation of that of the Fourier series expansion. For the discrete-time Fourier transform the vector space **V** is formed by all sequences, and the vector space **W** contains all periodic functions with period 2π. The associated parameters n (integer index) and ω are defined in the domains A and B, respectively. Here,

$$A = \{n: -\infty < n < \infty\},$$
$$B = \{\omega: -\pi \leqslant \omega < \pi\}.$$

The basis functions exist in both **V** and **W** and are given by $\phi(n, \omega) = \exp(j\omega n)$. Similar to the Fourier series expansion, we have one inner product defined in the discrete form for **V** and the other defined in the continuous form for **W**:

$$\langle x(n), y(n) \rangle_n = \sum_{n=-\infty}^{\infty} x(n)\bar{y}(n),$$
$$\langle X(\omega), Y(\omega) \rangle_\omega = \frac{1}{2\pi} \int_{-\pi}^{\pi} X(\omega)\bar{Y}(\omega) \, d\omega.$$

The basis functions are orthogonal in both vector spaces **V** and **W** with respect to both inner product definitions:

$$\langle \phi(n, \omega), \phi(n, \omega') \rangle_n = \sum_{n=-\infty}^{\infty} \exp[jn(\omega - \omega')] = 2\pi\, \delta(\omega - \omega'),$$

$$\langle \phi(n, \omega), \phi(n', \omega) \rangle_\omega = \frac{1}{2\pi} \int_{-\pi}^{\pi} \exp[j\omega(n - n')]\, d\omega = \delta(n - n').$$

Here $\delta(\omega - \omega')$ is the Dirac delta function and $\delta(n - n')$ denotes the Kronecker delta function. The discrete-time Fourier transform and its inverse can be formulated as

$$\begin{aligned} X(\omega) &= \langle x(n), \phi(n, \omega) \rangle_n \\ &= \sum_{n=-\infty}^{\infty} x(n) \exp(-j\omega n), \\ x(n) &= \langle X(\omega), \bar{\phi}(n, \omega) \rangle_\omega \\ &= \frac{1}{2\pi} \int_{-\pi}^{\pi} X(\omega) \exp(j\omega n)\, d\omega. \end{aligned}$$

The Parseval's relation for the discrete-time Fourier transform can be written as

$$\begin{aligned} \frac{1}{2\pi} \int_{-\pi}^{\pi} |X(\omega)|^2\, d\omega &= \langle X(\omega), X(\omega) \rangle_\omega \\ &= \langle x(n), x(n) \rangle_n \\ &= \sum_{n=-\infty}^{\infty} |x(n)|^2. \end{aligned}$$

The discrete-time Fourier transformation illustrates an example of discrete-to-continuous mapping. We can show that the discrete-time Fourier transform represents the dual operation of the Fourier series expansion by exchanging the vector spaces **V** and **W**.

10.5. DISCRETE FOURIER TRANSFORM (DFT)

The discrete Fourier transform or DFT represents a typical case of a finite-dimensional orthogonal mapping. Both **V** and **W** are vector spaces of N-point sequences defined over the domain $\{0, 1, \ldots, N - 1\}$. The basis sequences for the DFT are $\phi(n, k) = \exp(j2\pi nk/N)$ where n and k are integers (Oppenheim and Schafer, 1975). For any two sequences in **V**, say $x(n)$ and $y(n)$, the inner product is defined as

$$\langle x(n), y(n) \rangle_n = \sum_{n=0}^{N-1} x(n)\bar{y}(n).$$

In the space **W**, the inner product for arbitrary sequences $X(k)$ and $Y(k)$ is formulated as

$$\langle X(k), Y(k)\rangle_k = \frac{1}{N}\sum_{k=0}^{N-1} X(k)\bar{Y}(k).$$

The basis sequences are orthogonal in both vector spaces with respect to the corresponding inner products,

$$\langle \phi(n, k), \phi(n, k')\rangle_n = \sum_{n=0}^{N-1} \exp\left[\frac{j2\pi n(k-k')}{N}\right] = N\,\delta(k-k')$$

$$\langle \phi(n, k), \phi(n', k)\rangle_k = \frac{1}{N}\sum_{k=0}^{N-1} \exp\left[\frac{j2\pi(n-n')k}{N}\right] = \delta(n-n').$$

These are Kronecker delta functions because both **V** and **W** have discrete domains. It can be seen that the inner product in **W** has a uniform weighting of $1/N$ in order to satisfy the properties of the general model.

The DFT mapping then can be defined as

$$\begin{aligned} X(k) &= \mathrm{DFT}\{x(n)\} \\ &= \langle x(n), \phi(n, k)\rangle_n \\ &= \sum_{n=0}^{N-1} x(n)\bar{\phi}(n, k) \\ &= \sum_{n=0}^{N-1} x(n)\exp\left(-\frac{j2\pi nk}{N}\right). \end{aligned}$$

Similarly, the inverse DFT is formulated as

$$\begin{aligned} x(n) &= \mathrm{IDFT}\{X(k)\} \\ &= \langle X(k), \bar{\phi}(n, k)\rangle_k \\ &= \frac{1}{N}\sum_{k=0}^{N-1} X(k)\exp\left(\frac{j2\pi nk}{N}\right). \end{aligned}$$

In this case we obtain the associated Parseval's relation as follows:

$$\begin{aligned} \sum_{n=0}^{N-1} |x(n)|^2 &= \langle x(n), x(n)\rangle_n \\ &= \langle X(k), X(k)\rangle_k \\ &= \frac{1}{N}\sum_{k=0}^{N-1} |X(k)|^2. \end{aligned}$$

The Fourier transform, Fourier series expansion, discrete-time Fourier transform, and DFT are closely related and are often considered as a family of transformations. These transformations have very similar transform properties such as shift, modulation, scaling, duality, and convolution. This is because the basis functions of these transformations all have exponential form and involve the product of the associated parameters.

10.6. HANKEL TRANSFORM

The Hankel transform is also a degenerate special case of the Fourier transform. The vector spaces **V** and **W** are identical and are formed by causal signals. The Hankel transform is often used to analyze two-dimensional signals with radial symmetry (Karni and Byatt, 1982). The associated domains are also the same, namely,

$$A = \{r : r \geqslant 0\},$$

$$B = \{R : R \geqslant 0\}.$$

The orthogonal basis functions are the zero-order Bessel functions $\phi(r, R) = J_0(rR)$. These basis functions are orthogonal under the following definitions of the inner products:

$$\langle x(r), y(r) \rangle_r = \int_0^\infty r x(r) \bar{y}(r) dr.$$

$$\langle X(R), Y(R) \rangle_R = \int_0^\infty R X(R) \bar{Y}(R)\, dR.$$

It should be noted that these inner products each have linear weighting. These factors are caused by the change of variables from two-dimensional radially symmetric functions into one-dimensional causal signals. We can formulate the Hankel transform pair in terms of inner product operations as follows:

$$\begin{aligned} X(R) &= \text{Hankel transform } \{x(r)\} \\ &= \int_0^\infty r x(r) J_0(rR)\, dr, \\ x(r) &= \text{inverse Hankel transform } \{X(R)\} \\ &= \int_0^\infty R X(R) J_0(rR)\, dR. \end{aligned}$$

The norm of a signal is invariant under the Hankel transform, and the corresponding Parseval's relation can be written as

$$\begin{aligned}\int_0^{\infty} r|x(r)|^2\, dr &= \langle x(r), x(r)\rangle_r \\ &= \langle X(R), X(R)\rangle_R \\ &= \int_0^{\infty} R|X(R)|^2\, dR.\end{aligned}$$

The Hankel transform is an example of an orthogonal transformation that is similar to the Fourier transform in the sense that it operates in the same vector space, in the same domains, and with the same weighting for the inner products.

10.7. HILBERT TRANSFORM

The Hilbert transformation has long been known as a time-domain convolution operation and is frequently used in the generation of analytic signals (Whalen, 1971). In this section, we also describe the Hilbert transform as an orthogonal transform. In this case, the vector spaces **V** and **W** both consist of functions defined over all time. We use t and τ to distinguish the time parameters before and after the Hilbert transform. That is,

$$\begin{aligned}A &= \{t: -\infty < t < \infty\}, \\ B &= \{\tau: -\infty < \tau < \infty\}.\end{aligned}$$

The basis functions for the Hilbert transform are real. In this case $\phi(t, \tau) = 1/\{\pi(\tau - t)\}$. These basis functions are orthogonal with respect to the inner products:

$$\begin{aligned}\langle \phi(t, \tau), \phi(t, \tau')\rangle_t &= \delta(\tau - \tau'), \\ \langle \phi(t, \tau), \phi(t', \tau)\rangle_\tau &= \delta(t - t').\end{aligned}$$

The Hilbert transform pair can be defined in terms of the inner product operations as follows:

$$\begin{aligned}X(\tau) &= \text{Hilbert transform } \{x(t)\} \\ &= \int_{-\infty}^{\infty} x(t) \frac{1}{\pi(\tau - t)}\, dt, \\ x(t) &= \text{inverse Hilbert transform } \{X(t)\} \\ &= \int_{-\infty}^{\infty} X(\tau) \frac{1}{\pi(\tau - t)}\, d\tau.\end{aligned}$$

The associated Parseval's relation in this case is

$$\int_{-\infty}^{\infty} |x(t)|^2\, dt = \int_{-\infty}^{\infty} |X(\tau)|^2\, d\tau.$$

10.8. SAMPLING THEOREM

When the sampling theorem is introduced, the key objective is to be able to reconstruct the signal from its discrete samples without aliasing (Papoulis 1977; McGillem and Cooper, 1974). We can perform distortion-free conversion between a band-limited signal and its data samples because the conversion is basically an orthogonal transformation. For the case of band-limited signal sampling and reconstruction, the vector space **V** consists of sample sequences with sample spacing T, and the vector space **W** is formed by band-limited signals with cutoff frequency f_c where $f_c = 1/(2T)$. The two associated parameters are n (sample index) and f (frequency), and they are defined in the domains A and B, respectively, where,

$$A = \{n: -\infty < n < \infty\},$$
$$B = \{f: -f_c \leqslant f \leqslant f_c\}.$$

The orthogonal basis functions for the mapping are $\phi(n,f) = \exp(jn\pi f/f_c)$. We define the two inner products in **V** and **W** with uniform weighting as follows:

$$\langle x(n), y(n)\rangle_n = \sum_{n=-\infty}^{\infty} x(n)\bar{y}(n),$$

$$\langle X(f), Y(f)\rangle_f = \frac{1}{2f_c}\int_{-f_c}^{f_c} X(f)\bar{Y}(f)\, df.$$

The basis functions are orthogonal with respect to both inner products. The orthogonal transformation pairs for sampling and reconstruction can be represented in the following manner:

$$\begin{aligned} X(f) &= \langle x(n), \phi(n,f)\rangle_n \\ &= \sum_{n=-\infty}^{\infty} x(n)\exp\left(\frac{-jn\pi f}{f_c}\right). \\ x(n) &= \langle X(f), \bar{\phi}(n,f)\rangle_f \\ &= \frac{1}{2f_c}\int_{-f_c}^{f_c} X(f)\exp\left(\frac{jn\pi f}{f_c}\right) df. \end{aligned}$$

The structure of the mapping is very similar to the Fourier series expansion and the discrete-time Fourier transformation. The first equation represents the reconstruction of the band-limited spectrum from the data samples, while the second is the signal sampling in terms of the signal's spectral content. The corresponding Parseval's relation in this case can be written as

$$\sum_{n=-\infty}^{\infty} |x(n)|^2 = \frac{1}{2f_c} \int_{-f_c}^{f_c} |X(f)|^2 \, df.$$

This is a description of the sampling theorem as a samples-spectrum orthogonal transformation. We can also describe the orthogonal mapping strictly in the time domain, which is the more conventional version of the sampling theorem.

The vector space formulation of orthogonal transformation pairs can be also applied to the Legendre, Laguerre, Hermite, and Chebyshev decomposition techniques discussed in Chapter 3. In all but the first case, we observe nonuniform weighting in the orthogonality and Parseval's relations.

10.9. AMPLITUDE ESTIMATION

We now change our focus from orthogonal transformations to a different but related topic commonly known as amplitude estimation, a fundamental technique in signal detection. Amplitude estimation has many engineering applications in signal processing, control, communication, and imaging (Whalen 1971; Lee 1985b). In fact, orthogonal transformations can be described as special cases of amplitude estimation. In this section, we first present the amplitude estimation technique as a statistical approach and then generalize it by using vector space methods for the multidimensional cases. Consider a received signal $y(t)$ of the following form:

$$y(t) = ax(t) + n(t), \qquad 0 \leqslant t \leqslant T.$$

Here $x(t)$ is a *known signal* and $n(t)$ denotes the *noise* term. The signal $y(t)$ is observed within the time interval $[0, T]$ and the amplitude a is the estimation objective. Suppose that the additive noise is white and Gaussian. The likelihood function of $y(t)$ is

$$p(y|a) = K \exp\left\{-\frac{1}{N_0}\int_0^T |y(t) - ax(t)|^2 \, dt\right\}.$$

Here K is the normalization constant for the Gaussian distribution. The maximum likelihood estimate of the amplitude maximizes $p(y|a)$, which is equivalent to maximizing

$$\ln\{p(y|a)\} = \ln(K) - \frac{1}{N_0}\int_0^T |y(t) - ax(t)|^2 \, dt.$$

The optimal solution must satisfy

$$\begin{aligned}\frac{\partial}{\partial a}\ln\{p(y|a)\} &= \frac{2}{N_0}\int_0^T [y(t) - ax(t)]x(t)\,dt \\ &= \frac{2}{N_0}\left[\int_0^T y(t)x(t)\,dt - a\int_0^T x^2(t)\,dt\right] \\ &= 0.\end{aligned}$$

It follows that the *maximum likelihood* estimation of the amplitude can be written as

$$\hat{a} = \frac{\displaystyle\int_0^T y(t)x(t)\,dt}{\displaystyle\int_0^T x^2(t)\,dt}.$$

Alternatively, we can formulate the analysis of the amplitude estimation problem by using vector space methods. We first define a valid inner product of two arbitrary real functions $w(t)$ and $v(t)$ in the interval $[0, T]$:

$$\langle w, v\rangle \triangleq \int_0^T w(t)v(t)\,dt.$$

Then we take the inner product of the received signal $y(t)$ with the known signal $x(t)$. This yields

$$\begin{aligned}\langle y, x\rangle &= \langle ax + n, x\rangle \\ &= \langle ax, x\rangle + \langle n, x\rangle \\ &= a\langle x, x\rangle \\ &= a\|x\|^2.\end{aligned}$$

The term $\langle n, x\rangle$ is zero because the additive noise is assumed to be uncorrelated with the signal $x(t)$. This approach leads to the identical maximum likelihood estimate

$$\begin{aligned}\hat{a} &= \frac{\langle y, x\rangle}{\|x\|^2} \\ &= \frac{\displaystyle\int_0^T y(t)x(t)\,dt}{\displaystyle\int_0^T x^2(t)\,dt}.\end{aligned}$$

We can easily extend the vector space method to multidimensional amplitude estimation problems. Consider the case when we have N known independent signal components with unknown amplitude coefficients a_n. That is, the received signal is

$$y(t) = \sum_{n=1}^{N} a_n x_n(t) + n(t).$$

We take inner products of the received signal $y(t)$ and the N known signal components and obtain N independent equations

$$\begin{aligned}\langle y, x_k \rangle &= \left\langle \sum_{n=1}^{N} a_n x_n + n, x_k \right\rangle \\ &= \sum_{n=1}^{N} a_n \langle x_n, x_k \rangle + \langle n, x_k \rangle \\ &= \sum_{n=1}^{N} a_n \int_0^T x_n(t) x_k(t)\, dt.\end{aligned}$$

The white and Gaussian noise term $n(t)$ is again assumed to be uncorrelated with the known signals, and therefore the inner products $\langle n, x_k \rangle$ will vanish. We now define two N-dimensional column vectors:

$$a = [a_1, a_2, \ldots, a_N]^{\mathrm{T}},$$

$$r = [r_1, r_2, \ldots, r_N]^{\mathrm{T}}.$$

Here $r_k = \langle y, x_k \rangle$. In addition, an $N \times N$ square matrix can be formed as follows:

$$R = \begin{bmatrix} R_{11} & R_{12} & \cdots & R_{1N} \\ \vdots & \vdots & & \vdots \\ R_{N1} & R_{N2} & \cdots & R_{NN} \end{bmatrix}.$$

Here the elements of R are defined

$$\begin{aligned}R_{n,k} &= \langle x_n, x_k \rangle \\ &= \int_0^T x_n(t) x_k(t)\, dt.\end{aligned}$$

The N independent equations can then be written in the compact form as, $Ra = r$. This is commonly referred to as the normal equation, and the matrix R is known as the Gram matrix (Luenberger, 1969). We now obtain the optimal solution as

$$\hat{a} = R^{-1} r$$

It can be seen that the finite-dimensional amplitude estimation process has been transformed into a linear matrix operation. The key step of this matrix operation is the accurate formation of the inverse of the Gram matrix R. There are two ideal cases which make the matrix inversion very simple. The first case is when the known signals $x_k(t)$ are mutually orthogonal, in which case the R matrix becomes diagonal. The second case occurs when the R matrix is not diagonal but orthonormal. For this case, the inverse is simply the transpose. For these two special cases, the amplitude estimation operations can be characterized as orthogonal transformations, and there is no computation needed for matrix inversion. To achieve computationally efficient amplitude detection, it is therefore desirable for us to use certain types of signaling components such that the amplitude estimation operation can be reduced to an orthogonal mapping. In general, however, the known signals are mutually independent but correlated due to the relatively short observation period T. This implies that the R matrix may have full rank but is often ill-conditioned. We now present three methods that are widely used to implement the estimation process when direct Gaussian elimination is not appropriate for the matrix inversion.

10.10. GRAM–SCHMIDT DECOMPOSITION

The Gram–Schmidt procedure is an orthogonalization process which was introduced in Chapter 3. In this section, we utilize this orthogonalization process to implement matrix inversion for the multidimensional amplitude estimation problem. We begin by partitioning the Gram matrix R into N column vectors,

$$R = [r^1, r^2, \ldots, r^N]$$

As indicated earlier, these column vectors are usually independent and correlated. Therefore, a set of orthonormal vectors, $q^1, q^2, \ldots, q^N$ can be generated from this linearly independent set using the Gram–Schmidt algorithm (Bachman and Narici, 1966). We then form an $N \times N$ matrix Q by collecting the ordered orthonormal column vectors, $Q = [q^1, q^2, \ldots, q^N]$. The linear relationship between these two vector sets can be written in matrix form as

$$Q^{\mathrm{T}} = CR^{\mathrm{T}}.$$

Here C is a lower-triangular matrix, and the elements of this matrix are obtained during the Gram–Schmidt orthogonalization procedure. The two sets of column vectors span the same vector space. Therefore a vector r can be represented uniquely by two sets of vectors:

$$\begin{aligned} r &= Ra \\ &= \sum_{n=1}^{N} a_n r^n \\ &= \sum_{n=1}^{N} a'_n q^n. \end{aligned}$$

The coefficients a'_k can be directly obtained by utilizing the orthogonality of the columns of Q. In particular,

$$\begin{aligned}\langle r, q^k\rangle &= \left\langle \sum_{n=1}^{N} a'_n q^n, q^k \right\rangle \\ &= \sum_{n=1}^{N} a'_n \langle q^n, q^k\rangle \\ &= a'_k \|q^k\|^2 \\ &= a'_k .\end{aligned}$$

Here we have made use of the fact that the orthogonal vectors $\{q^k\}$ have been normalized so that $\|q^k\| = 1$. Next, we define a column vector a' as $a' = [a'_1, a'_2, \ldots, a'_N]^T$. We can then write

$$\begin{aligned}r &= Qa' \\ &= RC^T a' \\ &= Ra.\end{aligned}$$

Since the columns of R are linearly independent, it can be seen that the vectors a and a' are related by the C matrix. The vector a' can be obtained by taking the inner products of the r vector and the orthonormal columns of Q. The C matrix is formed by recording the coefficients computed during the orthogonalization process. Therefore the amplitude coefficients can be obtained by linearly transforming the vector a' by a matrix operator as follows:

$$a = C^T a'.$$

10.11. SINGULAR VALUE DECOMPOSITION

Singular value decomposition is also an orthogonalization technique widely used for system identification and sensitivity analysis (Noble and Deniel, 1977; Pratt, 1978). The objective here is to represent the Gram matrix R by a unique singular value decomposition

$$R = U\Lambda V^T.$$

Here U and V are orthonormal matrices and Λ is a diagonal matrix with its diagonal elements being the *singular values* associated with the R matrix

$$\Lambda = \text{diag}\{\lambda_1, \lambda_2, \ldots, \lambda_N\}.$$

Subroutines for obtaining singular value decompositions are available for most computer systems. The solution of the amplitude estimation problem can be written in terms of the singular value decomposition as

$$a = R^{-1}r$$
$$= V\Lambda^{-1}U^{\mathrm{T}}r.$$

Here the inverse of the diagonal matrix is again a diagonal matrix with the diagonal elements inverted.

$$\Lambda^{-1} = \operatorname{diag}\{1/\lambda_1, 1/\lambda_2, \ldots, 1/\lambda_N\}.$$

Because R has full rank but is often ill-conditioned, some singular values may be very small. To avoid amplification of noise, we select a threshold value for the small singular values. The singular values greater than the threshold, say the first M of them, are then used for the inversion. Therefore we approximate the inverse of Λ as

$$\Lambda^{-1} \cong \operatorname{diag}\{1/\lambda_1, 1/\lambda_2, \ldots, 1/\lambda_M, 0, \ldots, 0\}.$$

Let u^i and v^i be the ith column vectors of U and V respectively. We can then approximate the amplitude vector as

$$\hat{a} \cong \sum_{n=1}^{M} \left[\frac{(u^n)^{\mathrm{T}}r}{\lambda_n}\right] v^n$$
$$= \sum_{n=1}^{M} b_n v^n.$$

Here the coefficients are defined as $b_n = (u^n)^{\mathrm{T}}r/\lambda_n$ for $1 \leqslant n \leqslant M$.

This process can be regarded as a two-step operation. First, we decompose the vector r with respect to the orthonormal vectors $\{u^i\}$ and normalize the components by a corresponding gain $1/\lambda_i$. We then reconstruct the amplitude vector with another set of orthonormal vectors using the normalized coefficients obtained from the decomposition. When the R matrix is symmetric, the orthonormal matrices U and V are identical. In this case the singular value decomposition becomes an eigenvalue decomposition.

10.12. STEEPEST DESCENT METHOD

The Gram–Schmidt orthogonalization procedure and the singular value decomposition method presented earlier for matrix inversion are basically one-step techniques. In this section, we show that we can also perform amplitude estimation by using the method of steepest descent. This method is a recursive algorithm and is often referred to as the gradient method (Luenberger, 1973). A sequence of increasingly accurate estimates $\{\hat{a}(k)\}$ is generated as follows:

$$\hat{a}(k+1) = \hat{a}(k) - \alpha(k)g(k).$$

Here $\hat{a}(k)$ denotes the kth estimate of the amplitude vector, $g(k)$ is the *gradient* of an error function evaluated at $\hat{a}(k)$, and the nonnegative scalar $\alpha(k)$ is the optimal *step size* along the gradient direction. To formulate the optimization process, we first define a positive-definite matrix W as

$$W = R^{\mathrm{T}}R.$$

Recall from Section 10 that $Ra = r$. Thus

$$\begin{aligned} Wa &= R^{\mathrm{T}}Ra \\ &= R^{\mathrm{T}}r \\ &\triangleq b. \end{aligned}$$

We define a quadratic error function $E(\hat{a})$ as follows:

$$\begin{aligned} E(\hat{a}) &= \tfrac{1}{2}[R(\hat{a} - a)]^{\mathrm{T}}[R(\hat{a} - a)] \\ &= \tfrac{1}{2}(\hat{a} - a)^{\mathrm{T}}W(\hat{a} - a). \end{aligned}$$

Then an optimization problem can be formulated with the following objective and associated constraint:

$$\begin{aligned} &\text{Minimize} \quad E(\hat{a}) = \tfrac{1}{2}(\hat{a} - a)^{\mathrm{T}}W(\hat{a} - a), \\ &\text{Subject to} \quad Wa = b. \end{aligned}$$

Each step of the iterative operation for solving the optimization problem requires two inputs to complete the updating: the gradient and the step size. The gradient of the error function can be obtained by differentiating $E(\hat{a})$ with respect to the amplitude estimate $\hat{a}$:

$$\begin{aligned} g(\hat{a}) &= \frac{\partial}{\partial \hat{a}} E(\hat{a}) \\ &= W(\hat{a} - a) \\ &= W\hat{a} - b. \end{aligned}$$

Thus the kth gradient can be written as

$$\begin{aligned} g(k) &\triangleq g[\hat{a}(k)] \\ &= W\hat{a}(k) - b. \end{aligned}$$

The kth optimal step size can be evaluated by minimizing the error function along the gradient direction. Thus we need an α such that

$$\frac{\partial}{\partial \alpha} E[\hat{a}(k) - \alpha g(k)] = 0.$$

This is also widely known as a line search, which can be regarded as a one-dimensional optimization process. The optimal step size can be obtained from the constraint on the derivative and written in closed form as

$$\alpha(k) = \frac{g^{\mathrm{T}}(k)g(k)}{g^{\mathrm{T}}(k)Wg(k)}.$$

The general equation for updating the estimate then becomes

$$\hat{a}(k+1) = \hat{a}(k) - \frac{g^{\mathrm{T}}(k)g(k)}{g^{\mathrm{T}}(k)Wg(k)}\, g(k)$$

where

$$g(k) = W\hat{a}(k) - b.$$

The estimation error decreases with every updating step and the amplitude vector converges to the optimal solution. That is,

$$\lim_{k \to \infty} \hat{a}(k) = a.$$

The gradient is zero when $\hat{a}(k)$ reaches the optimal solution and the algorithm will be terminated. In practice, the norm of the gradient is compared against a threshold at every updating step, and the algorithm stops when the norm is smaller than some threshold value. The convergence rate of the steepest descent method is relatively slow, and it is influenced by the eigenvalue distribution of the W matrix.

For amplitude estimation it is always desirable to have a long observation period to ensure that the correlations between the signal components and the noise term are zero. On the other hand, it is also preferable to have a sufficient observation period to reduce the correlations among the signal components themselves to avoid having ill-conditioned matrices. When the R matrix is seriously ill-conditioned, it degrades the amplitude estimation algorithms in different ways. For the Gram–Schmidt decomposition method, we may not be able to generate a complete set of orthonormal vectors. For the singular value decomposition method, only a portion of the orthonormal vectors can be utilized for matrix inversion because of the small singular values. Finally, for the steepest descent method, an ill-conditioned R reduces the convergence rate of the process. The solution we obtain will be the minimum-norm solution. The minimum-norm solution is essentially the orthogonal projection of the actual solution onto the subspace spanned by the subset of the vectors in the R matrix that are effective for the matrix inversion.

10.13. DISCRETE MODEL

Discrete data acquisition has become very common in recent years because of the use of digital computers. Instead of the continuous signal $y(t)$, we now detect a finite number of equally spaced data samples during the observation period. In this section, we extend the vector space method to formulate the amplitude estimation problem for the discrete model. Suppose that M data samples are detected and available for the estimation process. In this case

$$\begin{aligned} y(t)|_{t=m\Delta} &= y(m\Delta) \\ &= \sum_{n=1}^{N} a_n x_n(m\Delta) + n(m\Delta). \end{aligned}$$

Here Δ is the sample spacing. Now we modify the definition of the inner product to include a weighting function $w(t)$:

$$\langle u, v \rangle = \int_0^T u(t)v(t)w(t)\,dt.$$

For discrete detection systems, we let

$$w(t) = \sum_{m=1}^{M} \delta(t - m\Delta)$$

where $\delta(\,)$ denotes the Dirac delta. The weighing function is the sampling function associate with the analog-discrete conversion. Similarly, we take the inner product of $y(t)$ and $x_k(t)$ as we did in previous sections and obtain

$$\begin{aligned} r_k &= \langle y, x_k \rangle \\ &= \int_0^T y(t)x_k(t) \sum_{m=1}^{M} \delta(t - m\Delta)\,dt \\ &= \sum_{m=1}^{M} y(m\Delta)x_k(m\Delta). \end{aligned}$$

On the other hand, we also have

$$\begin{aligned}
r_k &= \langle y, x_k \rangle \\
&= \left\langle \sum_{n=1}^{N} a_n x_n + n, x_k \right\rangle \\
&= \sum_{n=1}^{N} a_n \langle x_n, x_k \rangle \\
&= \sum_{n=1}^{N} a_n \int_0^T x_n(t) x_k(t) w(t)\, dt \\
&= \sum_{n=1}^{N} a_n \sum_{m=1}^{M} x_n(m\Delta) x_k(m\Delta) \\
&= \sum_{n=1}^{N} a_n R(n, k).
\end{aligned}$$

Here the R matrix is now defined as

$$R(n, k) = \sum_{m=1}^{M} x_n(m\Delta) x_k(m\Delta).$$

The formulation of the optimal estimate of the amplitude remains the same.

$$\hat{a} = R^{-1} r.$$

The total number of data samples must be greater than the number of unknown amplitude parameters to make the R matrix invertible. For the discrete model, the matrix R and vector r can be regarded as approximate versions of their counterparts formulated for the continuous cases. This approach produces a least squares solution identical to that obtained conventionally. We point out that the analog-discrete conversion is easily achieved by inserting a sampling function as the weighing of the inner product.

When the total number of available data samples is not sufficient, $M < N$, the R matrix does not have full rank and hence is not invertible. For this case, we approach the problem with a different objective, namely, to obtain the minimum-norm solution among all feasible solutions. We first reformulate $y(m\Delta)$ in terms of inner products as follows:

$$\begin{aligned}
y(m\Delta) &= \sum_{n=1}^{N} a_n x_n(m\Delta) \\
&= \langle a, x(m\Delta) \rangle.
\end{aligned}$$

Here $a = [a_1, a_2, \ldots, a_N]^T$ and $x(m\Delta) = [x_1(m\Delta)\, x_2(m\Delta), \ldots, x_N(m\Delta)]^T$. Now the problem can be described as one of finding the orthogonal projection of an unknown N-dimensional amplitude vector onto an M-dimensional subspace. The optimal (minimum-norm) estimate will be in the M-dimensional subspace spanned by the M vectors. Hence the orthogonal projection of a and can be represented as a linear combination of the M vectors as follows:

$$\hat{a} = \sum_{m=1}^{M} b_m x(m\Delta).$$

Here the b_m's are the coefficients of the optimal combination. They must satisfy

$$\begin{aligned} y(k) &= \langle \hat{a}, x(k\Delta) \rangle \\ &= \left\langle \sum_{m=1}^{M} b_m x(m\Delta), x(k\Delta) \right\rangle \\ &= \sum_{m=1}^{M} b_m \langle x(m\Delta), x(k\Delta) \rangle \\ &= \sum_{m=1}^{M} b_m S(m, k). \end{aligned}$$

Here S is an $M \times M$ matrix defined as

$$\begin{aligned} S(m, k) &= \langle x(m\Delta), x(k\Delta) \rangle \\ &= \sum_{n=1}^{N} x_n(m\Delta) x_n(k\Delta). \end{aligned}$$

Thus in matrix form we can write

$$y = Sb$$

Here $y = [y(\Delta), (2\Delta), \ldots, y(M\Delta)]^T$ and $b = [b_1, b_2, \ldots, b_M]^T$. The matrix has full rank, and the optimal coefficients can be obtained as

$$b = S^{-1} y.$$

Finally, the minimum-norm solution to the amplitude parameters can be formed as

$$\hat{a}_n = \sum_{m=1}^{M} b_m x_n(m\Delta).$$

This result is identical to that obtained by conventional methods.

PROBLEMS

1. Verify the following requirement for orthogonal transformations for the case of the Fourier transform:

$$\|\phi(\alpha, \beta')\|_A \|\delta(\beta - \beta')\|_B = 1,$$

$$\|\phi(\alpha', \beta)\|_B \|\delta(\alpha - \alpha')\|_A = 1.$$

2. Show the orthogonality of the basis functions for the Fourier series expansion and Hilbert transform.

3. Repeat problem 1 for the case of the Fourier series expansion.

4. Repeat problem 1 for the case of the discrete Fourier transform (DFT).

5. Use the vector space model for the sampling theorem by describing the relation between a low-pass time signal and its data sample sequence as an orthogonal transformation.

6. Use the vector space method to formulate Legendre, Laguerre, Hermite, and Chebyshev decompositions as orthogonal transformations.

7. Modify the optimal solution of the N-dimensional amplitude estimation problem when the received signal and the component functions are complex.

8. Consider the received signal

$$y(t) = \int_{-\infty}^{\infty} a(\tau)x(t - \tau)\, d\tau. + \text{noise}$$

Given $x(t)$, extend the vector space method to estimate the unknown function $a(t)$. This is often called the deconvolution problem in signal processing.

9. Find the C matrix in Section 10 when the R matrix is

$$R = \begin{bmatrix} 1 & \rho & \rho^2 \\ \rho & 1 & \rho \\ \rho^2 & \rho & 1 \end{bmatrix}$$

where ρ is a real and positive number.

10. Derive the following formula in Section 12 for the optimal step size:

$$\alpha(k) = \frac{g^{\mathrm{T}}(k)g(k)}{g^{\mathrm{T}}(k)Wg(k)}.$$

REFERENCES

Bachman, G. and L. Narici, *Functional Analysis*, Academic, New York, 1966.

Frederick, D. K. and A. B. Carlson, *Linear Systems in Communication and Control*, Wiley, New York, 1971.

Gabel, R. A. and R. A. Roberts, *Signals and Linear Systems*, Wiley, New York, 1980.

Karni, S. and W. J. Byatt, *Mathematical Methods in Continuous and Discrete Systems*, Holt, Rinehart and Winston, New York, 1982.

Lee, H. "On Orthogonal Transformations," *IEEE Trans. Circuits Syst.*, **32**(11), 1169–1177 (1985).

Lee, H. "Resolution Enhancement of backward Propagated Images by Wavefield Orthogonalization," *J. Acoust. Soc. Amer.*, **77**(5), 1845–1848 (1985).

Luenberger, D. G. *Optimization by Vector Space Methods*, Wiley, New York, 1969.

Luenberger, D. G. *Introduction to Linear and Nonlinear Programming*, Addison-Wesley, Reading, 1973.

McGillem, C. D. and G. R. Cooper, *Continuous and Discrete Signal and System Analysis*, Holt, Rinehart and Winston, New York, 1974.

Noble, B. and J. W. Deniel, *Applied Linear Algebra*, McGraw-Hill, New York, 1977.

Oppenheim, A. V. and W. R. Schafer, *Digital Signal Processing*, Prentice-Hall, Englewood Cliffs, NJ, 1975.

Papoulis, A. *Signal Analysis*, McGraw-Hill, New York, 1977.

Pratt, W. K. *Digital Image Processing*, Wiley, New York, 1978.

Whalen, A. D. *Detection of Signals in Noise*, Academic, New York, 1971.

APPENDIX A LIST OF SYMBOLS

SCALARS, VECTORS, MATRICES

SYMBOL	NAME
$\alpha, \beta, \gamma, \ldots$	Scalars
$a, b, c, \ldots$	Vectors
$A, B, C, \ldots$	Matrices
a_k	k-th component of the vector a
A_{kj}	component of matrix A in row k, column j
a^j	j-th column of matrix A: $A = [a^1, a^2, \ldots, a^n]$
0	zero matrix: $0x = 0$
I	identity matrix: $Ix = x$
i^j	j-th column of I: $I = [i^1, i^2, \ldots, i^n]$
$\mathrm{Re}(A)$	real part of A
$\mathrm{Im}(A)$	imaginary part of A
$\bar{A}$	complex conjugate of A: $\bar{A} = \mathrm{Re}(A) - j\mathrm{Im}(A)$
A^T	transpose of A: $A^T_{kj} = A_{jk}$
A^H	Hermitian transpose of A: $A^H = \bar{A}^T$
A^{-1}	inverse of A: $A^{-1}A = AA^{-1} = I$
$\det(A)$	determinant of A
$\mathrm{tr}(A)$	trace of A: $\mathrm{tr}(A) = A_{11} + A_{22} + \cdots + A_{nn}$
$\Delta(\lambda)$	characteristic polynomial: $\Delta(\lambda) = \det(\lambda I - A)$
λ_k	kth eigenvalue: $\Delta(\lambda_k) = 0$
$m(k)$	Multiplicity of kth eigenvalue
$\sigma(A)$	Spectrum of A: $\sigma(A) = \{\lambda_1, \lambda_2, \ldots, \lambda_q\}$
$[x]^{\mathbf{B}}$	Coordinates of vector x with respect to basis **B**
$[\mathbf{A}]^{\mathbf{B,C}}$	Matrix representation of **A** with respect to **B,C**
$\|x\|$	Norm of vector x

SYMBOL	NAME
$\|x\|_p$	pth Holder norm of x: $1 \leqslant p \leqslant \infty$
$\langle x, y\rangle$	Inner product of vectors x and y
$\mathbf{I}$	Standard basis for $\mathbf{F}^n$: $\{\mathrm{i}^1, \mathrm{i}^2, \ldots, \mathrm{i}^n\}$
$\|\alpha\|$	Absolute value of α

FUNCTIONS

SYMBOL	NAME
$f: U \to V$	Function f mapping U into V
f	Entire function f
$f(x)$	Function f evaluated at x
$f(S)$	Image of S under f: $f(S) = \{f(x): x \in S\}$
f^{-1}	Inverse function: $f^{-1}\{f(x)\} = x$
$f^{(k)}$	kth derivative of f
$f * g$	Convolution of f with g
$L\{f(t)\}$	Laplace transform of $f: F(s) = L\{f(t)\}$
$Z\{f(k)\}$	Z-transform of $f: F(z) = Z\{f(k)\}$
$\delta(t)$, $\delta(k)$	Unit impulse, unit pulse
$1(t)$, $1(k)$	Unit step functions
$\exp(\lambda t)$	The exponential function: $e^{\lambda t}$

VECTOR SPACES

SYMBOL	NAME
$\mathbf{F}$	Field of scalars
$\mathbf{R}$, $\mathbf{C}$	Real, complex fields
$\mathbf{U}$, $\mathbf{V}$, $\mathbf{W}$	Vector spaces
$\mathbf{V} \times \mathbf{W}$	Product space: $\mathbf{V} \times \mathbf{W} = \{(v, w): v \in \mathbf{V}, w \in \mathbf{W}\}$
$\mathbf{span}(\mathbf{U})$	Subspace spanned by U
$\mathbf{U}^{\perp}$	Orthogonal complement of U
$\mathbf{V}^*$	Dual of $\mathbf{V}$: $\mathbf{V}^* = \mathbf{B}(\mathbf{V}, \mathbf{F})$
$\dim(\mathbf{V})$	Dimension of the space $\mathbf{V}$
$\mathbf{F}^{\mathbf{n}}$	Column vectors with components in $\mathbf{F}$
$\mathbf{F}^{\mathbf{m} \times \mathbf{n}}$	$m \times n$ matrices with components in $\mathbf{F}$
$\mathbf{F}(\mathbf{V}, \mathbf{W})$	Functions mapping $\mathbf{V}$ into $\mathbf{W}$
$\mathbf{C}(\mathbf{V}, \mathbf{W})$	Continuous functions mapping $\mathbf{V}$ into $\mathbf{W}$

SYMBOL	NAME
$\mathbf{L(V, W)}$	Linear transformations from $\mathbf{V}$ into $\mathbf{W}$
$\mathbf{B(V, W)}$	Bounded linear transformations from $\mathbf{V}$ into $\mathbf{W}$
$\mathbf{P(T)}$	Polynomials mapping T into $\mathbf{F}$
$\mathbf{P^n(T)}$	nth degree polynomials mapping T into $\mathbf{F}$
$\ell_\mathbf{p}$	Summable sequences with norm $\|x\|_p$
$\mathbf{L_p(T, w)}$	Integrable functions with norm $\|x\|_p$
$\mathbf{L_p(T)}$	Integrable functions, uniform weighting

TRANSFORMATIONS, OPERATORS

SYMBOL	NAME
$A: \mathbf{V} \to \mathbf{W}$	Linear transformation mapping $\mathbf{V}$ into $\mathbf{W}$
A^{-1}	Inverse of A: $A^{-1}A = AA^{-1} = I$
A^*	Adjoint of A: $\langle Ax, y\rangle \equiv \langle x, A^*y\rangle$
$\mathbf{N}(A)$	Null space of A: $\mathbf{N}(A) = \{x : Ax = 0\}$
$\mathbf{R}(A)$	Range space of A: $\mathbf{R}(A) = \{Ax : x \in \mathbf{V}\}$
0	Zero transformation: $0x = 0$
I	Identity transformation: $Ix = x$
nullity(A)	Nullity of A: nullity(A) = $\dim\{\mathbf{N}(A)\}$
rank(A)	Rank of A: rank(A) = $\dim\{\mathbf{R}(A)\}$
$\sigma(A)$	Spectrum of A
$\sigma_p(A)$	Point spectrum of A
$\sigma_r(A)$	Residual spectrum of A
$\sigma_c(A)$	Continuous spectrum of A
$\pi(A)$	Approximate spectrum of A
$r(A)$	Spectral radius of A
$W(A)$	Numerical range of A
$\rho(A)$	Resolvent set of A: $\rho(A) = \mathbf{C} - \sigma(A)$
$\mathbf{G}(A)$	Graph of A
$P_{\mathbf{kj}}(A)$	Spectral polynomial of A
$\mathbf{G}(\lambda_\mathbf{k})$	Eigenspace: $\mathbf{N}(\lambda_\mathbf{k}I - A)$
$\mathbf{H}(\lambda_\mathbf{k})$	Generalized eigenspace: $\mathbf{N}\{(\lambda_\mathbf{k}I - A)^{\mathbf{m(k)}}\}$
$A > 0$	A is positive-definite: $\langle Ax, x\rangle > 0$, $x \neq 0$
$A \geqslant 0$	A is positive-semidefinite: $\langle Ax, x\rangle \geqslant 0$
$A \geqslant B$	$A - B$ is positive-semidefinite
E	Orthogonal projection operator

SETS

SYMBOL	NAME	PAGE
$\{x, y, z\}$	Set containing elements x, y, z	
$\{x : p(x)\}$	Set of all x satisfying property $p(x)$	
$x \in S$	x is a member of S	
$x \notin S$	x is not a member of S	
$\mathring{S}$	Interior of S	
$\tilde{S}$	Closure of S	
δS	Boundary of S: $\delta S = \tilde{S} - \mathring{S}$	
$R - S$	Complement of S relative to R	
$R \oplus S$	Direct sum of R and S	
$R \times S$	Product set: $R \times S = \{(r, s) : r \in R, s \in S\}$	
$R \cap S$	Intersection of R and S	
$R \cup S$	Union of R and S	
$R \subseteq S$	R is a subset of S	
$\{x^k\}$	Sequence $\{x^1, x^2, \ldots\}$	
$[a, b]$	Closed interval: $\{x : a \leqslant x \leqslant b\}$	
$[a, b)$	Semiclosed interval: $\{x : a \leqslant x < b\}$	
$(a, b]$	Semiclosed interval: $\{x : a < x \leqslant b\}$	
(a, b)	Open interval: $\{x : a < x < b\}$	
$B(x, r)$	Open ball at x: $B(x, r) = \{y : \lVert y - x \rVert < r\}$	
$\varnothing$	Empty set	

RELATIONS, QUALIFIERS

SYMBOL	NAME
$=$	Is equal to
$\neq$	Is not equal to
$\cong$	Is approximately equal to
$\triangleq$	Equals by definition
$>$	Is greater than
$\geqslant$	Is greater than or equal to
$<$	Is less than
$\leqslant$	Is less than or equal to
$\rightarrow$	Tends to
$\Rightarrow$	Implies

SYMBOL	NAME
$\Leftrightarrow$	If and only if
min	Minimum
max	Maximum
inf	Infimum: largest lower bound
sup	Supremum: smallest upper bound
lim	Limit
$k!$	k factorial: $(1)(2) \cdots (k)$
$\binom{k}{j}$	Binomial coefficient, $k!/\{j!(k-j)!\}$
$[k/2]$	Integer part of $k/2$
■	End of remark

APPENDIX B THE LOGIC OF PROOFS

The basic building blocks of a mathematical theory are logical propositions that are more commonly called theorems. Typically, these propositions take the form of one or more premises implying one or more conclusions. A proof of a proposition is a sequence of statements that, taken together, demonstrate the validity of the proposition. As such, the conclusions of a proposition are true if the premises are true and if the proposition has been validated with a proof. To develop a strategy for construction proofs, we begin by examining the structure of logical propositions.

Let P and Q denote *logical expressions*. That is, P and Q are expressions or statements whose values are either true or false. There are three basic *logical operators* we can use with such expressions called the NOT, AND, and OR operators. Refer to Table B.1 for a truth table summary of the basic logical operators.

Here T and F denote *true* and *false*, respectively. The NOT operator is a unary operator that simply negates the truth value of its argument. That is, $\bar{P}$ is true whenever P is *not* true. The AND and OR operators, on the other hand, are binary operators. Here $P \cdot Q$ is true whenever both P *and* Q are true, while $P + Q$ is true whenever either P *or* Q is true. Note from Table B.1 that the AND and OR operators are examples of *commutative* binary operators. That is, the order

TABLE B.1. Basic Logical Operators

P	Q	NOT P	P AND Q	P OR Q
F	F	T	F	F
F	T	T	F	T
T	F	F	F	T
T	T	F	T	T
P	Q	$\bar{P}$	$P \cdot Q$	$P + Q$

of the arguments can be changed or commuted without changing the truth value of the result.

There are a host of identities one can develop which interrelate the NOT, AND, and OR operators. The easiest way to establish these identities is to use a truth table and thereby enumerate all possible cases. This is called the method of *perfect induction.* Refer to Table B.2 for a summary of three simple identities which will be useful for our purposes. The last two are complementary forms of DeMorgan's law.

Next, we consider several noncommutative binary operators. We distinguish these operators from the others by referring to them as *statement connectives.* Table B.3 provides a truth table description of three fundamental statement connectives.

A *logical proposition* is simply a group of logical expressions interconnected by statement connectives. The letters V and I in Table B.3 denote *valid* and *invalid*, respectively. Thus the proposition $P \Rightarrow Q$ is valid if whenever P is true, Q is also true. In this case we say P *only if* Q, or P *implies* Q, or for P to be true it is *necessary* that Q be true. Similarly, the proposition $P \Leftarrow Q$ is valid if whenever Q is true, P is also true. In this case we say P *if* Q, or P is implied by Q, or for P to be true it is *sufficient* that Q be true. Finally, the proposition $P \Leftrightarrow Q$ is valid if whenever P is true, Q is also true and conversely. That is

$$(P \Leftrightarrow Q) = (P \Leftarrow Q) \cdot (P \Rightarrow Q).$$

In this case we say P *if and only if* Q, or P *implies and is implied by* Q, or for P to be true it is both *necessary and sufficient* that Q be true. Another way of stating that the proposition $P \Leftrightarrow Q$ is valid is to say that the logical expressions P and Q are *equivalent.* Note that the equivalence operator is commutative, whereas the IF and ONLY IF operators are not. Indeed, the IF operator is obtained from the ONLY IF operator by commuting the arguments:

$$(P \Leftarrow Q) = (Q \Rightarrow P).$$

It is evident that all three statement connectives in Table B.3 can be synthesized from the ONLY IF operator. It is for this reason that we shall adopt the following as the *standard form* for a logical proposition:

$$P \Rightarrow Q.$$

TABLE B.2. Logical Identities

Identity	Name
$\bar{\bar{P}} = P$	Double negation
$\overline{P \cdot Q} = \bar{P} + \bar{Q}$	DeMorgan's law
$\overline{P + Q} = \bar{P} \cdot \bar{Q}$	DeMorgan's law

TABLE B.3. Statement Connectives

P	Q	$P \Rightarrow Q$	$P \Leftarrow Q$	$P \Leftrightarrow Q$
F	F	V	V	V
F	T	V	I	I
T	F	I	V	I
T	T	V	V	V
P	Q	ONLY IF	IF	IF AND ONLY IF

Given the standard form for a logical proposition, we now consider techniques that can be used to demonstrate its validity. The first three techniques are general, whereas the last two apply only to special types of propositions.

B.1. DIRECT METHOD

The *direct method* of validating the proposition $P \Rightarrow Q$ is to assume that the expression P is true and show that the expression Q must therefore also be true. Typically, this involves going back to the definitions of the terms appearing in the expressions P and Q and making use of the defining properties. In the early stages of the development of a theory, relying exclusively on the basic definitions is often the only approach available. However, as a theory becomes refined, more and more interrelated results or validated propositions become available. Often one can use certain of the results, in conjunction with the definitions, to demonstrate the validity of a new proposition. Whenever these applicable results are available, they should be put to full use. Although all verifiable results can be derived directly from definitions, this brute force approach typically leads to long, cumbersome, less than elegant proofs. Much of the beauty of a highly refined theory lies in its hierarchical structure. Failure to fully exploit this structure in further extending the theory is tantamount to reinventing the wheel every time one wants to travel.

B.2. CONTRAPOSITIVE METHOD

A second method for establishing the validity of the proposition $P \Rightarrow Q$ is to prove the following equivalent proposition:

$$\bar{Q} \Rightarrow \bar{P}$$

TABLE B.4. Contrapositive Form

P	Q	$P \Rightarrow Q$	$\bar{Q} \Rightarrow \bar{P}$
F	F	V	V
F	T	V	V
T	F	I	I
T	T	V	V

We call this the *contrapositive form* of $P \Rightarrow Q$. The equivalence of the two follows directly from the definitions of the NOT and the ONLY IF operators as can be seen from the two truth tables in Table B.4.

It is evident from Table B.4 that the proposition $P \Rightarrow Q$ and its contrapositive form $\bar{Q} \Rightarrow \bar{P}$ are both valid except when the expression $P \cdot \bar{Q}$ is true. Proof by the contrapositive method is therefore simply proof by the direct method, but applied to the contrapositive form. We assume that the conclusion Q is false, and then we show that this implies that the premise P must also be false. The merit of this approach lies in the observation that it is often easier to show that an expression is false since a single counterexample will suffice.

B.3. CONTRADICTION

A third method for establishing the validity of the proposition $P \Rightarrow Q$ is to use proof by contradiction. Here we assume that the proposition $P \Rightarrow Q$ is not valid, and then show that this leads to a *contradiction.* This is accomplished by establishing the validity of the following proposition:

$$P \cdot \bar{Q} \Rightarrow \bar{P} + Q.$$

Note from Table B.3 that assuming the premise $P \cdot \bar{Q}$ is true is equivalent to assuming that the proposition $P \Rightarrow Q$ is not valid. Moreover, by applying the identities in Table B.2, one can show that the conclusion $\bar{P} + Q$ is simply the negation of the premise $P \cdot \bar{Q}$. Thus proving that the proposition $P \cdot \bar{Q} \Rightarrow \bar{P} + Q$ is valid is tantamount to establishing the following contraction: if the premise $P \cdot \bar{Q}$ is true, then it is also false. Since this is impossible, it follows that $P \cdot \bar{Q}$ must not be true in which case $P \Rightarrow Q$ is valid.

B.4. MATHEMATICAL INDUCTION

A fourth method for establishing the validity of the proposition $P \Rightarrow Q$ is called proof by mathematical induction. This technique is applicable only in certain special cases. In particular, suppose that the conclusion is of the form $Q(k)$ where

$Q(k)$ is a logical expression which depends on an *integer index* k. Let j be a fixed integer and consider the following proposition:

$$k \geqslant j \Rightarrow Q(k).$$

Here the premise is simply $P = (k \geqslant j)$. Since this proposition is in the standard form, $P \Rightarrow Q$, we could apply any of the methods discussed thus far. However, for this class of propositions it is often easier to take a fourth approach. Suppose we break the original proposition into the following two parts:

$$i = j \Rightarrow Q(i),$$
$$(i \geqslant j) \cdot Q(i) \Rightarrow Q(i + 1).$$

If the validity of both of these is established, then the original proposition, $k \geqslant j \Rightarrow Q(k)$, is valid. We call this method proof by *mathematical induction*. Here we first show that $Q(j)$ is true, and then we show that if $Q(i)$ is true for an arbitrary $i \geqslant j$, then $Q(i + 1)$ is also true.

B.5. EQUIVALENCE

Many logical propositions are not in the standard form $P \Rightarrow Q$, but instead take the form $P \Leftrightarrow Q$. This IF AND ONLY IF type of proposition is in fact a special case of the following more general proposition where $n \geqslant 2$:

$$P_1 \Leftrightarrow P_2 \Leftrightarrow \cdots \Leftrightarrow P_n.$$

Here if any one of the logical expressions $\{P_1, P_2, \ldots, P_n\}$ is true, then they are all true and conversely. Alternatively, we say that the expressions $\{P_1, P_2, \ldots, P_n\}$ are logically equivalent. A straightforward approach to establishing equivalence is to consider all possible pairs of expressions $\{P_k, P_j\}$ and show that $P_k \Rightarrow P_j$ and $P_j \Rightarrow P_k$. However, this entails proving $2(n - 1)!$ propositions of the standard form $P \Rightarrow Q$. A considerable savings for the cases when $n \geqslant 3$ can be made if we instead establish the validity of the following equivalent proposition:

$$P_1 \Rightarrow P_2 \Rightarrow \cdots \Rightarrow P_n \Rightarrow P_1.$$

The fact that this chain of ONLY IF operators is *closed* or circular means that each expression implies every other expression in the list. Note that this reduces the problem to one of establishing the validity of n standard propositions of the form $P \Rightarrow Q$. Since $n \geqslant 2$, this method will never be longer than the straightforward approach and, when $n > 2$, it will be shorter.

INDEX